D0869286

BICYCLES & TRICYCLES

Bicycles & Tricycles

AN ELEMENTARY TREATISE ON THEIR
DESIGN AND CONSTRUCTION

WITH EXAMPLES AND TABLES

BY

ARCHIBALD SHARP, B.Sc.

WHITWORTH SCHOLAR
ASSOCIATE MEMBER OF THE INSTITUTION OF CIVIL ENGINEERS
MITGLIED DES VEREINS DEUTSCHER INGENIEURE
INSTRUCTOR IN ENGINEERING DESIGN AT THE CENTRAL TECHNICAL COLLEGE
SOUTH KENSINGTON

WITH NUMEROUS ILLUSTRATIONS

THE MIT PRESS
CAMBRIDGE, MASSACHUSETTS, AND LONDON, ENGLAND

Fifth printing, 1993

First MIT Press paperback edition, 1979
Copyright © 1977 by
The Massachusetts Institute of Technology

This book was reproduced from an original in the Massachusetts
Institute of Technology Libraries.

This book was printed and bound by Braun-Brumfield Company
in the United States of America.

Library of Congress Cataloging in Publication Data
Sharp, Archibald, 1862–1934.
Bicycles and tricycles.

Reprint of the 1896 ed. published by Longmans, Green,
London, New York.
Includes index.
1. Bicycles and tricycles. I. Title.
TL410.S5 1977   629.22′72   77–4928
ISBN 0–262–19156–3
ISBN 0–262–19156–3 (hard)
ISBN 0–262–69066–7 (paper)

FOREWORD TO REPRINT EDITION

DAVID GORDON WILSON

Archibald Sharp's definitive work on bicycles and tricycles marked, and helped to bring about, the end of an exciting period in mechanical engineering. The invention of the pedaled bicycle bore fruit after steam power had been applied successfully to road, rail, and water. But the excitement engendered by the bicycle was in some ways more widespread and intense than was inspired by the triumphs of steam. For on the one hand it emancipated large numbers of ordinary people, and particularly women, from lifetimes spent mainly within walking distances of their homes. And on the other hand it gave scope to tinkerers, blacksmiths, and engineers, allowing their imaginations full reign in conceiving new designs of machines and of components.

The result was that in the two decades following Starley's 1876 invention of the tangent-spoked tension wheel, whose light weight and high strength opened up broad areas of design freedom, an extraordinary variety of bicycles was produced. Many of these showed "utter ignorance of mechanical science," as Sharp states in his preface. One of his aims in writing this book was to make it more difficult for manufacturers to produce such "mechanical monstrosities."

And he seems to have been supremely successful. His was not the first book on bicycle design. He acknowledges his indebtedness in his preface to R. P. Scott's *Cycling Art, Energy, and Locomotion* and C. Bourlet's *Traité des Bicycles et Bicyclettes*. But Sharp's work was more thorough, complete, and authoritative — and was almost the last book as well as the last word on bicycle design. The designs he scorns here virtually disappeared after the publication of his book (it is true that many were already becoming extinct because of their own all-too-obvious inadequacies), while the simple diamond-frame chain-driven "safety" bicycle, which he praised, ruled supreme and almost unchallenged as a utility and recreation vehicle during the following five decades of bicycle popularity all over the world.

The mass-produced, low-priced automobile running on inexpensive fuel relegated bicycles to the role of children's playthings, first in the United States in the twenties and thirties and later in Europe and other places in the fifties. But the very success of the automobile led to the end of the cheap-fuel era and to concern over the pollution and congestion resulting from over-use of automobiles in cities. A second great bicycle boom started in the United States in the late sixties and appears to be spreading to other countries. And with it has come a new outpouring of creative energy on the part of skilled engineers and of less skilled enthusiasts. Many have already wasted much effort repeating some of the mistakes Sharp pointed out in 1896. The reprinting of his book will have practical as well as historical and educational value.

His book can also be read for sheer enjoyment. He was no dry academic. It is true that at the time his book was published he was instructor in mechanical design in a prominent London college and that thereafter he was often irreverently and inaccurately referred to as "Professor" Sharp. But he spent much of his professional life as an independent consultant, inventing and developing improved machines and components for himself and for others. His salty and down-to-earth nature is very evident in the acrimonious letters he wrote to the correspondence columns of bicycling journals,* in which he attacked poor mechanical-engineering science wherever he saw it, regardless of how important, or how self-important were the perpetrators. He was almost always right.

The same forthrightness, and concern for the accuracy of engineering fundamentals, shines through every page of this book. The MIT Press will earn our gratitude for bringing back to us the complete Archibald Sharp: educator, engineer, entrepreneur, and concerned human being.

*Frank Rowland Whitt, senior author of *Bicycling Science,* sent me copies of some of Archibald Sharp's correspondence "wars" with contemporary bicycle manufacturers.

PREFACE

A BICYCLE or a tricycle is a more or less complex machine, and for a thorough appreciation of the stresses and strains to which it is subjected in ordinary use, and for its efficient design, an extensive knowledge of the mechanical sciences is necessary. Though an extensive literature on nearly all other types of machines exists, there is, strange to say, very little on the subject of cycle design; periodical cycling literature being almost entirely confined to racing and personal matters. In the present work an attempt is made to give a rational account of the stresses and strains to which the various parts of a cycle are subjected; only a knowledge of the most elementary portions of algebra, geometry, and trigonometry being assumed, while graphical methods of demonstration are used as far as possible. It is hoped that the work will be of use to cycle riders who take an intelligent interest in their machines, and also to those engaged in their manufacture.

The present type of rear-driving bicycle is the outcome of about ten years' practical experience. The old 'Ordinary,' with its large front driving-wheel, straight fork, and curved backbone, was a model of simplicity of construction, but with the introduction of a smaller driving-wheel, driven by gearing from the pedals, and the consequent greater complexity of the frame, there was more scope for variation of form of the machine. Accordingly, till a few years ago, a great variety of bicycles were on the market, many of them utterly wanting in scientific design. Out of these, the present-day rear-driving bicycle, with diamond-frame, extended wheel-base, and long socket steering-head—the fittest—has survived. A better technical education on the part of bicycle manufacturers and their customers might have saved them a great amount of trouble and expense. Two or three years ago, when there seemed a chance of the dwarf front-driving bicycle coming into popular favour, the same variety in design of frame was to be seen ; and even now with tandem bicycles there are many frames on the market which evince on the part of their designers utter ignorance of mechanical science. If the present work is the means of influencing makers, or purchasers, to such an extent as to make the manufacture and sale of such mechanical monstrosities in the future more difficult than it has been in the past, the author will regard his labours as having been entirely successful.

The work is divided into three parts. Part I. is on Mechanics and the Strength of Materials, the illustrations and examples being taken with special reference to bicycles and tricycles ; Part II. treats of the cycle as a complete machine ; and Part III. treats in detail of the design of its various portions.

The descriptive portions are not so complete as might be wished ; however, the 'Cyclist Year Books,' published early in each year, enable anyone interested in this part of the subject to be well informed as to the latest novelties and improvements.

The author would like to express his indebtedness to the following works :

The 'Cyclist Year Books' ;

'Bicycles and Tricycles of the Year,' by H. H. Griffin, a valuable series historically, which extends from 1878 to 1889 ;

'Cycling Art, Energy, and Locomotion,' by R. P. Scott ;

'Traité des Bicycles et Bicyclettes,' par C. Bourlet ;

The 'Cyclist' weekly newspaper ;

and to the various cycle manufacturers mentioned in the text, who have, without exception, always afforded information and assistance when asked. He has also to thank Messrs. Ackermann and Farmer for assistance in preparing drawings, and Messrs. Ackermann and Hummel for reading the proofs.

In a work like the present, containing many numerical examples, it is improbable that the first issue will be entirely free from error ; corrections, arithmetical and otherwise, will therefore be gladly received by the author.

CONTENTS

———◦◦◦———

PART I

PRINCIPLES OF MECHANICS

CHAPTER I

FUNDAMENTAL CONCEPTIONS OF MECHANICS

CHAPTER II

SPEED, RATE OF CHANGE OF SPEED, VELOCITY, ACCELERATION, FORCE, MOMENTUM

CHAPTER III

KINEMATICS : ADDITION OF VELOCITIES

Contents

PART II

CYCLES IN GENERAL

Contents

xiii

CHAPTER XVII

STABILITY OF CYCLES

CHAPTER XVIII

STEERING OF CYCLES

CHAPTER XIX

MOTION OVER UNEVEN SURFACES

CHAPTER XX

RESISTANCE OF CYCLES

PART III

DETAILS

CHAPTER XXII

THE FRAME : DESCRIPTIVE

CHAPTER XXIII

THE FRAME : STRESSES

CHAPTER XXVII

TOOTHED-WHEEL GEARING

CHAPTER XXVIII

LEVER-AND-CRANK GEAR

CHAPTER XXIX

TYRES

CHAPTER XXX

PEDALS, CRANKS, AND BOTTOM BRACKETS

LIST OF TABLES

PART I
PRINCIPLES OF MECHANICS

PART I

PRINCIPLES OF MECHANICS

CHAPTER I

FUNDAMENTAL CONCEPTIONS OF MECHANICS

1. **Division of the Subject.**—*Geometry* is the science which treats of relations in space. *Kinematics* treats of space and time, and may be called the geometry of motion. *Dynamics* is the science which deals with force, and is usually divided into two parts—statics, dealing with the forces acting on bodies which are at rest ; kinetics, dealing with forces acting on bodies in motion. *Mechanics* includes kinematics, statics, kinetics, and the application of these sciences to actual structures and machines.

2. **Space.**—The fundamental ideas of time and space form part of the foundation of the science of mechanics, and their accurate measurement is of great importance. The British unit of length is the *imperial yard*, defined by Act of Parliament to be the length between two marks on a certain metal bar kept in the office of the Exchequer, when the whole bar is at a temperature of 60° Fahrenheit. Several authorised copies of this standard of length are deposited in various places. The original standard is only disturbed at very distant intervals, the authorised copies serving for actual comparison for purposes of trade and commerce. The yard is divided into three *feet*, and the foot again into twelve *inches*. Feet and inches are the working units in most general use by engineers. The inch is further subdivided by engineers, by a process of repeated division by two, so that $\frac{1}{2}''$, $\frac{1}{4}''$, $\frac{1}{8}''$, $\frac{1}{16}''$, &c., are the fractions generally used by them. A more convenient

subdivision is the decimal system into $\frac{1}{10}$, $\frac{1}{100}$, $\frac{1}{1000}$, &c. ; this is the subdivision generally used for scientific purposes.

The unit of length generally used in dynamics is the *foot*.

Metric System. — The metric system of measurement in general use on the Continent is founded on the *metre*, originally defined as the $\frac{1}{10,000,000}$ part of a quadrant of the earth from the pole to the equator. This length was estimated, and a standard constructed and kept in France. The metre is subdivided into ten parts called decimetres, a decimetre into ten centimetres, and a centimetre into ten millimetres. For great lengths a kilometre, equal to a thousand metres, is the unit employed.

1 metre	= 39·371 inches = 3·2809 feet.
1 kilometre	= 0·62138 miles.
1 inch	= 25·3995 millimetres.
1 mile	= 1·60931 kilometres.

3. **Time.**—The measurement of time is more difficult theoretically than that of space. Two different rods may be placed alongside each other, and a comparison made as to their lengths, but two different portions of time cannot be compared in this way. 'Time passed cannot be recalled.'

The measurement of time is effected by taking a series of events which occur at certain intervals. If the time between any two consecutive events leaves the same impression as to duration on the mind as that between any other two consecutive events, we may consider, tentatively at least, that the two times are equal. The standard of time is the *sidereal day*, which is the time the earth takes to make one complete revolution about its own axis, and which is determined by observing the time from the apparent motion of a fixed star across the meridian of any place to the same apparent motion on the following day. The intervals of time so measured are as nearly equal as our means of measurement can determine.

The *solar day* is the interval of time between two consecutive apparent movements of the sun across the meridian of any place. This interval of time varies slightly from day to day, so that for purposes of everyday life an average is taken, called the *mean solar day*. The mean solar day is about four minutes longer than the sidereal day, owing to the nature of the earth's motion round the sun.

The mean solar day is subdivided into twenty-four *hours*, one hour into sixty *minutes*, and one minute into sixty *seconds*. The second is the unit of time generally used in dynamics.

4. **Matter.**—Another of our fundamental ideas is that relating to the existence of matter. The question of the measurement of quantity of matter is inextricably mixed up with the measurement of force. The *mass*, or quantity of matter, in one body is said to be greater or less than that in another body, according as the force required to produce the same effect is greater or less. The mass of a body is practically estimated by its weight, which is, strictly speaking, the force with which the earth attracts it. This force varies slightly from place to place on the earth's surface at sea level, and again as the body is moved above the sea level. Thus, the mass and the weight of a body are two totally different things ; and many of the difficulties encountered by the student of mechanics are due to want of proper appreciation of this. The difficulty arises from the fact that the *pound* is the unit of matter, and that the *weight* of this quantity of matter, *i.e.* the force by which the earth attracts it, is used often as a unit of force. A certain quantity of lead will have a certain weight, as shown by a spring-balance, in London at high level water-mark, and quite a different weight if taken twenty thousand feet above sea level, although the mass is the same in both places.

The British unit of mass is the imperial *pound*, defined by Act of Parliament to be the quantity of matter equal to that of a certain piece of platinum kept in the office of the Exchequer.

: The unit of mass in the metrical system of measurement is the *gramme*, originally defined to be equal to the mass of a cubic centimetre of distilled water of maximum density. This is, however, defined practically, like the British unit, as that of a certain piece of platinum kept in Paris.

CHAPTER II

5. **Speed.**—A body in relation to its surroundings may either be at rest or in motion. *Linear speed* is the rate at which a body moves along its path.

Speed may be either *uniform* or *variable*. With uniform speed the body passes over equal spaces in equal times; with variable speed the spaces passed over in equal times are unequal. The motion may be either in a straight or curved path, but in both cases we may still speak of the *speed* of a point as the rate at which it moves along its path.

6. **Uniform Speed** is measured by the space passed over in the unit of time. The unit of speed is one foot per second. Let s be the space moved over by the body moving with uniform speed in the time t, then if v be the speed, we have by the above definition.

$$v = \frac{s}{t}. \quad . \quad . \quad . \quad . \quad . \quad . \quad . \quad . \quad (1)$$

Example.—If a bicycle move through a space of one mile in four minutes we have, reducing to feet and seconds,

$$v = \frac{3 \times 1760}{4 \times 60} = 22 \text{ feet per second.}$$

It will be seen that the unit of speed is a compound one, involving two of the fundamental units, space and time.

In the above example, the same speed is obtained whatever be the time over which we make the observations of the space described. For example, in one minute the bicycle will move through a distance of a quarter of a mile, that is 440 yards, or 3 × 440 feet.   Using formula (1) we get

$$v = \frac{1320}{60} = 22 \text{ feet per second,}$$

the same result as before.

Now, consider the space described by the bicycle in a small fraction of a second, say $\frac{1}{10}$th, if the speed is uniform, this will be 2·2 ft.   Using formula (1) again, we have

$$v = \frac{2 \cdot 2}{\frac{1}{10}} = 22 \text{ feet per second.}$$

Proceeding to a still smaller fraction of a second, say $\frac{1}{1000}$th, if our means of observation were sufficiently refined, the distance passed over in the time would evidently be found to be the $\frac{22}{1000}$th part of a foot, *i.e.* = ·022 feet.   Again using formula (1) we have

$$v = \frac{·022}{\frac{1}{1000}} = 22 \text{ feet per second.}$$

Uniform Motion in a Circle.—Another familiar example of uniform motion is that of a point moving in a circular path ; a point on the rim of a bicycle wheel has, relative to the frame of the bicycle, such a motion, uniform when the speed of the bicycle is uniform.   The linear speed, relative to the frame, of a point on the extreme outside of the tyre will be the same as the linear speed of the bicycle along and relative to the road, while that of any point nearer the centre of the wheel will be less.

7. **Angular Speed.**—When a wheel is rotating about its axis, the linear speed of any point on it depends on its distance from the centre, is greatest when the point is on the circumference of the wheel, and is zero for a point on the axis.   The number of complete turns the wheel, as a whole, makes in a second gives a convenient means of estimating the rotation.   Let O (fig. 1) be the centre of a wheel, and A a point on its circumference ; OA may thus represent the position of a spoke of the wheel at a certain instant.   At the end of one second, suppose the spoke which was initially in the position OA_1 to occupy the position OA_2 ; if the motion of rotation of the wheel is uniform, the linear

s peed of the point *A* on the rim is measured by the arc $A_1 A_2$, while the angular speed of the wheel is measured by the angle $A_1 O A_2$. Generally, the angular speed of a body rotating uniformly is the angle turned through in unit of time.

The angular speed may be expressed in various ways. For example, the number of degrees in the angle $A_1 O A_2$ swept out per second may be expressed ; this method, however, is little used practically. The method of expressing angular velocity most in use by engineers, is to give the number of revolutions per minute, *n*.

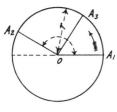

FIG. 1.

One revolution $= 360°$; revolutions per minute can be converted into degrees per second by multiplying by 360 and dividing by 60, that is, by multiplying by 6.

For scientific purposes another method is used. Mathematicians find that the most convenient unit angle to adopt is not obtained by dividing a right angle into an arbitrary number of parts ; they define the unit angle as that which subtends a circular arc of length equal to the radius. Thus, in figure 1, if the arc $A_1 A_3$ be measured off equal to the radius $O A_1$, the angle $A_1 O A_3$ will be the *unit angle*. This is called a *radian*.

The ratio of the length of the circumference of a circle to its diameter is usually denoted in works on mathematics and mechanics by the Greek letter π (pronounced like the English word 'pie'), and is $3.14159 \ldots$. This number is 'incommensurable,' which means that it cannot be expressed exactly in our ordinary system of numeration. It may, however, be expressed with as great a degree of accuracy as is desired ; a very rough value often used for caculations is $3\frac{1}{7}$. It is easily seen that there are π radians in an angle of half a revolution, and therefore the angle of one revolution, that is, four right angles, is 2π radians. Therefore, 1 radian $= \dfrac{360}{2\pi} = 57.28°$. The angular speed ω of a rotating body is expressed in radians turned through per second, and

$$\omega = \frac{2\pi n}{60} \qquad \ldots \ldots \quad (2)$$

8. **Relation between Linear and Angular Speeds.**—The connection between the angular speed of a rotating body and the linear speed of any point in it may now be easily expressed. Let O (fig. 1) be the centre of the rotating body, and A a point on it, distant r from the centre, which moves in unit of time from A_1 to A_2, the number of units in the linear speed of A is equal to the number of units in the length of the arc $A_1 A_2$, similarly the angular speed of the rotating body is numerically equal to the angle $A_1 O A_2$ in radians. But this by definition must be equal to the arc $A_1 A_2$ divided by the radius $O A_1$, hence if ω (omega) be the angular speed of a rotating body, v the linear speed of any point on it distant r from the centre, we have

$$\omega = \frac{v}{r} \quad . \quad . \quad . \quad . \quad . \quad . \quad . \quad (3)$$

The speed of a bicycle is conveniently expressed in miles per hour, and the angular speed of the driving-wheel in revolutions per minute. Let V be the speed in miles per hour, D the diameter of the driving-wheel in inches, and n the number of revolutions of the driving-wheel per minute ; then feet and seconds being the units in (3),

$$\omega = \frac{2\pi n}{60}, \quad v = \frac{V \times 5280}{3600}, \quad r = \frac{D}{2 \times 12}.$$

Substituting in (3) we get

$$\frac{2\pi n}{60} = \frac{V \times 5280 \times 2 \times 12}{3600 \times D},$$

from which $\qquad V = \dfrac{n D}{336 \cdot 13} \quad . \quad . \quad . \quad . \quad . \quad . \quad . \quad (4)$

that is, the speed of the bicycle in miles per hour is equal to the number of revolutions per minute of the driving-wheel, multiplied by the diameter of the driving-wheel in inches, and divided by $336 \cdot 13$.

A more convenient rule than the above for finding the speed of a bicycle can be deduced. Let N be the number of revolutions of the driving-wheel made in t seconds ; then

$$N = \frac{n \times t}{60}, \text{ and } n = \frac{60 N}{t}.$$

Substituting in (4), we get

$$V = \frac{60 \ N \ D}{336 \cdot 13 \ t}.$$

Now, suppose that N be chosen equal to V; that is, t is chosen such that the number of revolutions in t seconds is equal to the number of miles travelled in one hour. Substituting above we get

$$t = \frac{D}{5 \cdot 502}, \quad . \quad . \quad . \quad . \quad . \quad . \quad . \quad . \quad (5)$$

which is equivalent to the following convenient rule. Divide the diameter of the driving-wheel in inches by $5 \cdot 502$, the number of revolutions of the driving-wheel made in the number of seconds equal to this quotient is equal to the speed of the cycle in miles per hour.

If, in a geared-up cycle, D be taken as the diameter to which the driving-wheel is geared, N will be the number of revolutions per minute of the crank-axle, and formula (5) will still apply.

9. **Variable Speed.**—The numerical example in section 6 may help towards a clear understanding of the measurement of variable speed. When the speed of a moving body is changing from instant to instant, if we want to know the speed at a certain point, it would be quite incorrect to observe the space described by the body in, say, one hour or one minute after passing the point in question ; but the smaller the interval of time chosen, the more closely will the *average* speed during that interval approximate to the speed *at the instant* of passing the point of observation.

Now, suppose the body after passing the point to move with exactly the same speed it had at the point, and that in t seconds it moves over s feet, its speed at the point of observation would be $\frac{s}{t}$ feet per second. In a very small fraction of a second, say $\frac{1}{1000}$th, the amount of change in the speed of the body is very small, and by taking a sufficiently small period of time the average speed during the period may be considered equal to the speed at the beginning of the period, without any appreciable error. The speed at any instant will thus be expressed by equation (1), *provided t be chosen small enough.*

Suppose a bicyclist just starting to race, and that we wish to observe his speed at a point 5 feet from the starting-point. We observe the instant he passes the point, and the distance he travels in a period of time reckoned from that instant. If in a minute he travel 2,400 feet, his *average* speed during that time $= \frac{2400}{60} = 40$ feet per second. But in a quarter-minute, reckoned from the same instant, he may only travel 420 feet, giving an average speed of $\frac{420}{15} = 28$ feet per second; while in five seconds he may only have travelled 110 feet, in one second 15 feet, in one-tenth of a second 1·05 foot, in one-hundredth part of a second one-tenth part of a foot, with average speeds during these periods of 22, 15, 10·5, and 10 feet per second. The last of these values may be taken as a very close approximation to his speed when passing the point in question.

10. **Velocity.**—If the speed of a point and the direction of its motion be known, its *velocity* is defined : thus, in the conception 'velocity,' those of 'speed' and 'direction' are involved. Velocity has been defined as 'speed directed,' or 'rate of change of position.' Again, speed may be defined as the *magnitude* of velocity.

Velocity, involving as it does the idea of direction, can therefore be represented by a straight line, the direction of which indicates the direction of the motion, and, by choosing a suitable scale, the length of the line may represent the speed, or the magnitude of the velocity. A quantity which has not only magnitude and algebraical sign, but also direction, is called a *vector* quantity. Thus, velocity is a vector quantity. A quantity which has magnitude and sign, but is independent of direction, is called a *scalar* quantity. Speed is a scalar quantity.

Velocity may be *linear* or *angular* ; it may also be uniform or variable. A point on a body rotating with uniform angular speed about a fixed axis has its linear speed uniform, but since the direction of its motion is continually changing, its linear velocity is variable, its angular velocity is uniform. *Angular velocity* can also be represented by a vector, the direction of the vector being parallel to the axis of rotation, and the length of the vector being equal to the angular speed.

11. **Rate of Change of Speed.**—If a moving body at a certain instant has a speed of 3 feet per second, and a second later a speed of 4 feet per second, two seconds later a speed of 5 feet per second, three seconds later a speed of 6 feet per second, and so on ; in one second the speed increases by 1 foot per second. In other words, its rate of change of speed is 1 foot per second per second.

Rate of change of speed may be either uniform or variable. An important example of uniform rate of change of speed is that of a body falling freely under the action of gravity. If a stone be dropped from a height, its speed at the instant of dropping is zero ; at the end of one second, as determined by experiment, 32·2 feet per second approximately ; at the end of two seconds, 64·4 feet per second ; at the end of three seconds, 96·6 feet per second —at least, these would be the speeds if the air offered no resistance to the motion. Thus the rate of change of speed of a body falling freely under the action of gravity is 32·2 feet per second per second.

If a be the rate of change of speed of a body starting from rest, at the end of t seconds its speed will be

$$v = a\,t \quad . \quad . \quad . \quad . \quad . \quad . \quad . \quad (6)$$

Its *average* speed during the time will be $\frac{1}{2}\,a\,t$, and therefore the space it passes over in time t is

$$s = \frac{1}{2}\,a\,t \times t = \frac{1}{2}\,a\,t^2 \quad . \quad . \quad . \quad . \quad (7)$$

A cyclist starting in a race affords a good example of variable rate of change of speed. At the instant of starting the speed of the machine and rider is zero ; at the end of two seconds it may be five miles an hour ; at the end of three seconds, nine miles an hour ; at the end of four seconds, thirteen ; at the end of five, seventeen ; at the end of six, twenty ; at the end of seven, twenty-two ; at the end of eight, twenty-three ; at the end of nine, twenty-three and threequarters—the increase in the speed with each second becoming smaller and smaller until, say fifteen or twenty seconds from the start, the maximum speed is reached, the speed remains constant, and the rate of change becomes zero. In this case not only the speed, but also its rate of change, is variable. The rate of change probably increases at first, and reaches its maximum soon after the start, then diminishes, and ultimately reaches the value zero. If the speed of a body diminish, its rate of change of speed is negative. A cyclist while pulling up previous to stopping is moving with negative rate of change of speed.

The unit of rate of change of speed, like that of speed, is a compound one, into which the fundamental units of time and space enter. In expressing rate of change of speed we have used the phrase 'feet per second per second'; this deserves careful study on the part of the beginner, as a proper understanding of the ideas involved in these units is absolutely necessary for satisfactory progress in mechanics. This rate of change is often loosely spoken of in some of the earlier text books as so many 'feet per second'; this method of expression is quite wrong. For instance, considering the rate of change of speed due to gravity, we have stated above that it is 32 feet per second per second. This means that at the end of one second the speed of a freely falling body is increased by an additional speed of 32 feet per second, or 1,920 feet per minute. In one minute the speed would be increased by sixty times the above additional speed—that is, by 1,920 feet per second, or 115,200 feet per minute. This rate of change of speed may therefore be expressed either as '32 feet per second per second,' '1,920 feet per minute per second,' or '115,200 feet per minute per minute.'

The relation between the units of rate of change of speed, space, and time is expressed by the formula (7), which may be written

$$a = \frac{2\,s}{t^2},$$

which shows that the magnitude of the unit rate of change of speed is proportional to that of unit space, and inversely proportional to the square of that of unit time.

12. **Rate of Change of Angular Speed.**—The angular speed of a rotating body may be either constant or variable; in the

latter case the *rate of change of angular speed* is the increment in one unit of time of the angular speed. Let θ be the rate of change of angular speed, a the rate of change of linear speed of any point on the body distant r from the centre, then

$$\theta = \frac{a}{r} \qquad . \quad . \quad . \quad . \quad . \quad . \quad (8)$$

13. **Acceleration** is rate of change of velocity; it may be zero, uniform, or variable. When it is zero the velocity remains constant, and the motion takes place in a straight line.

When a point is moving with uniform speed in a circle, though its speed does not change, the direction of its motion changes, and therefore its velocity also changes. It must therefore be subjected to acceleration. An acceleration which does not change the speed of the body on which it acts must be in a direction at right angles to that of the motion, and is called *radial* acceleration. An acceleration which does not change the direction of a moving body must act in the direction of motion, and is called *tangential* acceleration. The *magnitude* of the tangential acceleration is the rate of change of speed.

14. **Force.**—The definition and measurement of force has afforded scope for endless metaphysical disquisitions. Force has been defined as ' that which produces or tends to produce motion in a body.' The unit of force is defined as ' that force which, acting for one unit of time on a body initially at rest, produces at the end of the unit of time a motion of one unit speed.' If the units of space, mass, and time be one foot, one pound, and one second respectively, the unit of force is called a *poundal.* In the centimetre-gramme-second system of units, the unit of force is called a *dyne.* The measurement of the unit of mass involves the idea of force, so that perhaps no satisfactory logical definition can be given.

The unit of force above defined is called the *absolute unit.* The magnitude of a force in absolute units is measured by the acceleration it would produce in unit of time on a body of unit mass. The force with which the earth attracts one pound of matter is equal to 32·2 poundals, since in one second it produces an acceleration of 32·2 feet per second per second. Generally, if a force f acting on a mass m produces an acceleration a, we have

$$f = m a \quad . \quad . \quad . \quad . \quad . \quad . \quad . \quad (9)$$

The unit of force used for practical purposes is the *weight* of one *pound* of matter ; this is called the gravitation unit of force. If f be the number of absolute, and F the number of gravitation units in a force, $f = gF$, or

$$F = \frac{f}{g} \quad . \quad . \quad . \quad . \quad . \quad . \quad . \quad (10)$$

The acceleration due to gravity is usually denoted by the letter g. The value of g, or, in other words, the weight of unit mass in absolute units of force, as has already been stated above, varies from place to place on the earth's surface. For Britain its value is approximately 32·2, the foot-pound-second system of units being used.

Great care must be exercised in distinguishing between one pound quantity of matter and 1 lb. weight, the former being a unit of mass, the latter an arbitrary unit of force.

15. **Momentum.**—The product of the mass of a body into its velocity is called its *quantity of motion* or *momentum*. The momentum of a body of mass one pound moving with a velocity of ten feet per second, is thus the same as that of a body of mass ten pounds moving with a velocity of one foot per second.

16. **Impulse.**—Multiply both sides of equation (9) by t, we then get

$$f t = m a t.$$

But if the body start from rest, $a t = v$, its velocity at the end of t seconds, therefore

$$f t = m v \quad . \quad . \quad . \quad . \quad . \quad . \quad (11)$$

Equation (11) asserts therefore that the momentum, $m v$, of a body initially at rest is equal to the product of the force acting on it and the time during which the force acts. The product $f t$ is called the *impulse* of the force.

Equation (11) is true, however small t, the time during which the force acts, may be. Now a momentum of 10 foot-pounds per second may be generated by the application of a force of 1 lb. acting for ten seconds, or a force of ten poundals for one second, or a force of 1000 poundals acting for $\frac{1}{100}$th part of a second ; and so on. When two moving bodies collide, or when a blow is struck by a hammer, the surfaces are in contact for a very small fraction of a second, and the mutual force between the bodies is very great. Neither the force nor the time during which it acts can be directly measured, but the momentum of the bodies before and after collision can be easily measured. Such forces of great magnitude acting for a very short space of time are called *impulsive* forces ; they differ only in degree, but not in kind, from forces acting for appreciably long periods.

17. **Moments of Force, of Momentum, &c.**—Let figure 2 represent a body fastened by a pin at O, so that it is free to turn

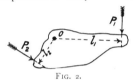

FIG. 2.

about O as a centre, but is otherwise constrained. Let it be acted on by the forces P_1 and P_2. Now, it is a matter of every day experience that the turning effect of such a force as P_1 depends not only on its magnitude, but also, in popu-

lar language, on its leverage, that is, on the length of the perpendicular from the centre of rotation to the line of action of the force. For example, in screwing up a nut, if a long spanner be used the force required to be exerted at its end is much smaller than if a short spanner be used. The product $P_1 l_1$ of the force into this distance is called the *moment* of the force about the given centre. The force P_1 tends to turn the body in the direction of the hands of a watch, while P_2 tends to turn the body in the opposite direction. Therefore, if the moment $P_2 l_2$ be considered positive, the moment $P_1 l_1$ must be considered negative. The positive direction is usually taken *contra-clockwise*.

If the body be at rest under the action of the forces P_1 and P_2 their moments must be equal in magnitude but of opposite sign ; that is, their algebraic sum must be zero.

The *moment of momentum* about a given point O of a body of mass m moving with velocity v is the product of its momentum $m v$, and the length of the perpendicular l from the given point to the direction of motion—*i.e.*, $m v l$. In the same way the moment of an impulse $f t$ is the product of the impulse and the length of the perpendicular from the given point to the line of action of the impulse—*i.e.*, $f t l$.

18. **Graphic Representation of Velocity.**—For the complete specification of a velocity two elements—its *magnitude* and *direction*—are necessary. If a body be moving at any instant with a certain velocity, the direction of the motion may be represented by the direction of a straight line drawn on the paper, and the speed of the body by the length of the straight line. For this purpose the unit of speed is supposed to be represented by any convenient length on the paper; the number of times this length is contained in the straight line drawn will be numerically equal to the speed of the body. For example, the line *a b* (fig. 3) represents a velocity of three feet per second in the direction of the arrow, while the line *c d* represents a velocity of two feet per second in a direction at right angles to that of the former. The scale of velocity in this diagram has been taken 1 foot per

FIG. 3.

second = $\frac{1}{4}$ inch. In the same way, any quantity which involves direction as well as magnitude can be represented by a straight line having the same direction and its length proportional to some scale to the magnitude of the quantity. Such a straight line is called a *vector*.

Example.—If a wheel be turning about its axis with uniform speed, the velocities of all points on its rim are numerically equal, but have all different directions. Thus, the velocities of the points *A*, *B*, and *C* on the rim (fig. 4) are represented by the three equal lines, *A a*, *B b*, and *C c* respectively at right angles to the radii *O A*, *O B*, and *O C*.

19. **Addition of Velocities.**—A body may be subjected at

the same instant to two or more velocities, and its aggregate velocity may be required. For example, take a man climbing the mast of a ship. Let the ship move horizontally in the direction $a\,b$ (fig. 5), and let the length $a\,b$ indicate the space passed over by it in one second. Let $a\,c$ be the mast, and as it passes the point a let the man commence climbing. At the end of one second suppose he has climbed the distance $a\,d$. The line $a\,d$ will represent the velocity of the man climbing up the mast, the line $a\,b$ the velocity of the ship. But if $a\,c$ be the position of the mast at the beginning of the second, at the end of the second it will be in the position $b\,c^1$, and the man will be at d^1, the length $b\,d^1$ being, of course, the same as $a\,d$. The actual velocity, in

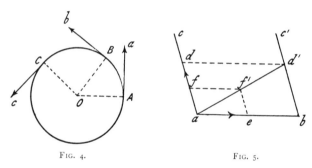

F<small>IG</small>. 4.　　　　　　　　　　F<small>IG</small>. 5.

magnitude and direction, of the man is represented by the line $a\,d^1$. At the end of half a second the foot of the mast would be at e, $a\,e$ being equal to $\frac{1}{2}\,a\,b$, and the man would have ascended the mast a distance $a\,f$; the actual position of the man would be f^1, midway between a and d^1. Thus his actual motion in space will be along the line $a\,d^1$.

The two velocities $a\,b$ and $a\,d$ above are called the *component* velocities, and the velocity $a\,d^1$ the *resultant* velocity of the man.

20. **Relative Velocity.**—We have spoken above of the motion of a body, meaning thereby the motion of the body in relation to the objects in its immediate neighbourhood, but these objects themselves may be in motion in relation to some other objects. For example, a man walking from window to window of

a railway carriage in rapid motion has a motion of a certain velocity relative to the carriage. But the carriage itself is in motion relative to the earth, and the motion of the man relative to the earth is quite different from that relative to the carriage. Again, the earth itself is not at rest, but rotates on its own axis, so that the man's motion relative to a plane of fixed direction passing through the earth's axis is still more complex. But besides a motion round its own axis, the earth has a motion round the sun. The sun itself, and with it the whole solar system, has a motion of translation relative to the visible universe ; in fact, there is no such thing as absolute rest in nature. Therefore, having no body at rest to which we can refer the motion of any body, we know nothing of absolute motion. The motions we deal with, therefore, are all relative, and the velocities are also relative. It will thus often be necessary, in specifying a velocity to express the body in relation to which it is measured.

21. **Parallelogram of Velocities.**—Given two component uniform velocities to which a body is subjected, the resultant velocity of the body may therefore be found as follows :— Draw a parallelogram with two adjacent sides, *o a* and *o b* (fig. 6),

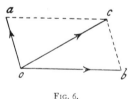

Fig. 6.

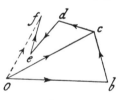

Fig. 7.

representing in magnitude and direction the component velocities The resultant velocity is represented in magnitude and direction by the diagonal *o c* of the parallelogram. This proposition is known as *the parallelogram of velocities*. Since velocity involves the two ideas of speed and direction, but not position, the resultant of two velocities may also be found by the following method :—Let *o b* (fig. 7) be one of the given velocities ; from *b* draw *b c* equal and parallel to the other ; *o c* will represent the resultant velocity.

Vector Addition.—The geometrical process used above is

called 'vector addition,' and is used in compounding any physical quantities that can be represented by, and are subject to the same laws as, vectors. Accelerations, forces acting at a point, rotations about intersecting axes, are treated in this way. In general, the sum of any number of vectors is obtained by placing at the final point of one the initial point of another, and so forming an unclosed irregular polygon ; the vector formed by joining the initial point of the first to the final point of the last is the required sum, the result being independent of the order in which the component vectors are taken. Thus, the sum of the vectors $o\,b$, $b\,c$, $c\,d$, $d\,e$, and $e\,f$ (fig. 7) is the vector $o\,f$.

22. Velocity of any Point on a Rolling Wheel.—Let a wheel roll along the ground, its centre having the velocity v. The wheel as a whole partakes of this velocity, which may be

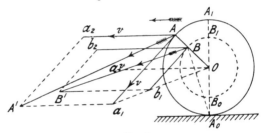

FIG. 8.

represented by the line $O\,a$ (fig. 8). The relative motion of the wheel and ground will be the same if we consider the centre of the wheel fixed and the ground to move backwards with velocity v. In this way it is seen that the linear speed of any point on the rim of the wheel relative to the frame is equal to v. We can now find the velocity, relative to the earth, of any point A on the rim of the wheel. The point A is subjected to the horizontal forward velocity $A\,a_2$ with speed v, and to the velocity with speed v, in a direction $A\,a_1$ at right angles to $O\,A$, due to the rotation of A round O. The resultant velocity is obtained by completing the parallelogram $A\,a_1\,A^1\,a_2$. The diagonal $A\,A^1$ represents the velocity of A in magnitude and direction. If the point on the rim be taken at A_0, the point of contact with the ground, it will be seen that the parallelogram in the above construction reduces to two coincident straight lines. In this case, however, it is easily seen that the velocity of A, due to the rotation of the wheel, is backwards, and, therefore, when added to the forward velocity due to the translation of the wheel, the resultant velocity is zero. On the other hand, if the point be taken at A_1, the top of the wheel, the velocity due to rotation is in the forward direction. Thus, the velocity of the uppermost point of the wheel is $2\,v$—that is, twice the velocity of the centre.

In the same way the velocity of any point B on one of the spokes may be found. Join $O\,a_1$ and draw $B\,b_1$ parallel to $A\,a_1$, meeting $O\,a_1$ at b_1. The velocity of B, due to rotation, is represented by $B\,b_1$. Draw $B\,b_2$ equal and parallel to $A\,a_2$, and complete the parallelogram $B\,b_2\,B^1\,b_1$. The velocity of B is represented by $B\,B^1$. It will be found that the velocity of B is greatest when passing its topmost position B_1, and least when passing its lowest position B_0.

The above problem can be dealt with by another method. The motion of the wheel has been compounded of two motions, the linear motion of the bicycle and the motion of rotation of the wheel about its axis. But the resultant motion of the wheel— that is, its motion relative to the ground—can be more simply expressed. If the wheel rolls on the ground without slipping, its point of contact A_0 is, at the moment in question, at rest. The linear velocity of the wheel's centre O is evidently the same as that of the bicycle v, and is in a horizontal direction. The centre of the wheel, therefore, may be considered to rotate about the point A_0. But as the wheel is a rigid structure, every point on it must be rotating about the same centre. The point A_0 is called the *instantaneous centre of rotation* of the wheel. The linear velocity of any point on the wheel is, by (3) (chap. ii.), equal to $\omega\,r$, where r is the distance of the point from the centre of rotation A_0. But ω is equal to $\dfrac{v}{r_0}$, where r_0 is the radius of the wheel ; therefore, the linear velocity of any point B on the wheel is equal to $\dfrac{v}{r_0}.A_0B$, and is in the direction $B\,B^1$ at right angles to $A_0\,B$.

The centre of rotation A_0 of the wheel is called an *instan-*

t͞ineous centre of rotation, as distinguished from a *fixed* centre of rotation, since when the wheel is rolled through any distance however small, its point of contact with the ground, and therefore its centre of rotation, is changed.

23. **Resolution of Velocities** is the converse of the addition of velocities, and has for its object the finding of components in two given directions, whose resultant motion shall be equal to the given motion. If *o c* (fig. 6) be the given velocity, *o b* and *o a* the directions of the required components, the latter are found by drawing from *c* straight lines, *c b* and *c a*, parallel respectively to *o a* and *o b*, cutting them at *b* and *a* : *o b* and *o a* represent the required components.

Example.—Suppose a cyclist to ride up an incline of one in ten at the rate of ten miles an hour. To find at what rate he

rises vertically, draw a horizontal line *A B* (fig. 9) ten inches long, and a vertical line *B C* one inch long ; join *A C*.

FIG. 9.

Along this line to any convenient scale mark off *A D*, the velocity ten miles an hour (14⅔ feet per second). Draw *D E* at right angles to *A B*, meeting *A B*, produced, if necessary, at *E*. *D E* is the required vertical velocity of the cyclist. By measurement this is found to be 1·46 feet per second (less than 1 mile per hour).

Example.—A cyclist is riding along the road with a velocity indicated in direction and magnitude by *O A* (fig. 10). The wind is blowing with velocity *O B*, and is therefore partially in the direction

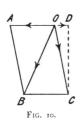

FIG. 10.

in which the cyclist is riding. To find the apparent direction of the wind, that is, its direction relative to the moving bicycle, join *A B* and draw *O C* equal and parallel to *A B* ; *O C* will be the velocity of the wind apparent to the cyclist, which is thus apparently blowing partially against him. The velocity *O C* can be resolved into two, *O D* dead against the cyclist, and *D C* sideways, *C D* being drawn at right angles to *A O*. For, from the parallelogram of velocities it is seen that the actual velocity, *O B*, of the wind relative to the earth is compounded of its velocity relative to the bicycle $O C$, and the velocity of the bicycle, $O A$, relative to the earth.

The above figure may explain why cyclists seldom seem to feel a back wind, while head winds seem always to be present.

24. **Addition and Resolution of Accelerations.**—An acceleration involves the idea of magnitude and direction, but not position; it may, therefore, be represented by a vector. Figs. 6 and 7 are, therefore, directly applicable to the compounding and resolving of accelerations.

25. **Hodograph** —If a body move in any path, its velocity at any instant, both as to direction and magnitude, can be conveniently represented by a vector drawn from a fixed point; the curve formed by the ends of such vector is called the *hodograph* of the motion.

26. **Uniform Circular Motion.**—The hodograph for uniform circular motion can easily be found as follows:—When the body is at A (fig. 11), its velocity is in the direction $A A^1$. From a fixed point o (fig. 12) set off $o a$ equal and parallel to $A A^1$. When the body is at B its velocity is represented by $B B^1$, equal in length to $A A^1$; the corresponding line $o b$ on the hodograph (fig. 12) is equal and parallel to $B B^1$. Repeating this construction for a number of positions of the moving body, it is seen that the hodograph $a b c$ is a circle.

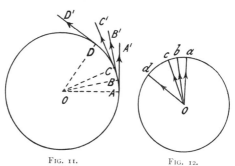

FIG. 11.   FIG. 12.

Since the direction of motion changes from instant to instant, the moving body must be subjected to an acceleration, which can be determined as follows:—When the body is at A, its velocity is represented by $o a$, and when at B by $o b$; therefore, in the interval of passing from A to B an additional velocity, represented by $a b$, has been impressed on it. If the point B be taken very close to A, *i.e.* if a very short interval of time be taken, b will be very close to a, and therefore $a\,b$, the direction of the impressed velocity, will be parallel to $A\,O$, *i.e.* directed towards the centre of rotation.   If the interval of time is taken suffi- ciently small, the additional velocity $a\,b$ is also very small, and the resultant velocity $o\,b$ does not sensibly differ in magnitude from $o\,a$; thus the only effect of the additional velocity is to change the direction of motion from $o\,a$ to $o\,b$ (fig. 12).

When at B suppose the body to undergo the same operation, at the end of it the direction of the motion will be $o\,c$.   After a number of such operations the body will be at D (fig. 11), and its velocity will be represented by $o\,d$ (fig. 12).   The total additional velocity imparted to it between the positions A and D has only had the effect of changing the direction, but not the magnitude of its velocity.   This total additional velocity is represented by the arc $a\,d$.

Now, suppose the body to take one second to pass from A to D, then $a\,d$ represents the increase of velocity in unit of time, and is, therefore, numerically equal to the acceleration a.   Let v be the linear speed of the body, and r the radius of the circle in which it moves; then the arc $A\,D$ is numerically equal to v, $o\,a$ is by definition equal to v, and since $o\,a$ and $o\,d$ are respec- tively parallel to the tangents at A and D, the angle $a\,o\,d$ is equal to the angle $A\,O\,D$; therefore,

$$\frac{a}{v} = \frac{\text{arc } a\,d}{\text{radius } a\,o} = \frac{\text{arc } A\,D}{\text{radius } A\,O} = \frac{v}{r}.$$

$$i.e. \qquad a = \frac{v^2}{r} \quad . \quad . \quad . \quad . \quad . \quad . \quad (1)$$

That is, in uniform circular motion, the radial acceleration is proportional to the square of the speed, and inversely propor- tional to the radius.

CHAPTER IV

27. **Definitions of Plane Motion**.—If a body move in such a manner that each point of it remains always in the same plane, it is said to have *plane motion*. Plane motion can be perfectly represented on a flat sheet of paper ; and, fortunately for the engineer, most moving parts of machines have only plane motion. In cycling mechanics there are more examples of motion in three dimensions. The motion of the wheels as the machine is moving in a curve and the motion of a ball in its bearing are examples of non-plane motion.

Each particle of a body having plane motion will describe a plane curve, which is called a *point-path*.

28. **General Plane Motion of a Rigid Body**.—The plane motion of a rigid body may be—

(1) *Simple translation*, without rotation. In this case any straight line drawn on the rigid body always remains in the same direction. The motion of the body will be completely determined if that one point of it is known.

(2) *Rotation about a fixed axis.*—In this case the path of any point is a circle of radius equal to the distance of the point from the axis of rotation.

(3) *Translation combined with a motion of rotation.*—We shall see later that in this case it is possible to represent the motion *at any instant* by a rotation of the body about an axis perpendicular to the plane of motion, the position of the axis, however, changing from instant to instant.

If the paths of two points of a rigid body be known, the path of any other point on the rigid body is determined. Let A, B, and C (fig. 13) be three points rigidly connected, A moving on the curve $a\,a$, B on the curve $b\,b$. The path of the point C can evidently be found as follows :—Let A_1 be any position of the

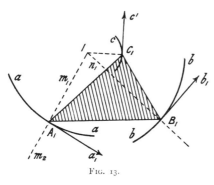

FIG. 13.

point on the curve $a\,a$; the corresponding position B_1 is found by drawing an arc with centre A_1 and radius $A\,B$, cutting the curve $b\,b$ in B_1. With centres A_1 and B_1, and radii $A\,C$ and $B\,C$ respectively, draw two arcs intersecting at C_1. C will be a point on the path described by C.

29. **Instantaneous Centre.**—Let A and B (fig. 13) be two points of a rigid body, $a\,a$ and $b\,b$ their respective point-paths. In the position shown the direction of the motion of A is a tangent at the point A_1 to the curve $a\,a$. The point A may therefore be considered to rotate about any point, m, lying on the normal at A_1 to the curve $a\,a$. For, if A be considered to rotate either about m_1 or m_2, the direction of the motion at the instant is in either case the same tangent, $A_1\,a_1$. In the same way, since the tangent $B_1\,b_1$ is also the tangent to any circle through B having its centre on the normal $B_1\,n_1$, the point B may be considered to rotate about any point in the normal at B_1 to the curve $b\,b$. If the normals $A_1\,m_1$ and $B_1\,n_1$ intersect at I, A and B may be both considered to be rotating at the instant about the centre I. No other point in the plane satisfies this condition, I is therefore called the *instantaneous centre of rotation* of the rigid body. Every point on the rigid body is at the instant rotating about the centre I, therefore the tangent at C_1 to the point-path $c\,c$ is at right angles to $C_1\,I$.

30. **Point-paths, Cycloidal Curves.**—A few point-paths described in simple mechanisms are of great importance in mechanics. We will briefly notice the most important.

Cycloid.—If a circle roll, without slipping, along a straight line, the curve described by a point on its circumference is called a *cycloid*. Let a circle roll along the straight line XX (fig. 14). The curve described by the point C on its circumference can be readily drawn as follows :—Divide the circumference of the circle into a number of equal parts (twenty-four will be convenient, as this division can be effected by the use of the 45° and 60° set squares), and number the divisions as shown. Through the centre draw a straight line parallel to XX; this will be the path of the centre of the circle. Along this line mark off a number of divisions, each equal in length to those on the circumference of the circle, and number them correspondingly. When any point,

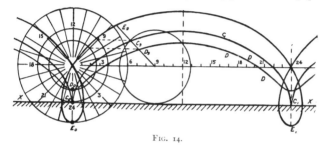

FIG. 14.

say 9, on the circumference of the circle is rolled into contact with the line XX, the centre of the circle will be on the corresponding point, 9, of the straight line. Draw the circle in this position. The corresponding position C_9 of C is evidently obtained by projecting over from the point 9 of the circumference. By repeating this process for each of the points of division, twenty four points on the cycloid will be obtained ; through these a fair curve may be drawn freehand. The curve $C_0 C C_1$ shows one portion of the cycloid. The point-path is a repetition, time after time, of this curve.

Prolate and Curtate Cycloid.—The path described by a point, D, inside the rolling circle is called a *prolate* cycloid. $D_0 D D_1$ shows one complete portion of the curve. The method of drawing it is exhibited in figure 14, and hardly requires any further explanation.

The curve described by a point lying outside the rolling circle is called a *curtate* cycloid. $E_0 E E_1$ (fig. 14) shows one complete portion.

A point on the circumference of a bicycle wheel describes a cycloid as the machine moves in a straight line. Any point on the spokes, or any point on the crank, describes a prolate cycloid.

Epicycloid and Hypocycloid.—If one circle roll on the circumference of another, the curve described by a point on the circumference of the rolling circle is called an *epicycloid* or a *hypocycloid*, according as the rolling circle lies outside or inside the fixed circle. These curves are of great importance in the theory of toothed-wheels.

In figure 15, *E E* is an epicycloid and *H H* a hypocycloid, in each of which the diameter of the rolling circle is one-third that of

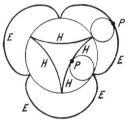

the fixed circle. The method of drawing these curves is similar to that of drawing the cycloid, the only difference being that the divisions along the path of the centre of the rolling circle will not be equal to those along the circumference of the rolling circle, but the divisions along the fixed and rolling circles will correspond.

FIG. 15.

A particular case occurs when the diameter of the rolling circle is equal to the radius of the fixed

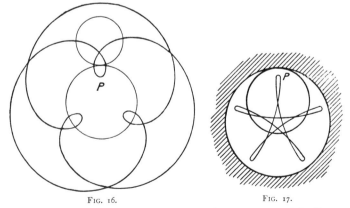

FIG. 16.   FIG. 17.

circle ; the hypocycloid in this case reduces to a straight line, a diameter of the fixed circle.

Epitrochoids and Hypotrochoids.—If the tracing point does not lie on the circumference of the rolling circle, the curve traced is called an *epitrochoid* or a *hypotrochoid*. Figures 16, 17, and 18 show some examples of epitrochoids and hypotrochoids.

Involute.—Let a string be wrapped round a circle and have a pencil attached at some point ; as it is unwound from the circle

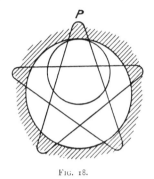

FIG. 18.

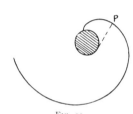

FIG. 19.

the pencil will describe a curve on the paper, called an *involute* (fig. 19).   This curve is also of great importance in the theory of toothed-wheels.

The involute is a particular case of an epicycloid.   If the rolling circle be of infinitely great radius its circumference will become a straight line.   The curve traced out by a point P (fig. 19) of a straight line, which rolls without slipping on a circle, is an involute.

31. Point-paths in Link Mechanisms.—We have already shown how to find the path described by any point of a rigid body of which two point-paths are known.   If the paths $a\,a$ and $b\,b$ (fig. 13) be circular arcs, the bar $A\,B$ may be considered as the coupling link between two cranks.   The variety of curves described by points rigidly connected

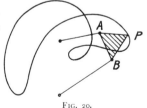

FIG. 20.

to such a coupling link is very great ; some of them have been of great practical use.   Figure 20 shows a point-path described by a tracing point, P, which does not lie on the axis of the link $A B$.

In Singer's 'Xtraordinary' bicycle the motion given to the pedal was such a curve. The mechanism and the path described by the pedal are discussed in chapter xxix.

32. **Speeds in Link Mechanisms.**—If the speed of any point in a mechanism be known, it will in general be possible to determine that of any other point. In a four-link mechanism, $A B C D$ (fig. 21), in which $C D$ is the fixed link, the nature of the motion will depend on the relative length of the links. If $D A$

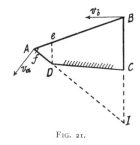

FIG. 21.

be the shortest, $A B + D A$ be less than $C D + B C$, and $A B - D A$ be greater than $C D - B C$, $D A$ will rotate continuously, and $C B$ oscillate. The speeds of points on the lever $C B$ are proportional to their distances from the fixed centre of rotation C; similarly for points on the lever $D A$. Now in any position of the mechanism the link $A B$ may be considered to have a rotation about the instantaneous centre I, the point of intersection of $A D$ and $C B$, produced if necessary; and thus the linear speed of any point of the link is proportional to its distance from I. If the point A rotates with uniform speed, the point B will oscillate in a circular arc with a variable speed. Let v_a be the uniform speed of A, and v_b the corresponding speed of B. Then, since the body $A B$ is rotating at the instant about the centre I,

$$\frac{v_a}{v_b} = \frac{I A}{I B}.$$

Draw $D e$ parallel to $C B$, meeting $A B$, produced if necessary, at e. Then the triangles $A D e$, $A I B$ are similar, and therefore

$$\frac{I A}{I B} = \frac{D A}{D e},$$

and

$$\frac{v_b}{v_a} = \frac{D e}{D A},$$

or

$$v_b = \frac{v_a}{D A} D e \quad . \quad . \quad . \quad . \quad . \quad (1)$$

Now DA is constant whatever be the position of the mechanism, and therefore if v_a, the speed of A, be constant, the speed of the point B is proportional to the intercept De.

Mark off Df along DA equal to De. The length Df is thus proportional to the speed of the point B when the crank DA is in the corresponding position. If this construction be repeated for all positions of the crank DA, the locus of the point f will be the *polar curve* of the speed of the point B.

33. **Speed of Knee-joint when Pedalling a Crank.**—In pedalling a crank-driven cycle, the motion of the leg from the hip

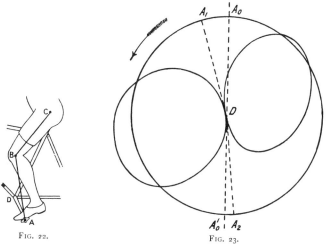

FIG. 22.

FIG. 23.

to the knee is one of oscillation about the hip-joint. If the ankle be kept quite stiff during the motion, as, unfortunately, is too often the case with beginners, the leg from the knee-joint downwards practically constitutes the coupling-link of a four-link mechanism. The pedal-pin (fig. 22) rotating with uniform speed, figure 23 shows the curve of speed of the knee-joint. It may be noticed that the maximum speed of the knee during the up-stroke is less than during the down-stroke. Also, the point B is at the upper end of its path when the pedal-pin is in the position A_1, some considerable distance after the vertical position DA_0 of the crank ; while B is in its lowest position when the pedal pin is at

A_2. The angle $A_1 D A_2$, passed through by the crank during the down-stroke of the knee is less than the angle passed through during the up-stroke ; consequently, since the speed of the pedal-pin is uniform, the average speed of the knee during the down-stroke is less than during the up-stroke. If the rider can just barely reach the pedal when at its lowest point, the speed of the knee-joint is very great immediately before and after coming to rest at the lowest point of its path.

34. **Simple Harmonic Motion.**—If P be a point moving with uniform speed in a circle of radius r of which $a\,b$ is a diameter, and $P\,p$ be a perpendicular let fall on $a\,b$ (fig. 24), while P moves in the circle, the point p will move backwards and forwards along the straight line $a\,b$. The point p is then said to have *simple harmonic motion*. The motion of a point on a

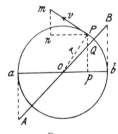

FIG. 24.

vibrating string, and of a particle of air in an organ-pipe when the simplest possible, is of this character. The speed of p will vary with its varying position. At any instant the velocity of the point P is in the direction $P\,m$, the tangent at P. Setting off $v = P\,m$ to scale along this line it may be resolved into two components $P\,n$ and $n\,m$ respectively parallel, and at right angles, to $a\,b$. The parallel component $P\,n$ is, of course, equal to the speed of the point p. If the scale of v be taken such that $P\,m$ is equal to r, the triangles $P\,m\,n$ and $P\,o\,p$ are equal, and therefore $P\,p$ is equal to $P\,n$. That is, in any position of the point p moving along $a\,b$ with simple harmonic motion, its speed may be represented by the ordinate $p\,P$ to the circle on $a\,b$ as diameter.

If P moves uniformly in the circle, its acceleration is constant in magnitude and equal to $\dfrac{v^2}{r}$, and is in the direction of the radius $P\,o$. The scale of acceleration may be chosen such that the vector $P\,o$ represents the acceleration of P, which may be decomposed into $P\,p$ and $p\,o$ respectively at right angles, and parallel to, $a\,b$. The parallel component $p\,o$ is, of course, equal to the acceleration of the point p along $a\,b$—that is, in simple harmonic motion the acceleration is proportional to the distance of the moving point from the centre of its motion. If an ordinate $p\,Q$ be set off equal to $o\,p$, the locus of Q will be the acceleration diagram of the motion ; this locus is a straight line $A\,B$ passing through o, the centre of the motion.

The motion of the knee-joint when pedalling approximates to simple harmonic motion, the approximation being closer the shorter the crank $D\,A$ (fig. 22) is in comparison with the lever $C\,B$ and connecting-link $A\,B$. If the motion were exactly simple harmonic motion, the polar curve of speed of knee-joint (fig. 23) would consist of two circles passing through D.

35. **Resultant Plane Motion.**—*Resultant of Two Transla-tions.*—If a rigid body be subjected to two motions of translation simultaneously, the resultant motion will evidently be a motion of simple translation, which can be found by an application of the parallelogram of velocities.

Resultant of Two Rotations about Parallel Axis.—Let a body be subjected to two rotations, ω_1 and ω_2, about the axes A and B

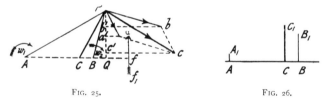

FIG. 25.   FIG. 26.

(fig. 25). If the motion be plane, the axes must be parallel, and at right angles to the plane of the motion. Let $\varGamma$ be any point in the body. Join P to A and B, and draw $P\,a$ and $P\,b$ at right angles to $P\,A$ and $P\,B$ respectively. The resultant linear velocity of P will be the resultant of the velocity $\omega_1 \times A\ P$ in the direction $P\,a$, and of $\omega_2 \times \overline{B\ P}$ in the direction $P\,b$. If $P\,a$ and $P\,b$ be marked off respectively equal to these velocities to any convenient scale, the resultant $P\,c$ can be found by the parallelogram of velocities.

From P draw a perpendicular $P\,Q$ on the line, produced if necessary, joining the centres A and B. Draw $a\,a^1$ and $b\,b^1$ per-pendicular to $P\,Q$. Then, the velocity of P due to the rotation ω_1 about A may be resolved into the velocity $a^1\,a$ parallel to, and the velocity $P\,a^1$ at right angles to, $A\,B$. Similarly, the velocity of P

due to the rotation ω_2 about B may be resolved into the two components $P\,b^1$ and $b^1\,b$. The triangles $A\,Q\,P$ and $P\,a^1\,a$ are similar; so, also, are the triangles $B\,Q\,P$ and $P\,b^1\,b$. It is, therefore, easy to show that the components of P's velocity due to ω_1, at right angles, and parallel, to $A\,B$, are respectively $(\omega_1 \times A\,Q)$ and $(\omega_1 \times Q\,P)$. Similarly, the components due to ω_2 are $(\omega_2 \times B\,Q)$ and $(\omega_2 \times Q\,P)$. Therefore, the components of P's resultant velocity at right angles, and parallel, to $A\,B$ are respectively : —

$$v_1 = (\omega_1 \times A\,Q) + (\omega_2 \times B\,Q) \quad . \quad . \quad . \quad . \quad (2)$$

and

$$v_2 = (\omega_2 + \omega_1)\,P\,Q \quad . \quad . \quad . \quad . \quad . \quad . \quad (3)$$

Let C be a point on the straight line $A\,B$, dividing it in the inverse ratio of the angular speeds ω_1 and ω_2, then

$$\frac{A\,C}{C\,B} = \frac{\omega_2}{\omega_1}$$

and

$$A\,C = \frac{\omega_2}{\omega_1 + \omega_2} A\,B, \quad C\,B = \frac{\omega_1}{\omega_1 + \omega_2} A\,B$$

Substituting $A\,Q = A\,C + C\,Q$, and $B\,Q = C\,Q - C\,B$ in (2), it is easily deduced that

$$v_1 = (\omega_1 + \omega_2)\,C\,Q \quad . \quad . \quad . \quad . \quad . \quad (4)$$

From (3) and (4) it is evident that the resultant velocity of P is $(\omega_1 + \omega_2)\,C\,P$. That is, any point P, and therefore the whole body, is rotating with angular speed equal to the sum of the component angular speeds, about a parallel axis in the same plane, and distant from the axis of the component rotations inversely as the component angular speeds.

The above result can be more simply attained by an application of the principle of 'addition of vectors.' Let ρ be the vector $A\,P$, from the axis A to any point P of the rotating body, and let a be the vector $A\,B$. Then $P\,a$ is a vector of magnitude $\omega_1\,\rho$, at right angles to ρ; $B\,P$ is the vector $(\rho - a)$; and $P\,b$ is a vector of magnitude $\omega_2(\rho - a)$, at right angles to $(\rho - a)$.

Vector Pc = vector Pa + vector Pb

$$= \omega_1 \rho + \omega_2 (\rho - a)$$
$$= (\omega_1 + \omega_2) \rho - \omega_2 a$$
$$= (\omega_1 + \omega_2) \left(\rho - \frac{\omega_2}{\omega_1 + \omega_2} a \right)$$
$$= (\omega_1 + \omega_2) (\rho - AC)$$
$$= (\omega_1 + \omega_2) \, CP, \text{ and at right angles to } CP.$$

That is, any point P rotates about the axis C (where $AC : CB = \omega_2 : \omega_1$) with angular speed equal to the sum of the component angular speeds.

Let figure 26 be a view of the body taken in a direction at right angles to that of figure 25, AB now representing the plane of the motion. The rotation ω_1 may be represented by a line AA_1 at right angles to AB, its length representing, to some scale, the magnitude of the rotation ω_1. In the same way BB_1 may represent the rotation ω_2. The resultant rotation, CC, is equal to the sum of the rotations ω_1 and ω_2, and takes place about an axis whose distances from A and B are inversely proportional to the rotations ω_1 and ω_2.

Thus, rotations about parallel axes can be represented in the same way as parallel forces, and their resultant found by the methods used to find the resultant of parallel forces (*see* chapter vi.).

Example.—Find the instantaneous centre of rotation of the crank of a front-driver geared two to one. Let n be the number of revolutions the cranks make in a minute, the wheel makes $2n$ revolutions, and the crank must make n revolutions backward relative to the wheel—*i.e.* makes $-n$ revolutions per minute. The crank's motion may be considered as the resultant of a rotation $2n$ about B, the point of contact of the wheel with the ground, and a rotation $-n$ about the wheel centre A (fig. 27). Applying the preceding results, the instantaneous centre is on the line AB, and

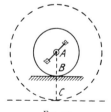

FIG. 27.

$$\frac{A\,C}{C\,B} = \frac{2\,n}{-\,n} = -\,2.$$

That is, $A\,C = -\,2\,C\,B$

or $A\,C = 2\,B\,C = 2\,A\,B.$

The motion of the cranks relative to the ground is, therefore, the same as if they were fixed to a wheel twice the size of the driving-wheel, and running on a flat surface below the ground.

Translation and Rotation.—Let a body be subjected to a rotation ω_1 about an axis A (fig. 25), and to a translation with velocity v in a direction f_1f in the same plane as that of the motion. From A draw Af at right angles to ff_1. Let P be any point on the body. From P draw $P\,Q$ at right angles to Af. Then proceeding as before, the components of $P's$ resultant velocity at right angles and parallel to Af are respectively

$$v_1 = (\omega_1 \times A\,Q) - v \quad . \quad . \quad . \quad . \quad . \quad . \quad . \quad (5)$$

$$v_2 = \omega_1 \times Q\,P \quad . \quad . \quad . \quad . \quad . \quad . \quad . \quad . \quad . \quad (6)$$

Let C be a point on Af such that $(\omega_1 \times A\,C) = v$; then (5) becomes

$$v_1 = \omega_1 \times (A\,Q - A\,C) = \omega_1 \times C\,Q \quad . \quad . \quad . \quad (7)$$

By comparing (6) and (7), it is evident that the resultant velocity of P is one of rotation about the centre C with angular velocity ω_1. Thus, the resultant of a rotation and a translation is a rotation of the same magnitude about a parallel axis, the plane of the two axes being at right angles to the direction of translation.

Example.—A cycle wheel, relative to the frame, has a motion of rotation about the axle ; the frame, and therefore the axle, has a motion of translation. The instantaneous motion of the wheel is the resultant of these two motions. The resultant axis of rotation of the wheel is the point of contact with the ground.

36. **Simple Cases of Relative Motion of Two Bodies in Contact.**—In the theory of bearings it is important to know the relative motion of the portions of two bodies in the immediate neighbourhood of the point of contact, the motion of the bodies

being such that they remain always in contact. Before discussing the general case we will notice a few simpler examples. It will be convenient to consider one of the bodies as fixed, we will then have to speak only of the motion of one of the bodies ; this may be done without in any way altering the relative motion.

Sliding.—If the motion of the body can be expressed as a simple translation, 'sliding' is said to take place at the point of contact. With this definition, pure sliding can only exist continuously when the surface of either the fixed or moving body is cylindrical ; the elements of the surfaces at the point of contact will constitute a 'sliding pair.' An example is afforded by the motion of a pump-plunger in its barrel.

Rolling.—If the instantaneous axis of rotation passes through, and lies in the tangent plane at, the point of contact of the fixed and moving bodies, the motion is said to be 'rolling' ; the rolling is therefore the same as the relative rotation. At the point of contact of a wheel rolling along the ground, the motion is pure rolling. The position of the instantaneous axis continually changes ; but in plane motion it always preserves the same direction.

Spinning.—If the instantaneous axis of rotation passes through, and it is at right angles to the tangent plane at, the point of contact, the motion is similar to the spinning of a top, and may be called *spinning*. An example of pure spinning is found at the centre of a pivot-bearing.

Rubbing.—In a turning pair, the motion can be expressed as a simple rotation about the axis of the pair. For example, the motion of a shaft of radius r in a plain cylindrical bearing is a rotation, ω, about the centre o of the bearing (fig. 28). The motion can also be expressed as an equal rotation, ω, about a parallel axis through P, a point on the surface of the bearing, and a translation with speed

FIG. 28.

$v = \omega r$ in the direction PT at right angles to OP. The motion at P is kinematically more complex than 'sliding,' as above defined, and yet there is nothing of what is commonly understood as *rolling* ; we may give it the name *rubbing*. Thus, rubbing at any point on the surface of contact of a cylindrical shaft of radius,

r is equivalent to a translation v and a rotation $\dfrac{v}{r}$ about an axis, parallel to that of the shaft, passing through the point in question.

More generally, let A and B be two bodies in contact at the point P (fig. 29), let r_a and r_b be their respective radii of curvature at P, and let I be the instantaneous axis of rotation of angular speed ω. I must lie on the common normal at P, since the bodies remain in contact during the motion. Suppose A fixed, and that the same point of the body B rubs along A with speed V for at least two consecutive instants. The motion of B on A may then be said to be pure rubbing. In this case I must

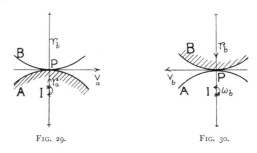

FIG. 29.   FIG. 30.

evidently coincide with the centre of curvature of the body A at the point P; then U_a, the rubbing of B on A, takes place with speed, $V_a = \omega\,r_a$, and is therefore equivalent to a translation V_a and a rotation $\dfrac{V_a}{r_a}$, or

$$U_a \equiv V_a \text{ and } \frac{V_a}{r_a}. \qquad \ldots \ldots \ldots \ldots (8)$$

Similarly, if the position of I be such that the same point of A rubs on B for at least two consecutive instants,

$$U_b \equiv V_b \text{ and } \frac{V_b}{r_b}. \qquad \ldots \ldots \ldots \ldots (9)$$

37. **Combined Rolling and Rubbing.**— In figure 29 let A be fixed, and let the motion of the body B be kinematically a translation $V_a = V$, and a rotation $\omega_a = \omega$ about the point of contact P. The motion at P is compounded of rubbing and

rolling. The rubbing has already been defined ; R_a, the rolling of B on A, will be the total motion less the rubbing, *i.e.*—

$$R_a = (V_a \text{ and } \omega_a) - \left(V_a \text{ and } \frac{V_a}{r_a} \right)$$

$$= \omega_a - \frac{V_a}{r_a} = \omega - \frac{V}{r_a} \quad . \quad . \quad . \quad . \quad . \quad (10)$$

In using the formula (10) the positive directions of the axis of ω_a, of r_a, and of V_a should be taken so that, in the order named, they form a right-handed system of rectangular axes. That is, looking along the positive direction of the axis of ω, a positive rotation, ω, will appear clock-wise, and the positive direction of r if rotated a right angle in the positive direction of ω, will come into the positive direction of V. r_a and r_b may be taken positive for convex surfaces, negative for concave surfaces. The positive directions of ω_a, r_a and V_a are shown in figure 29.

In figure 30 let the relative motion of the bodies be exactly the same as in figure 29, but let B be fixed. Then V_b and ω_b will be oppositely directed in space to V_a and ω_a respectively. But with the above conventions as to positive directions, taking r_b positive, V_b will be positive and equal to V, ω_b will be negative and equal to $-\omega$. Therefore

$$R_b = \omega_b - \frac{V_b}{r_b} = -\omega - \frac{V}{r_b} \quad . \quad . \quad . \quad . \quad . \quad (11)$$

From formulæ (10) and (11) it is seen that when rolling and rubbing combined take place, the 'rollings' of the two bodies are not reciprocal. The actions at the points of contact in the two bodies are not reciprocal, as may be illustrated by a few examples.

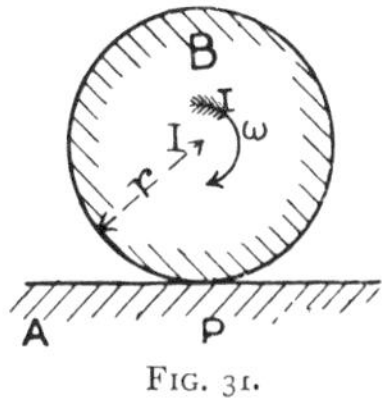

Example I.—Let the bodies A and B be a plane and cylinder of radius r respectively, in contact at P (fig. 31). Let the instantaneous

FIG. 31.

axis of rotation coincide with the axis of the cylinder, and let ω be the angular speed of B relative to A. Then at P:—$r_a = \infty$; $r_b = r$; the speed of rubbing $V = V_a$, $= -\omega r$.

$$R_a = \omega - \frac{V}{r_a} = \omega$$

$$R_b = -\omega - \frac{V}{r_b} = -\omega \frac{-\omega r}{r} = 0.$$

That is, the cylinder's motion on the plane is compounded of a

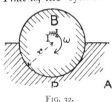

FIG. 32.

rubbing of speed ωr, and a rolling of angular speed ω. The plane's motion on the cylinder is one of pure rubbing with speed ωr.

Example II.—Let the bodies A and B be a circular bearing and shaft respectively, of the same radius r (fig. 32), ω being the angular speed of the shaft. Then at P, $r_a = -r, r_b = r$, $V = V_a = -\omega r$, and

$$R_a = \omega - \frac{V}{r_a} = \omega - \frac{-\omega r}{-r} = 0$$

$$R_b = -\omega - \frac{V}{r_b} = -\omega - \frac{-\omega r}{r} = 0.$$

Thus the definitions given in (10) and (11) of the magnitudes of the rollings of one body on the other are consistent with our usual conceptions in these simple cases.

CHAPTER V

KINEMATICS : MOTION IN THREE DIMENSIONS

38. **Resultant of Translations.**—If a body be subjected to a number of translations in different directions in space, the resultant velocity can be found by finding the resultant of any two of the given translations, which resultant must evidently lie in the same plane as the two given translations. The resultant of a third given translation with the resultant of the first two can again be found by the same method ; and so on for any number of given translations. Thus the resultant of any number of translations in space is a motion of translation.

39. **Resultant of Two Rotations about Intersecting Axes.**— Let the axes $O\,A$ and $O\,B$ of the rotations intersect at the point O (fig. 33). The rotations ω_1 and ω_2 may be represented by the length of the lines $O\,A$ and $O\,B$ respectively, and since rotations are resolved and compounded like forces, the resultant rotation will be represented by the diagonal $O\,C$ of the parallelogram of which $O\,A$ and $O\,B$ are adjacent sides. This proposition

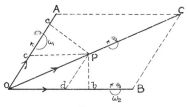

FIG. 33.

is called the *parallelogram of rotations*. In using this proposition, attention must be paid to the sense of the rotation. The lengths of the lines representing the magnitudes of the rotations must be set off along the axes of the rotations in such directions that when looking in the positive directions the motions both appear either in watch-hand direction, or both in contra watch-hand direction. In figure 33, the rotations are both in watch-hand direction ; the resultant rotation about the axis $O\,C$ will therefore be in the direction indicated by the arrow.

The above proposition is so important that a separate proof depending on first principles will be instructive. Let $O\,A$ and $O\,B$ be the axes of rotation, and let P be a point on the body lying in the plane $A\,O\,B$. Draw $P\,a$ and $P\,b$ perpendicular to $O\,A$ and $O\,B$ respectively. If P lie in the angle between the positive directions $O\,A$ and $O\,B$, the linear velocity of P, which is in a direction at right angles to the plane of the axes, will be

$$\omega_1\,\overline{a\,P} - \omega_2\,\overline{b\,P} \quad . \quad . \quad . \quad . \quad . \quad . \quad (1)$$

If P lie on the axis of resultant rotation its velocity is zero, and (3) becomes $\omega_1\,\overline{a\,P} - \omega_2\,\overline{b\,P} = O,$

or,
$$\frac{\omega_1}{\omega_2} = \frac{b\,P}{a\,P}.$$

Draw $P\,c$ and $P\,d$ parallel respectively to $O\,B$ and $O\,A$, meeting $O\,A$ and $O\,B$ at c and d respectively. Then, the triangles $P\,a\,c$ and $P\,b\,d$ are similar, and therefore—

$$\frac{b\,P}{a\,P} = \frac{P\,d}{P\,c} = \frac{O\,c}{O\,d} \quad . \quad . \quad . \quad . \quad . \quad (2)$$

That is, $O\,P$ is the diagonal of a parallelogram whose adjacent sides coincide with the direction of the axis of rotation, and are of lengths respectively proportional to the component angular velocities about these axes.

40. **Resultant of Two Rotations about Non-intersecting Axes.**—Let $A\,A$ and $B\,B$ (fig. 34) be the two axes, and let $g\,h$ be the common perpendicular to $A\,A$ and $B\,B$. Through h draw a line $C\,C$ parallel to $A\,A$. Then by section 35, the rotation ω_1 about

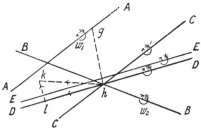

FIG. 34.

the axis $A\,A$ is equivalent to an equal rotation about the axis $C\,C$, together with a translation in the direction *h k* at right angles to the plane containing *A A* and *C C*. The resultant of the rotations about the axes *B B* and *C C* is, by section 39, a rotation about an axis *D D* passing through *h*. Thus, the given motion is equivalent to a rotation about an axis *D D*, and a translation in the direction *h k*. The translation in the direction *h k* may be resolved into two components, *h l* along *D D* and *l k* at right angles to *D D*. By section 35, the rotation about *D D* and the translation in the direction *l k* are equivalent to an equal rotation about a parallel axis *E E*. Thus, finally, the resultant motion is a rotation about an axis *E E* and a translation in the direction of that axis. Such a motion is called a *screw* motion.

41. **Most General Motion of a Rigid Body.**—In the same way it can be shown that the resultant of any number of translations and of any number of rotations about intersecting or non-intersecting axes may be reduced to a rotation about an axis and a translation in the direction of that axis. If a common screw bolt be fixed and its nut be moved, the motion imparted is of this character. The motion of the nut can be specified by giving the *pitch* of the screw and its angular speed of rotation about its axis. In the same way, the motion of a rigid body at any instant can be expressed by specifying the axis and pitch of its screw, and its angular speed.

42. **Most General Motion of Two Bodies in Contact.**—We have seen that the most general motion of a rigid body can be resolved into a rotation *ω* and a translation *v* in the direction of the axis of rotation. Also that a rotation about any axis is equivalent to an equal rotation about a parallel axis through any point, together with a translation at right angles to the plane of the parallel axes. Hence, if two bodies move in contact, the relative motion at any point of contact can be resolved into a translation, and a rotation about an axis passing through the point of contact. The direction of the translation must be in the tangent plane at the point ; since, if the two bodies move in contact, there can be no component of the translation in the direction of the normal.

Let figure 35 be a section of the two bodies *A* and *B* by a plane, passing through the point of contact *P*, at right angles to the instantaneous axis of rotation II. The body A may be considered fixed, the body B to have a rotation ω round, and a translation v along, II. If PI be perpendicular to II, the motion of B is equivalent to a rotation ω about the axis Pa, parallel to II, together with a translation $\omega \cdot \overline{IP}$ along Pb at right angles to the plane of PI and Pa, plus a translation v along Pa. The resultant of these two translations is a translation

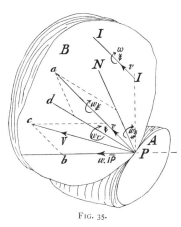

FIG. 35.

V along Pc. Pc must lie in the common tangent plane to the surfaces at P.

Let PN be the normal at P, and Pd the intersection of the tangent plane with the plane containing PN and Pa. Then, the rotation ω about Pa can be resolved into rotations ω_s and ω_r about PN and Pd respectively. Thus, the motion at P consists of translation with velocity V in the direction Pc, a spinning, ω_s, about the normal PN, and a rolling, ω_r, about the

axis Pd lying in the tangent plane. Therefore the most general relative motion of two bodies in contact is compounded of 'rubbing,' 'rolling,' and 'spinning.'

We have in the chapter on Plane Motion given examples of the pure motions just mentioned. We shall see, in the chapter on Bearings, that the motion of a ball on its path in the ordinary form of adjustable bearing is compounded of rolling and spinning ; while, in some special ball-bearings, the motion at the point of contact of a ball with its path is compounded of rubbing, rolling, and spinning.

CHAPTER VI

43. **Graphic Representation of Force.**—For the complete specification of a force acting on a body, its direction, line of application, and magnitude are required. A force can therefore be represented completely by a straight line drawn on a diagram, the length of the line representing to scale its magnitude, the direction and position of the line giving the direction and positions of application of the force. Thus a force can be represented by a *localised* vector.

44. **Parallelogram of Forces.**—When two or more forces are applied at the same point, a single force can be found which is equivalent to the original forces. This is called the *resultant* force, and the original forces are called the components. If the forces act in the same direction, the resultant is, of course, equal to the sum of the component forces. If two forces act in opposite directions, the resultant is the difference of the two. Generally, if a number of forces act along a straight line, some in one direction, others in the opposite direction, the resultant of the whole system is equal to the difference between the sum of the forces acting in one direction and that of the forces acting in the opposite direction.

Suppose two forces acting at a point in different directions are represented by *o a* and *o b* respectively (fig. 36), then it is evident that some force such as *o c* in a direction between *o a* and *o b* will be the resultant. The resultant *o c* is found by completing the parallelogram *o a c b* and drawing the diagonal *a c*, exactly as in the case of the parallelogram of velocities.

Want of space prevents a strict elementary mathematical proof of this proposition, but it can be easily verified experimentally as follows : Fasten two pulleys, *A* and *B* (fig. 37), on a wall, the pulleys turning with as little friction as possible on their spindles. Take three cords jointed together at *O* with weights W_1, W_2, W_3 at their ends. Let the heaviest weight hang vertically downwards from *O*, and let the other two cords be passed over the pulleys *A* and *B* respectively. Then, if the heaviest weight, W_3, underneath *O* be less than the sum of the other two, the whole system will come to rest in some particular

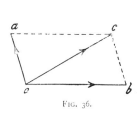

FIG. 36.

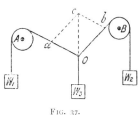

FIG. 37.

position. While in this position make a drawing on the wall of the three cords meeting at *o*. Produce the vertical cord upwards to any point *c*, and from *c* draw parallels *c a* and *c b* to the other two cords. It will be found on measurement that the lengths *O a*, *O b*, and *O c* are exactly proportional to the weights W_1, W_2, and W_3. Thus the resultant of the forces along *O a* and *O b* is given by the diagonal *O c* of the parallelogram whose sides represent the component forces.

Example.—The crank spindle of a bicycle is pressed vertically downwards by the rider with a force of 25 lbs., while the horizontal

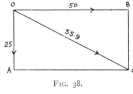

FIG. 38.

pull of the chain is 50 lbs. What is the magnitude and direction of the resultant pressure on the bearing? Set off *O A* (fig. 38) vertically downwards equal to 25 lbs. and *O B* horizontally equal to 50 lbs. Complete the parallelogram *O A B C*. The resultant is equal in magnitude and direction to the diagonal *O C*, which by measurement is found to be 55·9 lbs.

45. **Triangle of Forces.**—Suppose that in addition to the two forces *o a* and *o b* (fig. 36) a third force, *c o*, acts at the point ;

this third force being exactly equal, but opposite to, the resultant of the two forces. If these three forces act simultaneously no effect will be produced, and the body will remain at rest. *b c* is equal and parallel to *o a*, and may therefore represent in magnitude and direction the force *o a* acting at *A*. The three sides *o b*, *b c*, and *c o* of the triangle *o b c*, therefore, taken in order, represent the three forces acting at the point and producing equilibrium. The proposition of the parallelogram of forces may therefore be put in the following form, which is often convenient :

If three forces act at a point and produce equilibrium they can be represented in magnitude and direction by the three sides of a triangle taken in order round the triangle. The converse of this proposition is also true.

A very important proposition which can be deduced immediately from the triangle of forces is, that if three forces act on a body and produce equilibrium they must all act through the same point.

46. **Polygon of Forces.**—Since forces acting at a point can be represented by vectors, the resultant *R* of a number of forces,

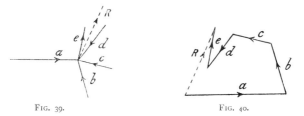

FIG. 39.　　　　　　　　　　FIG. 40.

a, *b*, *c*, *d*, and *e*, acting at the same point (fig. 39) can be found by drawing a vector polygon (fig. 40) whose sides represent the given forces ; the resultant vector *R* represents the resultant force. If a force equal, but oppositely directed, to *R* acted at the same point as the forces *a*, *b*, *c*, *d*, and *e*, they would be in equilibrium. Therefore, if a number of forces acting at a point are in equilibrium, they can be represented in magnitude and direction by the sides of a polygon, taken in order round the polygon. Conversely, if a number of forces acting at a point are represented in magnitude and direction by the sides of a polygon taken in order, they are in equilibrium.

In the preceding paragraph it must be clearly understood that the sides of the polygon represent the forces in magnitude and direction, but not in *position*. Thus the sides of the polygon *a*, *b*, *c*, *d*, *e* (fig. 40) represent in magnitude and direction the five forces acting at the same point. If a body were acted on by forces represented by the sides of a polygon, in *position* as well as in magnitude and direction, a turning motion would evidently be imparted to it.

47. **Resultant of any Number of Co-planar Forces.**—The resultant of any number of forces all lying in the same plane acting on a rigid body, and which do not necessarily all act at the same point, may be found by repeated applications of the principle of the parallelogram of forces. The resultant R_2 of any two of the given forces P_1 and P_2 passes through the point of intersection of the latter ; the resultant R_3 of R_2 and a third force, P_3, passes through the point of intersection of R_2 and P_3 ; and so on. This process is very tedious when a great number of forces have to be dealt with. The following method is more convenient :

Let figure 41 represent the position of the given forces, and figure 42 the corresponding force-polygon $P_1 P_2$. . . . The resultant R of all the given forces is evidently represented in magnitude and direction by the line $a f$ forming the closing side of the polygon ; for if a force of magnitude and direction $f a$ were added to the given forces, the resultant would be of zero magnitude. It only remains therefore to determine the *position* of the resultant R on figure 41.

No difference will be made if two equal and opposite forces be added to the system. We will add a force Q, represented by $O a$ in the force-polygon, which acts along any line a (fig. 41). The resultant of Q and P_1 is $O b$ (fig. 42), and it passes through p_1, the point of intersection of Q and P_1 (fig. 41). Draw from the point p_1 the line b parallel to $O b$ (fig. 42), cutting the line of action of P_2 at p_2. The resultant of Q, P_1, and P_2 is $O c$ (fig. 42), and it passes through p_2. Draw from the point p_2 the line c parallel to $O c$ (fig. 42). Continuing this process, the resultant of Q, P_1, P_2, P_3, P_4, and P_5 is Of (fig. 42), and acts through the point p_5. From p_5 draw the line f parallel to Of (fig. 42),

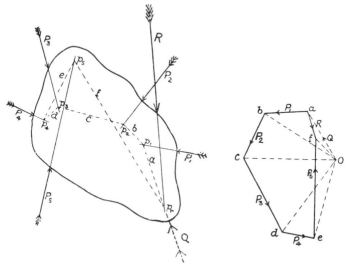

FIG. 41.   FIG. 42.

cutting the line a, the line of action of the added force Q, at p_r. The resultant of Of and $-Q$ is $af = R$ (fig. 42), and it acts through the point p_r.

The above construction may be expressed thus : Take any pole O and from it draw radius vectors to the corners of the force-polygon. Draw another polygon, which may be called the *link-polygon*, having its corners p_1, p_2 . . . on the lines of action of the given forces P_1, P_2, . . . and having its sides a, b, . . . parallel to the radius vectors Oa, Ob . . . of the force-polygon ; the sequence of sides and corners a, p_1, b, p_2 . . . in the link-polygon being the same as that of the corners and sides a, P_1, b, P_2, . . . of the force-polygon. The point of inter-section of the first and last sides of the link-polygon determines the position of the resultant R.

It is readily seen from the above, that if a system of forces acting on a rigid body are in equilibrium, both the force- and link-polygons must be closed.

48. **Resolution of Forces.**— A single force may be resolved into two components in given lines which intersect on the line of action of the given force. The principle of the parallelogram of

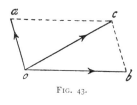

FIG. 43.

forces is, of course, used again here. Let *o c* (fig. 43) be the given force acting at *o*, and let its components in the directions *o a* and *o b* be required. From *c* draw *c a* and *c b* respectively parallel to *b o* and *a o*, meeting *o a* and *o b* in *a* and *b* respectively : *o a* and *o b* are the required components of the given force in the two given directions.

Example.—Given the vertical pressure on the hub of the driving-wheel of a Safety bicycle, to find the forces acting along the top and bottom forks, *O A* and *O B* (fig. 44).

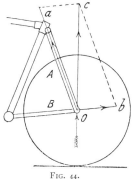

FIG. 44.

Draw *O c* vertical and equal to the given pressure on the hub. This is the direction and magnitude of the force with which the wheel presses on the hub spindle. From *c* draw *c a* and *c b* parallel to *O B* and *A O* respectively, meeting *O A* and *B O* produced in *a* and *b* respectively. *o a* and *o b* are the forces acting along the top and bottom forks respectively. It will be seen that the top fork *O A* is compressed and the bottom fork *O B* is in tension.

Resolution of a Force into Three Components in given Directions and Positions.—Let *R* be a force whose components acting along the given lines P_1, P_2, and P_3 (fig. 45) are required. Let *R* and P_1 intersect at *A*, P_2 and P_3 intersect at *B*. Then *R* may be resolved into two forces acting along P_1 and *A B* respectively, the latter into two forces acting along P_2 and P_3 respectively. The constructions necessary are indicated in fig. 46.

Any force, *R*, acting on a rigid body can be resolved into two, one acting along a given line P_1, the other passing through a given point *B*. The latter force must pass through *A*, the point of intersection of R and P_1. The construction is clearly shown in figures 45 and 46.

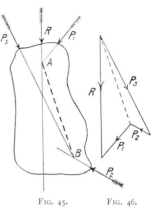

If the point of intersection A be inaccessible, as in figure 47, the link-polygon method may be used with advantage. In the force diagram (fig. 48) set off af equal to R to any convenient scale, draw fb parallel to P_1. Commence the link-polygon at B, by drawing the side a parallel to the vector Oa, then draw the side f parallel to the vector Of, cutting the line of action of P_1 at p_1. The closing side b of the link-polygon is the straight line p_1 B. Draw the vector Ob parallel to the side b

FIG. 45.   FIG. 46.

of the link-polygon, cutting the side P_1 of the force triangle at b. The force P_2 is represented in magnitude and direction by the

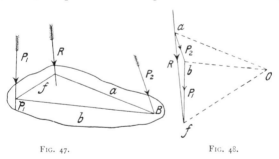

FIG. 47.   FIG. 48.

third side ab of the force triangle. Comparing with figures 41 and 42, the truth of the above construction is obvious.

49. **Parallel Forces.**—Let two parallel forces P_1 and P_2 act on a body (fig. 49). Required to find their resultant. It is evident that the resultant force R is equal to the sum $P_1 + P_2$; the only element to be found is the point at which it acts. Let AB be a line in the body at right angles to the directions of P_1 and P_2, and let C be the point at which the resultant R acts.

Let another force, Q, equal and opposite to R, be applied to the body ; then since it is equal and opposite to the resultant of P_1 and P_2, the body is in equilibrium under the action of the three

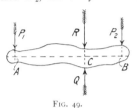

forces P_1, P_2, and Q. Consider the moments of the forces about the point C ; that of Q is zero, and, therefore, the algebraic sum of the moments of P_1 and P_2 must also be zero, since the body is in equilibrium. Therefore,

FIG. 49.

$$P_2 \times \overline{CB} = P_1 \times \overline{AC} \quad . \quad (1)$$

that is, the point C divides AB into two parts inversely proportionate to the forces P_1 and P_2.

If the forces P_1 and P_2 acted in opposite directions (fig. 50), paying attention to the sign of the moments, it is seen that the

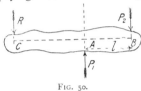

point C will lie beyond A, the point of application of the larger force. Here again

$$P_2 \times \overline{CB} = P_1 \times \overline{AC} \quad . \quad (1)$$

The above is often referred to

FIG. 50.

as the principle of the lever. The experimental verification is easy.

The resultant of any number of parallel forces P_1, P_2, ... can be found by the method of figures 41 and 42 ; the force-polygon (fig. 42) becoming in this case a straight line.

50. **Mass-centre.**—An important case of finding the resultant of a number of parallel forces is finding the centre of gravity of a body. The earth exerts an attraction on every part of the body, and therefore the resultant force of gravity on the body is the resultant of a great number of parallel forces.

Considering a body as made up of an indefinite number of small particles of equal mass, the *mass-centre* of the body is a point such that its distance from any plane is the mean distance of all the particles from that plane. If the body is subjected to gravitational attraction, every particle is acted on by a force, the total force acting on the body is the resultant of all such forces. The centre of gravity is a point at which the total mass of the body may be considered to be concentrated, in considering its attraction by other bodies. When the attractions on the particles of a body are proportional to their mass, as is practically the case on the surface of the earth, the mass-centre and the centre of gravity of a body are coincident.

If the density of the body is uniform, the mass centre will also be the geometrical centre of figure ; in fact, it is the geometrical centre of figure that is of importance in problems on mechanics.

The mass-centres for a few important cases may be given here.

Circular, Square, or Rectangular Disc.—If these discs be cut out of metal plate of uniform thickness, it is evident that the mass-centre will also be at the geometrical centre of the figure.

Triangle.—Let $A B C$ (fig. 51) be a triangle, which we may consider cut out of thin metal plate. Consider any narrow strip, $p\, p$, parallel to the side $B C$; the mass-centre p_1 of this strip is at the middle of its length. Dividing up the triangle into a number of such slips, their mass-centres will all lie on the line $A\, a$, joining A to the middle point of $B C$. In the same way, by dividing the triangle up into a number of strips parallel to $A B$, it may be

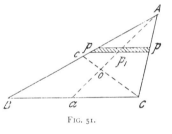

FIG. 51.

seen that the mass-centres of all the strips will lie on the line $C c$ joining C, the middle point of $A B$. The mass-centre of the whole triangle must lie somewhere on the line $A a$; it must also lie somewhere on the line $C c$; O, the point of intersection of these lines, is therefore the mass-centre. It can easily be proved that $a O$ is one-third of $a A$, and $C o$ one-third of $c C$.

Circular Arc.—Let $A B$ (fig. 52) be a portion of a circular arc with centre O. Consider the moment about any diameter $O X$. Let $P P^1$ be a portion of the arc so short that it does not

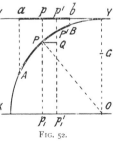

FIG. 52.

sensibly differ from the straight line $P P^1$, and its length is negligible in comparison with the radius. The mass may be considered proportional to the length of the line, and we may therefore say that the moment of $P P^{1}$ about $O X$ is $P P^{1} \times P p_{1}$; $P p_{1}$ being drawn perpendicular to $O X$; and $P^{1} Q$ being negligible compared with $P p_{1}$.

Draw $Y Y$ a tangent to the circle and parallel to the axis $O X$; from A, P, P^{1} and B project a, p, p^{1} and b on this tangent, the projectors being at right angles to it. Draw $P Q$ parallel to, and $P^{1} Q$ at right angles to $O X$, the two lines meeting at Q. Join $O P$. Then, since the triangles $P P^{1} Q$ and $O P p_{1}$ are similar,

$$\frac{P P^{1}}{P Q} = \frac{P O}{P p_{1}}.$$

Therefore, $P P^{1} \times P p_{1} = P Q \times P O = p p^{1} \times p p_{1}$—*i.e.* the moment of the arc $P P^{1}$ about the axis $O X$ is equal to the moment of the straight line $p p^{1}$ about the same axis.

This holds for all the elements of which the arc $A B$ may be considered made up. Therefore, by summing the moments of these elements we get the important result, that the moment of the arc $A B$ about the axis $O X$ is equal to the moment of the straight line $a b$, its projection on the tangent parallel to the axis.

If the arc under consideration be a semicircle of radius r, and G be its mass-centre, its length is πr, the length of its projection on the tangent is $2 r$, and we get

$$\pi r \times O G = 2 r \times r.$$

Therefore $$O G = \frac{2}{\pi} r \quad . \quad . \quad . \quad . \quad . \quad . \quad (2)$$

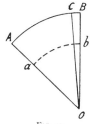

FIG. 53

Sector of a Circle.—The mass-centre of a sector of a circle $O A B$ (fig. 53) is found by dividing it up into a number of smaller sectors, $O C B$, the arc $B C$ being so short as not to differ sensibly from a straight line. The sector $O C B$ may then be considered a triangle, its mass-centre will be at a distance from O equal to two-thirds $O B$. Thus, the mass-centres of the small sectors into which $O A B$ can be divided all lie on the arc $a b$, whose radius is two-thirds that of the arc

$A\,B$; and therefore the mass-centre of the sector $O\,A\,B$ is the same as that of the arc $a\,b$.

In particular, the centre of area included between a semi-circle and its diameter is at a distance $\dfrac{4\,r}{3\,\pi}$ from the centre of the circle.

51. **Couples.**—If two parallel but opposite forces, P_1 and P_2 (fig. 54), are also equal, their resultant is zero, they tend to turn the body without giving it a motion of translation. Two equal, parallel, but oppositely directed forces constitute a *couple*, whose magnitude is measured by the product Pl of one of the equal forces into the perpendicular distance between their lines of action. A couple may be regarded as equivalent to a zero force acting at an infinite distance ; with this point of view they form no exception to the general case of finding the resultant of given forces.

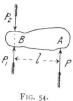

FIG. 54.

In the construction of figures 41 and 42, if the points a and f of the force-polygon coincide, the resultant of the given forces is zero. If, in addition, the line $p_5\,p_1$ is parallel to $O\,a$, the link-polygon is also closed, and the given forces are in equilibrium. If, however, $p_5\,p_1$ is not parallel to $O\,a$, the resultant of the given forces is a couple.

Let two parallel forces P_1 and P_2 (fig. 55), each equal to P, at a distance l apart, constitute a couple. The sum of the moments of the two forces about any point O in the plane of P_1 and P_2, distant x from P_1, is

$$P_2\,(x + l) - P_1\,x = Pl\,;$$

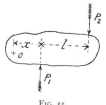

FIG. 55.

that is, the turning effect of a couple depends only on its moment Pl, and not on the position of its constituent forces relative to the axis of turning. The axis of the couple is at right angles to its plane.

Let a single force P act on a body at A (fig. 54). Introduce at B two opposite forces P_1 and P_2, each equal to, and distant l from, P. No change in the condition of the body is effected by this procedure, since P_1 and P_2 neutralise each other. But the system of forces may now be expressed as a single force P_1 acting at B, together with a couple Pl formed by the forces P and P_2. Thus, a force acting on a body at A is equivalent to an equal force acting at B, together with a *couple of transference* Pl.

A couple may be graphically represented by a vector parallel to its axis—*i.e.* at right angles to its plane ; the length of the vector being equal, to some scale, to the moment Pl of the couple.

52. **Stable, Unstable, and Neutral Equilibrium.**—If a heavy body be situated so that a vertical line through its mass-centre passes within its base it is in equilibrium. If the vertical line through

FIG. 56.

the mass-centre fall outside the base, the body is not in equilibrium, and will fall unless otherwise supported. If a body, supported in such a way that it is free to turn about an axis O (fig. 56), be left to itself it will come to rest in such a position that its mass-centre G will be vertically underneath the axis of suspension O. If the body be displaced slightly, so that its mass-centre is moved to G^1, when left to itself it will return to its original position. In fact, the forces now acting on the body are, its weight acting downwards through G^1, and the reaction at the support O acting vertically ; these two forces form a couple evidently tending to restore the body

FIG. 57.

to its original position. In this case the body is said to be in *stable* equilibrium.

If now the body be placed with its mass-centre above O (fig. 57), though in equilibrium, the smallest displacement will move G sideways, and the body will fall. The equilibrium in this case is said to be *unstable*.

If the mass-centre of the body coincide with the axis of suspension, the body will remain at rest in any position, and the equilibrium in this case is said to be *neutral*.

A body may have equilibrium of one kind in one direction, and of another kind in another direction : thus a bicycle resting on the ground in its usual position is in stable equilibrium in a longitudinal direction, and is in unstable equilibrium in a transverse direction. A bicycle wheel resting on the ground is in neutral equilibrium in a longitudinal direction, and in unstable equilibrium in a transverse direction.

53. **Resultant of any System of Forces.**—*Concurrent forces.*— If the given forces all pass through the same point, but do not all lie in the same plane, the method of section 46 can be extended to them ; their resultant will be represented as before, by the closing side of the vector-polygon, the only difference from the case of coplanar forces being that the vector-polygon is no longer plane. Thus, the resultant of a system of concurrent forces is either zero or a single concurrent force.

Non-concurrent, non-planar forces.—Let P_1, P_2, . . . be the given system of forces. Take any point O as origin and introduce two opposite forces, p_1 and $- p_1$, each equal and parallel to P_1. No change is made by this procedure, since p_1 and $- p_1$ neutralise each other. The force P_1 is therefore equivalent to a single force p_1 acting at O, and a couple of transference $P_1 l_1$; l_1 being the length of the perpendicular from O to P_1, and the axis of the couple being perpendicular to the plane of P_1 and p_1. Similarly, P_2 is equivalent to an equal and parallel force p_2 acting at O, together with a couple of transference $P_2 l_2$; and so on for all the given forces. The resultant of the concurrent forces p_1, p_2 . . . is either zero or a single concurrent force, p. Since the couples $P_1 l_1$, $P_2 l_2$, . . . are vector quantities, their resultant is also a similar vector quantity—*i.e.* a couple C. Hence the resultant of any system of forces can be expressed as the sum of a single force p and a couple C.

The magnitude of p does not depend on the position of the origin O, while that of C does. The couple C can be resolved into two couples C' and C'', having their axes respectively parallel to, and at right angles to, the direction of p. The resultant of p and C'' is a force p', equal to, parallel to, and at a distance $\dfrac{C''}{p}$ in a direction at right angles to the plane of p and the axis C'' from, p. Thus, finally, the resultant of any system of forces can be expressed as a single force p' and a couple C'' having its axis parallel to p'.

CHAPTER VII

DYNAMICS—GENERAL PRINCIPLES

54. **Laws of Motion.**—In section 13 we have seen that the measurement of force is closely associated with that of motion. The general phenomena of force and motion have been summed up by Newton in his well-known laws of motion :

I. Every body continues in its state of rest or of uniform motion in a straight line, except in so far as it may be compelled by applied forces to change that state.

II. Change of motion is proportional to the force applied, and takes place in the direction in which the force acts.

III. The mutual actions of any two bodies are always equal and oppositely directed in the same straight line ; or, action and reaction are equal and opposite.

These laws apply to forces acting in the direction of the motion, and also to forces acting in any other direction. A force like the latter will alter the direction of the body's motion, and may, or may not, increase or diminish its speed. It follows from Newton's first law that any body moving in a curved path must be continually acted on by some force so long as its motion in the curved path continues.

55. **Centrifugal Force.**—An important case of motion, especially to engineers and mechanicians, is uniform motion in a circle. If a stone at the end of a string be whirled round by hand, the string is drawn tight and a pull is exerted on the hand. This pull is called *centrifugal* force. At the other end the string exerts a pull on the stone tending to pull it inwards towards the hand. This pull is called the *centripetal* force, and it is the continual exercise of this force that gives the stone its circular path. If this force ceased to act at any instant the stone would continue

its motion, neglecting the influence of gravity, in a straight line in the direction it had at the instant the centripetal force ceased to act.

The distinction between the two forces must be carefully kept in mind.

Every point on the rim of a rapidly rotating bicycle wheel is acted on by a centripetal force which is supplied partially by the tension of the spokes. If the speed of rotation gets abnormally high, the centripetal force required to give the particles in the rim their curvilinear motion may be so great that the strength of the material is insufficient to transmit it, and the wheel bursts. The flywheels of steam engines are often run so near the speed limited by these considerations, that it is not uncommon for them to burst under the action of the centrifugal stress.

Let m be the mass in lbs. of the body moving with speed v feet per second in a circle of radius r. It has been shown (sec. 26) that the radial acceleration a is $\dfrac{v^2}{r}$. But if f be the radial force acting, by section 14,

$$f = ma = \frac{m\,v^2}{r} \text{ poundals, or } f = \frac{m\,v^2}{g\,r} \text{ lbs.} \quad . \quad . \ (1)$$

56. **Work.**—When a force acts on a body and produces motion it is said to do *work*. If a force acts on a body at rest, and no motion is produced, no work is done. The idea of *motion* is essential to work. If a man support a load without moving it, although he may become greatly fatigued, he cannot be said to have done mechanical work. The load, as regards its mechanical surroundings, might as well have been supported by a table. If the applied force be constant throughout the motion, the work done is measured by the product of the force into the distance through which it acts. The practical unit of work is the *foot-pound*, which is the work done in raising a weight of one pound through a vertical distance of one foot.

It should be noted particularly that the idea of *time* does not enter into work ; the work done in raising one ton ten feet high being the same whether a minute or a year be taken to perform it. In the same way, the work done by a cyclist in riding up a hill of a given height is the same whether he does it slowly or quickly.

The work done in raising a body through a definite height is quite independent of the manner or path of raising, neglecting frictional resistance and considering only the work done against gravity. The work a cyclist does against gravity in ascending a hill of a certain height is quite independent of the *gradient* of the road over which he travels.

Example.—Let the machine and rider weigh 200 lbs., then the work done by the rider in rising 100 feet vertically is 20,000 foot-lbs. If the gradient of the road be known, this can be calculated in another way, which, for the present purpose, is roundabout but instructive. Consider an extreme gradient of one vertical to two on the slope (fig. 58), the length of the hill

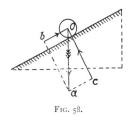

FIG. 58.

will be 200 feet. The work done in ascending the hill may be estimated by the product of the force required to push the machine and rider up the hill, into the length of the hill. The machine and rider weigh 200 lbs. ; this force acts vertically downwards, and can be resolved into two, one parallel to the road's surface, and one at right angles to it. If $O a$ be set off equal to 200 lbs., and the construction of section 48 be performed, it will be found that the component $b O$ required to push the machine and rider up the hill is 100 lbs. The work done will be the product of this force into the distance through which it acts, 200 feet ; the result, 20,000 foot-lbs., being the same as before.

This is only the work done against gravity. In riding along a level road there is no work done against gravity, any resistance being made up of the rolling friction of the wheels on the road, air resistance, and the friction of the bearings. These resistances will remain, to all intents and purposes, the same on an incline as on a level. The work done in riding along 200 feet of level road would have to be added to the 20,000 foot-lbs. of work done against gravity, in order to get the *total* work done by the cyclist in ascending the hill.

Generally, the work done by, or against, a force is the product of the force into the projection on the direction of action of the force of the path of the moving body. Thus, if a body move from *A* to *B*, and be acted on by the force *f*, which always retains the same direction, the work done is *f. A̅C̅* ; *B C* being perpendicular to *A C* (fig. 59).

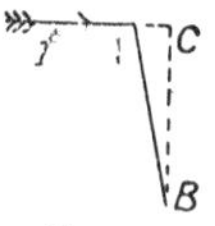

FIG. 59.

The centripetal force acting on a body moving in a circle is always at right angles to the direction of motion ; consequently in this case the projection of the path is zero, and no work is done.

In the Simpson lever-chain the pressure of the chain rollers on the teeth of the hub sprocket wheel is at right angles to the surface of the teeth, and consequently makes a considerable angle with the direction of motion of the rollers. In this case, therefore, the projection *A C* (fig. 59), on the line of action of the pressure, of the distance *A B* moved through, is very much less than *A B*. The claims of its promoters virtually amount to saying that the work done on the hub by the pull of the chain is *f . A̅B̅*, whereas the correct value is *f . A̅C̅*.

In driving a cycle up-hill, the work done against gravity by the rider at each stroke of the pedal is the product of the total weight and the vertical distance moved through during half a turn of the crank axle. Let the gradient be *x* parts vertical in 100 on the slope, *D* the diameter in inches to which the driving-wheel is geared, and *W* the total weight of machine and rider in lbs. The vertical distance passed through per stroke of pedal is

$$\frac{x}{100} \cdot \frac{\pi D}{2} \text{ inches.}$$

The work done per stroke of pedal is therefore

$$\frac{\pi x D}{200} W \text{ inch-lbs.}$$

$$= \cdot 001309 \, x \, D \, W \text{ foot-lbs. . . . . (2)}$$

Table I., on the following page, is calculated from equation (2).

TABLE I.—WORK DONE IN FOOT-LBS. PER STROKE OF
PEDAL, IN RAISING 100 LBS. WEIGHT AGAINST GRAVITY.

Diameter, to which driving-wheel is geared	Gradient, parts in 100							
	1	2	3	4	5	6	7	8
Inches								
40	5·24	10·47	15·70	20·94	26·18	31·41	46·65	41·89
45	5·89	11·78	17·67	23·56	29·45	35·34	41·23	47·12
50	6·55	13·09	19·63	26·18	32·72	39·27	47·97	52·36
55	7·20	14·40	21·60	28·79	36·00	43·20	50·39	57·59
60	7·86	15·71	23·56	31·42	39·27	47·12	54·97	62·83
65	8·51	17·02	25·52	34·04	42·54	51·05	59·56	68·08
70	9·16	18·32	27·49	36·65	45·81	54·98	64·14	73·30
75	9·82	19·64	29·45	39·27	49·09	58·90	68·72	78·54
80	10·47	20·94	31·41	41·89	52·36	62·83	73·30	83·78

57. **Power.**—The rate of doing work is called the *power* of an agent, and into its consideration time enters. The standard of power used by engineers is the *horse-power*. Any agent which performs 33,000 foot-lbs. of work in one minute is said to be of 1 H.P. This, Watt's estimate, is in excess of the average power of a horse, but it has been retained as the unit of power for engineering purposes. The average power of a man is about one-tenth that of a horse—that is, equal to 3,300 foot-lbs. per minute.

If V be the speed, in miles per hour, of a cyclist riding up a gradient of x parts in 100, the vertical distance moved through in one minute is

$$\frac{x}{100} \cdot \frac{V \times 5280}{60} = \cdot 88\, x\, V \text{ feet,}$$

and the power expended is

$$\cdot 88\, x\, V\, W \text{ foot-lbs. per minute} \quad . \quad . \quad . \quad . \quad (3)$$

Table II. is calculated from equation (3).

58. **Kinetic Energy.**—So far we have dealt with the work done by a force which gives motion to a body against a steady resistance, the speed of the body having no influence on the question, further than it must be the same at the end as at the beginning. If a body free to move be acted on by a force, the work done will be expended in increasing its speed. The work is

TABLE II.—WORK DONE, IN FOOT-LBS. PER MINUTE, IN
PUSHING 100 LBS. WEIGHT UP-HILL.

Speed. Miles per hour	Slope, parts in 100							
	1	2	3	4	5	6	7	8
4	352	704	1056	1408	1760	2112	2464	2816
5	440	880	1320	1760	2200	2640	3080	3520
6	528	1056	1584	2112	2640	3168	3696	4224
7	616	1232	1848	2464	3080	3696	4312	4928
8	704	1408	2112	2816	3520	4224	4928	5632
9	792	1584	2376	3168	3960	4752	5544	6336
10	880	1760	2640	3520	4400	5280	6160	7040
11	968	1936	2904	3872	4840	5808	6776	7744
12	1056	2112	3168	4224	5280	6336	7392	8448
13	1144	2288	3432	4576	5720	6864	8008	9152
14	1232	2464	3696	4928	6160	7392	8624	9856
15	1320	2640	3960	5280	6600	7920	9240	10560
16	1408	2816	4224	5632	7040	8448	9856	11264
17	1496	2992	4488	5984	7480	8976	10472	11968
18	1584	3168	4752	6336	7920	9504	11088	12672
19	1672	3344	5016	6688	8360	10032	11704	13376
20	1760	3520	5280	7040	8800	10560	12320	14080

stored in the moving body, and can be restored in bringing the body again to rest. This stored work is called *kinetic energy*.

59. **Potential Energy.**—Newton's first law of motion expresses the idea of permanence of motion of a body unless altered by applied forces. If the speed of a body on which no force acts remains constant, its kinetic energy must also remain constant. If a body free to move is acted on by a force, the work done by the force is stored up as kinetic energy. If work is done by moving the body against the resistance of a force which is constant in magnitude and direction, whatever be the direction of motion, the work is expended in changing the position of the body. For example, in raising a body from the ground, the resistance overcome is its weight, which always acts vertically downwards, whether the body be at rest or moving upwards or downwards. If the body be lowered by suitable means to the ground, the work done in raising it is again restored. The body

at rest a certain height above the ground possesses therefore an amount of energy due to its position ; this is called *potential energy*. If the body be allowed to fall freely under the action of gravity, at the instant of reaching the ground it possesses no potential energy, but kinetic energy due to its speed. Its initial store of potential energy has been all converted into kinetic energy.

60. **Conservation of Energy.**—The great principle of conservation of energy is an assertion that energy cannot be created or destroyed. This is one of the most comprehensive generalisa tions that has been deduced from our observations of natural phenomena. Applied to the case of a body moving under the action of force without any frictional resistance, it asserts that the sum of the kinetic and potential energies is constant. A cyclist riding down a short hill with his feet off the pedals and not using the brake, will have a greater speed at the bottom than at the top, part of the potential energy due to the high position at the top of the hill being converted into kinetic energy at the bottom. If another short hill of equal height has to be ascended immediately, the kinetic energy at the bottom gets partially converted into potential energy at the top ; the rider arriving at the top of the second hill with the same speed as he left the first. The friction of the air, tyres, and bearings has been neglected in the above discussion. If the rider just work hard enough to overcome these resistances as on a level road, the above statement will be strictly true.

Applied to mechanism used to transmit and modify power, the principle of the conservation of energy is sometimes quoted, ' No more work can be got out at one end of a machine than is put in at the other.' The work got out will be exactly equal to that put into the machine, provided the friction of the machine is zero, an ideal state of things sometimes closely approached, but never actually attained in practice. The chronic inventor of cycle driving-gears might save himself a great deal of trouble by mastering this principle.

61. **Frictional Resistance.**—It is a matter of every-day experience that a moving body left to itself will ultimately come to rest, thus apparently contradicting Newton's first law. A flat stone moved along the ground comes to rest very soon. If the

stone be round, it may roll along the ground a little longer, while a bicycle wheel with pneumatic tyre set off with the same speed will continue its motion for a still longer period. A wheel set rapidly rotating on its axis will gradually come to rest. If the wheel be supported on ball-bearings, the motion may continue for a considerable fraction of an hour, but ultimately the wheel will come to rest. In all these cases there is a force in action opposing the motion, the force of *friction*, which is always called into play when two bodies move in contact with each other. The amount of friction depends on the nature of the surfaces in contact. The friction is very great with the flat stone sliding along the ground, is less with the rolling stone, and still less with the pneumatic-tyred wheel. The friction of a ball-bearing may be reduced to a very small amount, but cannot be entirely abolished ; the less the friction, the longer the motion persists. The air also offers a considerable resistance to the motion, which varies with the speed. If a wheel with ball-bearings could be set in rapid rotation under a large bell-jar from which the air had been exhausted by an air-pump, the motion of the wheel might persist for several hours, and thus give a close approximation to an experimental verification of Newton's first law of motion. The movement of the planets through space affords the best illustration of the permanence of motion.

62. **Heat.**—The force of friction is thus seen to diminish the kinetic energy of a moving body, while if the body move in a horizontal plane, its potential energy remains the same throughout, and energy is said to be *dissipated*. The energy dissipated is not destroyed, but is converted into *heat*, the temperature of the bodies in contact being raised by friction. Heat is a form of energy, and the conversion of mechanical work by friction into heat is a matter of every-day experience ; conversely, heat can be converted into mechanical work. Steam engines, gas-engines, and oil-engines are machines in which this conversion is effected. Heat due to friction is energy in a form which cannot be utilised in the machine in which it arises ; hence popularly engineers speak of the work *lost* in friction, such energy being in a useless form.

In riding down-hill the potential energy of the machine and

rider gets less ; if the speed remains the same, the kinetic energy remains the same, and the potential energy is dissipated in the form of heat. If a brake be used, the heat appears at the brake-block and the wheel on which it rubs. If back-pedalling be employed, the same amount of heat is expended in heating the muscles of the legs, though the other physiological actions going on may be such as to render the detection or measurement of this heat difficult.

Mechanical Equivalent of Heat.—The conversion of heat into work, and work into heat, takes place at a certain definite rate. 780 foot-pounds of work are equivalent to one unit of heat ; the unit of heat being the quantity of heat required to raise the temperature of one pound of water one degree Fahrenheit. Thus, in descending a hill 100 feet high, a rider and machine weighing 200 lbs. would convert 20,000 foot-lbs. of work into $\frac{20000}{780} = 25\cdot6$ units of heat. If this could all be collected at the brake-block, it would be sufficient to raise the temperature of one pound of water 25·6 degrees.

CHAPTER VIII

63. **Dynamics of a Particle.**—A particle, an ideal conception in the Science of Mechanics, is a heavy body of such small dimensions that it may be considered a point. If a particle of mass m initially at rest, but free to move, be acted on for time t by a constant force f, we have seen (sec. 16) that the speed v imparted is such that

$$f t = m v \qquad . \qquad . \qquad . \qquad . \qquad . \qquad .$$

or

$$f = \frac{m v}{t} \qquad . \qquad . \qquad . \qquad . \qquad . \qquad (1)$$

$$f = m a \ . \qquad . \qquad . \qquad . \qquad . \qquad . \qquad (2)$$

a being the acceleration, or rate of change of speed, $m v$ is the momentum acquired in time t, hence $\frac{m v}{t}$ is the momentum acquired in unit of time, and (1) is equivalent to defining force as 'rate of change of momentum.'

Let s be the distance traversed in the time t; then since the *average* speed is half the speed at the end of the period,

$$s = \tfrac{1}{2} v t = \tfrac{1}{2} a t^2 \ . \qquad . \qquad . \qquad . \qquad . \qquad (3)$$

The work done during the period is $f s$, and

$$f s = \tfrac{1}{2} v f t = \tfrac{1}{2} m v^2 . \qquad . \qquad . \qquad . \qquad . \qquad (4)$$

If the particle has initially a speed v_0, equations (1), (3) and (4) become

$$f t = m (v - v_0) \ . \qquad . \qquad . \qquad . \qquad . \qquad (5)$$
$$s = \tfrac{1}{2} (v + v_0) t \ . \qquad . \qquad . \qquad . \qquad . \qquad (6)$$
$$f s = \tfrac{1}{2} m (v^2 - v^2_0) \ . \qquad . \qquad . \qquad . \qquad . \qquad (7)$$

Kinetic Energy.—The work done by the force has been expended in giving the body its speed v, and the body in coming to rest can restore exactly the same amount of work. The product $\frac{1}{2} m v^2$ is called the *kinetic* energy of the moving body ; it may be denoted by the symbol E.

The units employed above are all absolute units. The unit of kinetic energy in (4) is the foot-poundal ; in foot-pounds the kinetic energy is

$$E = \frac{m \, v^2}{2 \, g} \qquad \ldots \ldots \ldots (8)$$

Falling Bodies.—A body falling freely under the action of gravity is a special case of the above. Let the mass m be one pound, the force acting on the body is 1 lb. weight, *i.e.* g poundals. Writing g instead of f, and $m = 1$, in equations (1)–(4) the formulæ for falling bodies are obtained.

64. **Circular Motion of a Particle**.—Let the particle be constrained to move in a circle of radius r, and be acted on by a force of constant magnitude f, which is always in the direction of the tangent to the path of the particle ; then since the radial force does no work, equations (1) to (7) still hold. Multiply both sides of (1) by r, then

$$f r = \frac{m \, v \, r}{t} \qquad \ldots \ldots \ldots (9)$$

$f r$ is the moment of the applied force about the axis of rotation, $m v$ is the momentum, $m v r$ the *moment of momentum* or *angular momentum* ; hence the moment of a force is equal to the rate of change of angular momentum it produces.

If ω be the angular speed and θ the angular acceleration of the particle about the axis at the end of the time t, $v = \omega \, r$, $\theta = \frac{\omega}{t}$, and (9) may be written

$$f r = \frac{m \, \omega \, r^2}{t} = m \, r^2 \, \theta \qquad \ldots \ldots (10)$$

The product $m \, r^2$ is the *moment of inertia* of the particle about the axis of rotation, and may be denoted by i ; (10) may then be written

$$f r = i \, \theta \qquad \ldots \ldots \ldots (10)$$

That is, the moment of the force is equal to the product of the moment of inertia of the body on which it acts and the angular acceleration it produces.

Equation (4) becomes, for this case,

$$e = f s = \tfrac{1}{2}\, m\, v^2 = \tfrac{1}{2}\, m\, r^2\, \omega^2 = \tfrac{1}{2}\, i\, \omega^2 \quad . \quad . \quad . \quad (11)$$

That is, the kinetic energy of a particle moving in a circle is half the product of its moment of inertia about the centre and the square of its angular speed.

(9) may be written

$$f\, t\, r = m\, v\, r = m\, r^2\, \omega = i\, \omega \quad . \quad . \quad . \quad . \quad (12)$$

$f\, t$ is the impulse of the force ; therefore the moment of the impulse is equal to the product of the moment of inertia of the particle and the angular speed produced by the impulse.

65. Rotation of a Lamina about a Fixed Axis Perpendicular to its Plane.—A rigid body of homogeneous material may be considered to be made up of a great number of particles, all of equal mass uniformly distributed. A rigid lamina is a rigid body of uniform, but indefinitely small, thickness lying between two parallel planes ; a flat sheet of thin paper is a physical approximation to a lamina. Let O (fig. 60) be the fixed axis of rotation, perpendicular to the plane of the paper ; let A be any particle of the lamina distant r from O. Then using the same notation, equations (9) to (12) hold for the particle A, the acting force f being always at right angles to the radius $O\,A$. Now the rigid lamina may be considered made up of a number of heavy particles like A, embedded in a rigid weightless frame. Instead of the force f acting directly at A, suppose a force p act at a point B of the frame in a direction at right angles to $O\,B$. Let $O\,B = l$, then if

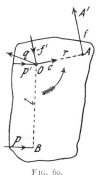

FIG. 60.

$$p\, l = f\, r \quad . \quad . \quad . \quad . \quad . \quad . \quad (13)$$

the effects of the forces f and p in turning the weightless frame and heavy particle A about the centre O are exactly the same ; the motion of A is unaltered by the substitution.

Also, if d be the space passed over during the period by the point B,

$$\frac{d}{l} = \frac{s}{r} \text{ or } s = \frac{d\,r}{l}, \text{ and}$$

therefore

$$f\,s = \frac{p\,l}{r} \cdot \frac{d\,r}{l} = p\,d.$$

Substituting in (10), (11) and (12) they may be written

$$p\,l = i\,\theta \quad . \quad . \quad . \quad . \quad . \quad . \quad . \quad . \quad (14)$$
$$e = p\,d = \tfrac{1}{2}\,i\,\omega^2 \quad . \quad . \quad . \quad . \quad . \quad (15)$$
$$p\,l\,t = i\,\omega \quad . \quad . \quad . \quad . \quad . \quad . \quad . \quad (16)$$

Let i_1, i_2 . . . . be the moments of inertia of the heavy particles A_1, A_2 . . . . of which the lamina is composed; p_1, p_2, ... the corresponding forces at the point B required to give them their actual motions; then for all the particles, (14), (15) and (16), may be written

$$(p_1 + p_2 + \quad . \quad . \quad . \quad) \, l = (i_1 + i_2 + \quad . \quad . \quad . \quad) \, \theta$$
$$(p_1 + p_2 + \quad . \quad \quad . \quad) \, d = \tfrac{1}{2}\,(i_1 + i_2 \quad . \quad . \quad . \quad) \, \omega^2$$
$$(p_1 + p_2 + \quad . \quad . \quad . \quad) l \, t = (i_1 + i_2 + \quad . \quad . \quad . \quad) \, \omega$$

l, t, θ and ω being the same for all the particles. Let $I = (i_1 + i_2 + \quad . \quad . \quad . \quad)$, then I is the moment of inertia of the lamina about the axis O; let $(p_1 + p_2 + \quad . \quad . \quad . \quad) = P$, then P is the actual force applied at the point B of the lamina; let $(e_1 + e_2 + \quad . \quad . \quad . \quad) = E$, then E is the kinetic energy of the lamina; and the above equations may be written

$$P\,l = I\,\theta \quad . \quad . \quad . \quad . \quad . \quad . \quad . \quad (17)$$
$$E = \tfrac{1}{2}\,I\,\omega^2 \quad . \quad . \quad . \quad . \quad . \quad . \quad (18)$$
$$P\,l\,t = I\,\omega \quad . \quad . \quad . \quad . \quad . \quad . \quad (19)$$

$P\,l$ is the magnitude of the applied turning couple.

66. **Pressure on the Fixed Axis.**—In the above investigation the pressure on the axis at O has been neglected, since whatever be its value, its moment about O is zero, and it does not, therefore, influence the speed of rotation. It is, however, desirable to know the pressure on the bearings of the rotating body; we therefore proceed to investigate it. Consider only the particle A, connected by the rigid weightless frame to B and O; if the force p at B

gives A its tangential acceleration, the weightless frame must press on the particle A with a force f, in the direction at right-angles to r, and the particle A must react on the frame with an equal and opposite force $-f$. But the particle A also presses on the frame with the centrifugal force $c = \dfrac{m\,v^2}{r} = m\,\omega^2\,r$, in the direction of the radius r. The frame being weightless must be in equilibrium under the forces acting on it ; since, by (2), a finite force, however small, acting on a body of zero mass would produce infinite acceleration. These forces are : $-f$ at A, p at B, the reaction q of the axis at O, and the centrifugal force c, which also acts through O. But the forces $-f$ at A, and p at B, are equivalent to equal and parallel forces at O, and the couples $-fr$ and pl. The couples equilibrate each other, therefore the four forces $-f^1$, p^1, q and c at O are in equilibrium. Therefore,

$$\text{vector } q = \text{vector } f - \text{vector } p - \text{vector } c \quad . \quad . \quad . \quad (20)$$

Let Q be the resultant reaction of the fixed axis on the lamina, due to the particles A_1, A_2, . . . of which it is composed, *i.e.*—

$$\text{vector } Q = \text{sum of vectors } q_1, q_2 \quad . \quad . \quad .$$

Similarly, let

$$\text{vector } P = \text{sum of vectors } p_1, p_2 \quad . \quad . \quad .$$
$$\text{vector } F = \text{sum of vectors } f_1, f_2 \quad . \quad . \quad .$$
$$\text{vector } C = \text{sum of vectors } c_1, c_2, c_3 \quad . \quad .$$

Then, adding equations (20) for all the particles A_1, A_2 . . .,

$$\text{vector } Q = \text{vector } F - \text{vector } P - \text{vector } C \quad . \quad . \quad (21)$$

But by (10)—

$$\text{vector } F = m\,\theta \times \text{vector sum } (r_1 + r_2 + \ldots .)$$

And the vector sum $(r_1 + r_2 + \ldots . r)$ is the vector $n \cdot \overline{OG}$; G being the mass-centre of the lamina (fig. 61), and n the number of particles, each of mass m, it contains.

Therefore,

$$\text{vector } F = M\,\theta \cdot \overline{OG} \quad . \quad . \quad . \quad . \quad . \quad . \quad . \quad (22)$$

M being the total mass of the lamina. The component forces f_1, f_2 . . . acting at right-angles to the corresponding vectors r_1,

r_2 . . ., the resultant force F will act at right angles to the re-sultant vector $O\,G$. Similarly,

$$\text{vector } C = M\,\omega^2 \,.\, \overline{O\,G} \quad . \quad . \quad . \quad . \quad . \quad . \quad . \quad . \quad (23)$$

the force C acting along $O\,G$.

Now, from (13) and (10) $p = \dfrac{f\,r}{l} = \dfrac{m\,r^2\,\theta}{l}.$

The vectors p are all in the same direction, at right angles to $O\,B$, and are therefore added like scalars. Therefore,

$$\text{vector } P = \frac{\theta}{l} \times \text{sum } (m_1\,r_1{}^2 + m_2\,r_2{}^2 \ldots) = \frac{I\,\theta}{l} \quad . \quad (24)$$

Substituting these values in (21), the reaction Q (fig. 61) of the fixed axis is the resultant of :—A force at O equal and parallel

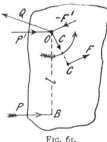

FIG. 61.

to that required to accelerate the mass M supposed concentrated at G ; a force at O equal, opposite and parallel to the applied force P ; the centripetal force $M \,.\, \omega^2 \,.\, \overline{O\,G}$, acting along $G\,O$.

From (21) many important results can be deduced. Let a couple act on a rigid lamina quite free to move in its plane ; then $P = 0$, $Q = 0$; and (21) becomes

vector F − vector $C = 0$.

But the vectors F and $-\,C$ are at right angles ; their sum can only be zero when each is zero. This is the case when $\overline{O\,G} = 0$ —see (22) and (23)—that is, when the mass-centre and the axis of rotation coincide. Hence a couple applied to a lamina free to move causes rotation about its mass-centre.

67. Dynamics of a Rigid Body.—Equations (17), (18) and (19) are applicable to the rotation of any rigid body about a fixed axis. Equations (21) to (24) are applicable if the rigid body is symmetrical about a plane perpendicular to the axis of rotation ; this includes most cases occurring in practical engineering. But in a non-symmetrical body, *e.g.* a pair of bicycle cranks and their axle, the resultant pressure on the bearings cannot be expressed as a single force, but is a couple. Thus, such a rigid body, if perfectly free, will turn about an axis, in general, not parallel to that of the acting couple.

From (23), the centrifugal pressure on the fixed axis of any rigid body is the same as if the whole mass were concentrated at the mass-centre G. If the mass-centre lies on the axis of rotation, the centrifugal pressure is zero. Hence the necessity of accurately balancing rapidly revolving wheels. In this case also (21) becomes $Q = -P$, *i.e.* the pressure on the bearing is equal and parallel to the applied force, provided Q can be expressed as a single force. If only a couple be applied, $P = 0$, and the pressure on the bearings is zero. In a rapidly rotating wheel with horizontal axis, P is the weight of the wheel; with vertical axis $P = 0$, the weight acting parallel to the axis.

The motion of a rigid body can be expressed (sec. 41) as a translation of its mass-centre, and a rotation about an axis passing through its mass-centre. Any applied force is equivalent to an equal parallel force at the mass-centre and a couple of transference. The rotation about the mass-centre is the effect of this couple. Hence, the turning effect of any system of forces acting on a free rigid body is the same as if its mass-centre were fixed. Since the resultant couple does not influence the motion of the mass-centre, the motion of the mass-centre of a rigid body under the action of any system of forces is the same as if equal parallel forces were applied at the mass-centre.

The kinetic energy of any moving body is the sum of the energy due to the speed of its mass-centre, and the energy due to its rotation about the mass-centre.

Moments of Inertia.—If M be the total mass of a rigid body, its moment of inertia may be expressed $I = Mk^2$; and k is called the radius of gyration. The I about an axis through the mass-centre is least : let it be denoted by I_0; that about any parallel axis distant h is

$$I = I_0 + Mh^2. \quad . \quad . \quad . \quad . \quad . \quad (25)$$

The values of I for a few forms may be given here. For a thin ring of radius r and mass M rotating about its geometric axis, $I_0 = Mr^2$. This is approximately the case of the rim and tyre of a bicycle wheel. For the same ring rotating about an axis at its circumference, as in rolling along the ground, $I = 2\,M\,r^2$.

For a bar of length l rotating about an axis through its end perpendicular to its own axis, $I = \dfrac{M\,l^2}{3}$. This is approximately the case of the spokes of a bicycle wheel.

For a circular disc of uniform thickness and radius r rotating about its geometric axis, $I_0 = \dfrac{M\,r^2}{2}$. For the same disc rolling along the ground, $I = \dfrac{3}{2}\,M\,r^2$.

68. **Starting in a Cycle Race.**—The work done by a rider at the beginning of a race is nearly all expended in giving himself and machine kinetic energy, the frictional resistances being small until a high speed is attained. If the winning-post be passed at top speed, the kinetic energy is practically not utilised. In a short distance race, this kinetic energy may be large in comparison to the energy employed in overcoming frictional resistances. The kinetic energy of translation of the machine and rider is $\dfrac{W\,v^2}{g\,r}$ foot-lbs., W being the total weight. Hence, a light machine, other things being equal, is better than a heavy one for short races. Further, there is the kinetic energy of rotation of the wheels and cranks. For the rims and tyres this is nearly equal to their translational kinetic energy ; therefore, at starting a race, one pound in the rim and tyres is equivalent to two pounds in the frame. In comparing racing machines for sprinting, the weight of the frame, added to twice that of the rims and tyres, would give a better standard than the weight of the complete machine. The pneumatic tyre, with its necessarily heavier rim, is, in this respect, inferior to the old narrow solid tyre. Of course, once the top speed is attained, the weight of the parts has no direct influence, but only so far as it affects frictional resistances.

69. **Impact and Collision.**—If two bodies moving in opposite directions collide, their directions of motions are apparently changed instantaneously ; but, as a matter of fact, the time during which the bodies are in contact, though extremely short, is still appreciable. The magnitude of the force required to generate

velocity in a body, or to destroy velocity already existing, is inversely proportional to the time of action ; if the time of action be very short, the acting force will be very large. Such forces are called *impulsive forces*.

Now in the case of colliding bodies, such as a pair of billiard balls, it is impossible either to measure f or t ; but the mass m of one of the balls, and its velocities v_0 and v before and after collision, may easily be measured. The expression on the right-hand side of (5) denotes the increase of momentum of the body due to the collision ; the product $f\,t$ on the left-hand side is called the *impulse* ; therefore, from (5), the impulse is equal to the change of momentum it produces.

We shall now have to examine more minutely the nature of the forces between two bodies in collision: At the instant that the bodies first come into contact they are approaching each other with a certain velocity. Suppose A (fig. 62) to be moving to the right, and B to the left ; immediately they touch, the equal impulsive forces f_1 and f_2 will be called into action, and will oppose the motions of A and B respectively. The parts of the bodies in the

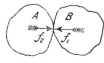

FIG. 62.

neighbourhood of the place of contact will be flattened, and this flattening will increase until the relative velocity of the bodies is zero. The time over which this action extends is called the *period of compression*. If the bodies are elastic, they will tend to recover their original shapes, and will therefore still press against each other ; the forces now tending to give the bodies a relative velocity in the direction opposite to their original relative velocity. These impulsive forces will be in action until the original shape has been recovered and the bodies leave each other. The time over which this action extends is called the *period of restitution* ; and the total impulse may be conveniently divided into two parts, the impulse of compression and the impulse of restitution.

Index of Elasticity.—Now it is an experimental fact that in bodies of given material the impulse of restitution bears a constant ratio to the impulse of compression ; this ratio is called the *index of elasticity*. A perfectly elastic material has its index of elasticity unity ; in an inelastic body the index of elasticity is zero ; if the index of elasticity lies between zero and unity, the body is imperfectly elastic. The index of elasticity e is, for balls of glass $\frac{15}{16}$, for balls of ivory $\frac{8}{9}$, and for balls of steel $\frac{5}{9}$. These are the values given by Newton, to whom the theory of collision of bodies is due.

Conservation of Momentum.—In figure 62, the force f_1 at any instant acting on A is exactly equal to the force f_2 acting on B ; the total impulse on A is therefore equal to the total impulse on B; and as they are in opposite directions their sum is zero. Thus, the momentum of the system is the same after collision as before it. This is true whether the bodies are inelastic, imperfectly elastic, or perfectly elastic. If two bodies of mass m_1 and m_2, moving with velocities v_0' and v_0'' respectively, collide, their velocities after collision can be easily determined, if the index of elasticity e is given. For cyclists, the most important case is when one of the bodies is rigidly fixed ; in other words, when m_2 is infinite and v_0'' zero. Let, as before, the mass of the finite body be m, its velocities before and after collision with the infinite body be v_0 and v ; then before collision its momentum is $m\,v_0$. Let C be the impulse of compression ; then since at the end of the compression period the velocity is zero, we get by substitution in (1)

$$C = m\,v_0 \quad . \quad . \quad . \quad . \quad . \quad . \quad (26)$$

The impulse of restitution, by definition, is $e\,C$; therefore, if v be the velocity of the body after collision, we have

$$e\,C = -\,m\,v_0$$

Substituting the value of C from (26), we get

$$v = -\,e\,v_0 \quad . \quad . \quad . \quad . \quad . \quad (27)$$

That is, the speed of rebound is equal to the speed of impact multiplied by the index of elasticity. The speed of rebound is therefore always less than the speed of impact.

This result at first sight seems to be contradictory to the principle of the conservation of momentum, but remembering that the mass of the fixed body may be considered infinite, and its velocity zero, its momentum is

$$\infty \times 0,$$

an expression which may represent *any* finite magnitude. We may say the fixed body gains the momentum lost by the moving body by the collision. For example, when a ball falls vertically and rebounds from the ground, the earth as a whole is displaced by the collision.

Loss of Energy.—The kinetic energy of the moving body before impact is

$$\frac{m v_0^2}{2 g} \text{ foot-lbs. ;}$$

the kinetic energy after impact is

$$E = \frac{e^2 m v_0^2}{2 g} = e^2 E_0 \quad . \quad . \quad . \quad . \quad (28)$$

The loss of energy due to collision is thus

$$(1 - e^2) E_0 \quad . \quad . \quad . \quad . \quad . \quad . \quad (29)$$

70. **Gyroscope.**—Let a wheel W (fig. 63), of moment of inertia I, be set in rapid rotation on a spindle S, which can be balanced by means of a counterweight w, on a pivot support T (fig. 63). If a couple C, formed by two equal and opposite vertical forces F_1 and F_2 acting at a distance l, be applied to the spindle, tending to make it turn about a horizontal axis, it is found that the axis of the spindle turns slowly in a horizontal plane. This motion is called 'precession.' This phenomenon, which,

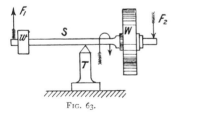

FIG. 63.

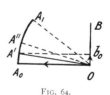

FIG. 64.

when observed for the first time, appears startling and paradoxical, can be strikingly exhibited by removing the counterweight w, so that statically the spindle is not balanced over its support. The explanation depends on the composition of rotations. Figure 64 is a plan showing the initial direction $O A_0$ of the axis of rotation of the wheel W. The initial angular momentum of the wheel can be represented to any convenient scale by the length $O A_0$. The couple C tends to give the wheel a rotation about the axis

$O B$ at right angles to $O A_0$. If this couple C acts for a very short period of time, t_1, the angular momentum it produces about the axis $O B$ is $C t_1$. This may be represented to scale by $O b_0$. The resultant angular momentum of the wheel at the end of the time, t_1, may therefore be represented in magnitude and direction by $O A^1$. If the time t_1 be taken very small, $O A^1$ is practically equal to $O A_0$, and the only effect of the couple C is to alter the *direction* of the axis of rotation. At the end of a second short interval of time, t_2, it may be shown in the same manner that the axis of rotation is $O A''$, $A' A''$ being at right angles to $O A'$. At the end of one second the increment of the angular momentum is numerically equal to C, and may be represented by the arc $A_0 A_1$; thus at the end of one second the axis of rotation is $O A_1$. Let θ be the angular speed of precession, then θ is numerically equal to the angle $A_0 O A_1$, *i.e.*,

$$\theta = \frac{\text{arc } A_0 A_1}{\text{radius } O A_0} = \frac{C}{I \omega} \quad . \quad . \quad . \quad (30)$$

or

$$C = I \omega\, \theta = M k^2\, \omega\, \theta$$
$$= M v k \theta \quad . \quad . \quad . \quad . \quad . \quad . \quad (31)$$

where M is the mass and k the radius of gyration of the wheel, and v the linear speed of a point on the wheel at radius k.

In drawing the diagram (fig. 64) care should be taken that the quantities $O A_0$ and $O b_0$ are marked off in the proper direction. If the rotation of the wheel when viewed in the direction $O A_0$ appear clock-wise, it may be considered positive; similarly, the rotation which the couple C tends to produce, appears clock-wise when measured in the direction $O b_0$, and is therefore also considered positive. If the couple C were of the opposite sign, the increment of angular momentum $O b_0$ would be set off in the opposite direction, and the precession would also be in the opposite direction.

The geometrical explanation of this phenomenon is almost the same as that given for centrifugal force in the case of uniform motion in a circle.

A cyclist can easily make an experiment on precession without any special apparatus as follows : Detach the front wheel from a bicycle, and, supporting the ends of the hub spindle between the thumb and first fingers of each hand, set it in rotation by striking the spokes with the second and third fingers of one hand. On withdrawing one hand the wheel will not fall to the ground, as it would do if at rest, but will slowly turn round, its axis moving in a horizontal plane. As the speed of rotation gradually gets less owing to friction of the air and bearings, the speed of precession gets greater, until the wheel begins to wobble and ultimately falls.

71. **Dynamics of any System of Bodies.**—The forces acting on any given system of bodies may be conveniently divided into 'external' and 'internal'; the former due to the action of bodies external to the given system, the latter made up of the mutual actions between the various pairs of bodies in the given system. The latter forces are in equilibrium among themselves; that is, the force which any body A exerts on any other body B of the system is equal and opposite to the force exerted by B on A. The motion of the mass-centre of the given system is therefore unaffected by the internal forces, and some of the results of section 67 can be extended to any system of bodies, thus :

The motion of the mass-centre of a system of bodies under the action of any system of forces is the same as if equal parallel forces were applied at the mass-centre.

The turning effect of a system of forces acting on any system of bodies is the same as if the mass-centre of the system were fixed.

The kinetic energy of any system of bodies is the sum of the kinetic energies due to : (*a*) the total mass collected at, and moving with the same speed as, the mass-centre of the system ; (*b*) the masses of the various bodies concentrated at their respective mass-centres, and moving round the mass-centre of the system ; (*c*) the rotations of the various bodies about their respective mass-centres.

Example.—If a retarding force be applied to the side wheel of a tricycle, the diminution of speed is the same as if the force were applied at the mass-centre of the machine and rider, while the turning effect on the system is the same as if the machine were at rest. (See chap. xviii.)

CHAPTER IX

72. Smooth and Rough Bodies.—If two perfectly *smooth* bodies are in contact, the mutual pressure is always in a direction at right angles to the surface of contact. Thus a smooth stone resting on the smooth frozen surface of a pond presses the ice vertically downwards, and the reaction from the ice is vertically upwards. If a horizontal force be applied to the stone it will move horizontally, the mutual pressure between it and the ice offering little resistance to this motion. A smooth surface may be defined as one which offers no resistance to the motion of a body upon it. No *perfectly* smooth surface exists in nature, but all are more or less rough, and offer resistance to the motion of a body upon them. This resistance is called *friction*.

Friction always acts in the direction opposed to the motion of a body, and thus tends to bring it to rest. In all machinery, therefore, great efforts are made to reduce the friction of the moving parts to the least possible value. In bearings of machinery friction is a most undesirable thing, but in other cases it may be a most useful agent. Without friction, no nut would remain tight after being screwed up on its bolt ; railways would be impossible ; and in cycling, not only would it be impossible to ride a bicycle upright on account of side-slip, but not even a tricycle could be driven by its rider along the ground, as the driving-wheels would simply skid.

73. Friction of Rest.—The greatest possible friction between two bodies is measured by the force parallel to the surface of contact which is just necessary to produce sliding. If a force acting parallel to the surface be less than this amount, the bodies will remain at rest.

It is found by experiment that friction varies with the nature of the surfaces of contact ; is proportional to the mutual normal pressure, and is independent of the area of the surface of contact so long as the pressure remains the same. When sliding motion actually takes place, the friction is often less than when the bodies are at rest in a state just bordering on motion.

74. **Coefficient of Friction.**—Let P be the force perpendicular to the surface of contact with which two bodies are pressed together, and F the force parallel to the surface which is just necessary to make one slide on the other. Then, as stated above, it is found experimentally that F is proportional to P. The ratio of F to P is called the *coefficient of friction* for the particular surfaces in contact ; this is usually denoted by the Greek letter μ. The coefficient of friction for iron on stone varies from ·3 to ·7 ; for wood on wood from ·3 to ·5 ; for metal on metal from ·15 to ·25 ; while for india-rubber on paper the author has observed values greater than 1·0.

Angle of Friction.—If two bodies be pressed together with a force P, making an angle θ with the normal to the surface, its components P_1, perpendicular to, and P_2, parallel to, the surface can be readily obtained by drawing. If $\dfrac{P_2}{P_1}$ be less than μ, no sliding will take place, but if $\dfrac{P_2}{P_1}$ be greater than μ, sliding will occur. The angle θ at which sliding just occurs is called the *angle of friction*.

If one of the bodies be an inclined plane and the other a body of weight W resting on it, the force P pressing them together is vertical, and therefore inclined at an angle θ to the normal to the surface ; the angle θ of the inclined plane at which the body will first slide down is evidently the same as the angle of friction, and is sometimes called the *angle of repose*.

75. **Journal Friction.**—It has been established by experiment that the friction of two bodies sliding on each other at moderate speeds, under moderate pressures, and with the surfaces either dry or very slightly lubricated, is independent of the speed of sliding and of the area of the surfaces of contact, and is simply

proportional to the mutual pressure. The experiments on which
the laws of friction rest were made by Morin in 1831. With well-
lubricated surfaces, such as in the bearings of machinery, the laws
of friction approximate to those relating to the friction of fluids.
Mr. Tower made experiments, for the Institution of Mechanical
Engineers, on the friction of cylindrical journals, which showed
that when the lubrication of the bearing was perfect, the total
friction remained constant for all loads within certain limits.
The coefficient of friction is therefore inversely proportional to the
load. The total friction also varies directly as the square root of
the speed. The coefficient of friction may therefore be represented by a formula

$$\mu = C \frac{\sqrt{v}}{P} \quad . \quad . \quad . \quad . \quad . \quad . \quad . \quad . \quad (1)$$

These experiments clearly show that with perfect lubrication the
journal does not actually touch the bearing, but floats on a thin
film of oil held between the two surfaces. The most perfect form
of lubrication is that in which the journal dips into a bath of oil.
The ascending surface drags with it a supply of oil, and so the
film between the journal and its bearing is constantly renewed.
If the lubrication is imperfect the coefficient of friction rises considerably, the conditions approaching then those which hold with
regard to solids.

The journal experimented on was 4 in. diameter by 6 in. long.
With oil-bath lubrication, running at 200 revolutions per minute,
and with a total load on the journal of 12,500 lbs., the total
friction at the surface of the journal was 12·5 lbs., giving a coefficient of friction of ·0010. With a total load of 2,400 lbs. the total
friction at the surface of the journal was 13·2 lbs., giving a coefficient of friction of ·0055.

76. **Collar Friction.**—The research committee of the Institution of Mechanical Engineers also carried out some experiments
on the friction of a collar bearing. The collar was a ring of mild
steel, 12 in. inside and 14 in. outside diameter, and bore against
gun-metal surfaces. The pressure per square inch which such a
bearing could safely carry was far less than in a cylindrical journal ;
the lowest coefficient of friction was ·031, corresponding to a

pressure of 90 lbs. per square inch, and a speed of 50 revolutions per minute. μ was practically constant, its average value being about ·036.

The much higher coefficient of friction in a collar than in a cylindrical bearing is no doubt due to the fact that a thin film of oil cannot be held between the surfaces, and be continually renewed.

77. **Pivot Friction.**—The relative motion of the surfaces of contact in a pivot bearing is one of rotation about an axis at right angles to the common surface of contact. Let figure 65 represent plan and elevation of a pivot bearing, o being the axis of rotation and ω the angular speed. The linear speed of rubbing of any point at a radius r from the centre will be ωr. Let W be the total load on the pivot, D its diameter, and R its radius. If we assume the pressure to be uniformly distributed over the surface of contact, the pressure per square inch will be,

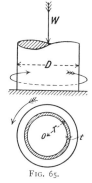

FIG. 65.

$$p = \frac{4 \ W}{\pi D^2} \cdot \ \cdot \ \cdot \ \cdot$$

The area of a ring of mean radius r and width t is $2 \pi r t$. The frictional resistance due to the pressure on this ring is $2 \mu \pi r t p$, and the moment about the centre O is $2 \mu \pi r^2 t p$. Summing the moments for all the rings into which the bearing surface of the pivot may be divided, the moment of the frictional resistance of the pivot is

$$\frac{2 \mu \pi R^3 p}{3} = \frac{\mu \ W D}{3} \quad \cdot \ \cdot \ \cdot \ \cdot \ \cdot \ \cdot \quad (2)$$

That is, the frictional resistance due to the load W may be supposed to act at a distance from the centre of one-third the diameter of the pivot.

If the diameter be very small, the average linear speed of rubbing, and therefore also the total work lost in friction, will be small. The work lost in friction is converted into heat, and the heat must be carried away as fast as it is generated, or the temperature of the bearing will rise and the surface will seize.

The pressure per square inch a bearing may safely carry will thus depend on the quantity of heat generated per unit of surface, and therefore on the speed of rubbing. This speed being small in pivot bearings, they may safely work under greater pressure than collar bearings.

It will be shown (chapter xxv.) that the motion of a ball in a ball bearing is compounded of rolling and spinning. Rolling friction is discussed in section 78.

Spinning friction of a ball on its path is analogous to pivot friction, with the exception that the surfaces have contact only

FIG. 66.

at a point when no load is applied. When the ball is pressed on its path by a force W (fig. 66) the surfaces in the immediate neighbourhood of the geometrical point of contact o are deformed, and contact takes place over an area $a\,o\,b$. The intensity of pressure is probably greatest at o, and diminishes to zero at a and b. The frictional resistance thus ultimately depends on the diameter of the ball, its hardness, the radius of curvature of its path, the load W as well as the coefficient of friction. No experiments on the spinning friction of balls have been made, to the author's knowledge, though they would be of great use in arriving at a true theory of ball-bearings.

78. **Rolling Friction.**—When a cylindrical roller rolls on a perfectly horizontal surface there is a resistance to its motion, called *rolling friction*. Professor Osborne Reynolds has investigated the nature of rolling resistance, and he finds that it is due

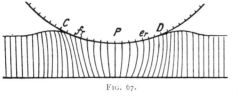

FIG. 67.

to actual sliding of the surfaces in contact. No material in nature is absolutely rigid, so that the roller will have an *area* of contact with the surface on which it rolls, the extent of which will vary with the material and with the curvature of the surfaces in contact. Figure 67 shows what takes place when an iron roller rests on

a flat thick sheet of india-rubber. The roller sinks into the
rubber and has contact with it from C to D. Lines drawn on the
india-rubber originally parallel and equidistant are distorted as
shown. The motion of the roller being from the left to the right,
contact begins at D and ceases at C. The surface of the
rubber is depressed at P, the lowest point of the wheel, and is
bulged upwards in front of, and behind, the roller. The vertical
compression of the layers of the rubber below P causes them to
bulge laterally, whilst the extension vertically of the layers in
front of D causes them to get thinner laterally. This creates a
tendency to a creeping motion of the rubber along the roller. If
the resistance to sliding friction between the surfaces be great, no
relative slipping may take place, but if the frictional resistance be
small, slipping will take place, and energy will be expended. $e\,r$
and $f\,r$ limit the surfaces over which there is no slipping ; between
$e\,r$ and D, and again between $f\,r$ and C, there is no relative slipping.

This action is such as to cause the distance actually travelled
by a roller in one revolution to be different from the geometric
distance. Thus, an iron roller rolled about two per cent. less
per revolution when rolling on rubber than when rolling on
wood or iron. The following table shows the actual slipping of
a rubber tyre three-quarters of an inch thick, glued to a roller.

Nature of surface	Distance travelled in one revolution	Circumference of the ring	Amount of slipping
Steel bar	22·55 in.	22·5 in.	— 0·05 in.
India-rubber 0·156 in. thick (clean)	22·55 ,,	22·5 ,,	— 0·05 ,,
Ditto (black-leaded) . .	22·55 ,,	22·5 ,,	— 0·05 ,,
Ditto 0·08 in. thick (clean) .	22·5 ,,	22·5 ,,	0·0 ,,
Ditto (black-leaded) . .	22·52 ,,	22·5 ,,	— 0·02 ,,
Ditto 0·36 in. thick (clean) .	22·39 ,,	22·5 ,,	0·11 ,,
Ditto (black-leaded) . .	22·42 ,,	22·5 ,,	0·08 ,,
Ditto 0·75 in. thick (clean) .	22·4 ,,	22·5 ,,	0·1 ,,
Ditto (black-leaded) . .	22·4 ,,	22·5 ,,	0·1 ,,

With regard to the work lost in rolling friction, a little consideration will show that a soft substance like rubber will waste
more work, and therefore have a greater rolling resistance than
a harder substance such as iron or steel. Professor Osborne

Reynolds has shown that the rolling resistance of rubber is about ten times that of iron. Experiments were made on a cast-iron roller and plane surfaces of different materials, the plane being inclined sufficiently to cause the roller to start from rest. The following table shows the mean of results for various conditions of surface and manner of starting, the figures tabulated giving the vertical rise in five thousand parts horizontal.

Nature of surface	Starts from rest		Starts from rest in the opposite direction		Mean
	Clean	Oiled or black-leaded	Clean	Oiled or black-leaded	
Cast-iron . .	5·66	5·61	2·57	2·36	4·05
Glass . . .	6·32	5·96	1·93	2·56	4·19
Brass . . .	7·75	6·53	2·07	2·587	4·73
Boxwood . .	10·05	9·25	5·71	2·34	7·09
India-rubber .	35·37	38·75	31·87	28·00	33·24

CHAPTER X

79. Action and Reaction.—Newton's third law of motion is thus enunciated :

"To every action there is always an equal and contrary re-action ; or, the mutual action of any two bodies are always equal, and oppositely directed in the same straight line ; or, action and reaction are equal and opposite."

We have in the preceding chapters spoken of single forces, but remembering that force can only be exerted by the mutual action of two bodies, the truth of Newton's third law is apparent. If a rider press his saddle downwards with a force of 150 lbs., the saddle presses him upwards with an equal force ; if he pull at his handles, the handles exert an equal force on his hands in the opposite direction. The passive forces thus called into existence are quite as real as what are apparently more active forces. For example, suppose a man to pull at the end of a rope with a force of 100 lbs., the other end of which is fastened to a hook in a wall, the hook exerts on the rope a contrary pull of 100 lbs. Suppose now that two men at opposite ends of the rope each exert a pull of 100 lbs., the 'active' pull of the second man in the second case is exactly equivalent to the 'passive' pull of the hook in the first case.

The different forces must be carefully distinguished in such cases. Thus, in figure 68 the

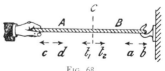

FIG. 68.

force exerted by the rope on the hook in the wall is in the direction a, the force exerted by the hook on the rope is in the direction b, the pull exerted by the man on the end of the rope is in the direction c, and the pull of the rope on the man is in the direction d.

80. **Stress and Strain.**—Consider the rope divided at C into two parts, A and B. The part A will exert a pull in the direction t_1 on B, and similarly the part B will exert a pull in the direction t_2 on A. The two forces t_1 and t_2 constitute a straining action at C.

In the case of a rope the forces b and c acting on its ends are directed outwards, and the straining action is called a *tension*.

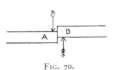

If a bar (fig. 69) be subjected to equal forces, a and b, at its ends acting inwards, the straining action is called a *compression*.

Fig. 69.

In figures 68 and 69 the parts A and B tend to separate from or approach each other in a direction at right angles to the plane C. If the parts A and B tend to slide relative to each

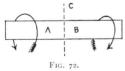

Fig. 70.

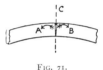

Fig. 71.

other in the direction of the plane (fig. 70), the straining action is called *shearing*.

If the parts A and B tend to rotate about an axis perpendicular to the axis of the bar (fig. 71), the straining action is called *bending*.

If the parts A and B tend to rotate in opposite directions

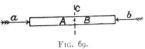

Fig. 72.

about the axis of the bar (fig. 72), the straining action is called *torsion*.

Compound straining actions consisting of all or any of the simple straining actions may take place.

These straining actions are resisted by the mutual action between the particles of the material, this mutual action constituting the *stress* at the point.

Tensile Stress.—If a bar be subjected to forces as in figure 68, every transverse section throughout its length is subject to a *tensile stress*. If P be the magnitude of the forces b and c (fig. 68), and A the area of the transverse section at C, the force acting on each unit of transverse section—that is, the tensile stress per unit of area—is

$$p = \frac{P}{A} \quad . \quad . \quad . \quad . \quad . \quad . \quad . \quad (1)$$

Compressive Stress.—In the same way, if the bar be subjected to forces directed inwards (fig. 69), every transverse section of it is subjected to a *compressive stress*. The compressive stress per unit of area will also be in this case

$$p = \frac{P}{A} \quad . \quad . \quad . \quad . \quad . \quad . \quad . \quad (1)$$

81. **Elasticity**.—If a bar of unit area (fig. 73) be fixed at one end, and subjected at the other end to a load, p, it is found that its length is increased by a small quantity. If the load does not exceed a certain limit, when it is removed the bar recovers its original length. It is found experimentally that with nearly all bodies, metals especially, this increase in length, x, is proportional to the load, and to the original length of the bar, so that we may write

$$x = \frac{p\,l}{E}$$

or,

$$\frac{p}{E} = \frac{x}{l} \quad . \quad . \quad . \quad . \quad . \quad . \quad (2)$$

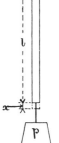

FIG. 73.

where E is a constant, the value of which depends on the nature of the material. The ratio of this elongation to the original length—that is, the extension per unit of length—is called the *extension*, and denoting it by e we have

$$e = \frac{x}{l} \quad . \quad . \quad . \quad . \quad . \quad . \quad (3)$$

substituting in (2) we have

$$\frac{p}{e} = E \quad . \quad . \quad . \quad . \quad . \quad . \quad (4)$$

E is called the *modulus of elasticity* of the material. A general idea of its nature may be had as follows : Conceive the material to be infinitely strong, and to stretch under heavy loads at the same rate as under small loads. Let the load be increased until the change of length, x, is equal to l, the original length of the bar. Substituting $x = l$ in (2) we have $p = E$. That is, the modulus of elasticity is the stress which would be required to extend the bar to twice its original length, provided it remained perfectly elastic up to this limit.

The value of E for cast iron varies from 14,000,000 to 23,000,000 lbs. per sq. in. ; for wrought-iron bars, from 27,000,000 to 31,000,000 lbs. per sq. in. ; for steel plate 31,000,000 lbs. per sq. in. ; for cast steel, tempered, 36,000,000 lbs. per sq. in.

Example.—The spokes of a wheel are No. 16 W.G., 12 inches long ; the nipples are screwed up till the spokes are stretched $\frac{1}{100}$ in. What is the pull on each spoke ?

Taking $E = 36,000,000$ lbs. per sq. in., and substituting in (2), we get

$$\frac{p}{36,000,000} = \frac{\frac{1}{100}}{12},$$

from which,

$$p = 30,000 \text{ lbs. per sq. in.}$$

A, the sectional area of each spoke (Table XII., p. 346), is ·00322 sq. in. ; P, the total pull on the spoke, is $p\,A$. Therefore,

$$P = 30,000 \times ·00322 = 96·6 \text{ lbs.}$$

82. **Work done in Stretching a Bar.**—In section 81 we have found the stress, p, corresponding to an extension, x, of a bar ; we can now find the work done in stretching the bar. It will be convenient to draw a diagram to represent graphically the relation between p and x. Let $A\,B_0$ (fig. 74) be the bar, fixed at A, and let B_0 be the position of the lower end when subjected to no load. Under the action of the load P let the lower end be stretched into position B, then $B_0\,B = x$. Let $B\,N$ be drawn at right angles to the axis of the bar, representing to any convenient scale the load P. If these processes be repeated for a

number of different values of P, the locus of the point N will be a straight line passing through B_0, and the area of the triangle $B_0 B N$ will represent the work done in stretching the bar the distance $B_0 B$. There-fore,

$$\text{Work done} = \tfrac{1}{2} P x \quad . \quad . \quad . \quad (5)$$

Substitute the value of x from (2) in (5), and remembering that $P = A p$, we get

$$\text{Work done} = \frac{p^2}{E} \cdot \frac{A l}{2} = \frac{p^2}{2E} \times \text{volume of}$$
$$\text{the bar} \quad . \quad . \quad . \quad . \quad . \quad (6)$$

Therefore the quantities of work done in producing a given stress, p, on different bars of the same material are proportional to the volumes of the bars. On bars of equal volume but of different materials the quan-tities of work done in producing a given

FIG. 74.

stress, p, are inversely proportional to the moduli of elasticity. The work done in stretching a given bar is proportional to the square of the stress produced.

If the bar be tested up to its elastic limit, f, the work done is

$$\frac{f^2}{2 E} \times \text{volume of bar.}$$

This gives a measure of the work that can be done on the bar without permanently stretching it. The quantity $\dfrac{f^2}{2 E}$ depends only on the material, is called its *modulus of resilience*, and gives a convenient measure of the value of the material for resisting im-pact or shock.

Example.—The work done in stretching the spoke in the example, section 81, is

$$\tfrac{1}{2} . \times 96 \cdot 6 \times \tfrac{1}{100} = \cdot 483 \text{ inch-lb. or } \cdot 04 \text{ foot-lb.}$$

83. **Framed Structures.**—A framed structure is formed by jointing together the ends of a number of bars by pins in such a manner that there can be no relative motion of the bars without

distorting one or more. If each bar be held at only two points, and the external forces be applied at the pins, the stress on any bar must be parallel to its axis, and there will be no bending. In

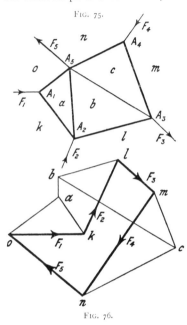

FIG. 75.

figure 75 let the external forces F_1, F_2 . . . be applied at the pins A_1, A_2 . . . Let the frame be in equilibrium under the forces, and let F_1, F_2 . . . (fig. 76) be the sides of the force-polygon. If all the forces F_1, F_2 . . . be known, it will be possible, in general, to find the stress on each bar of the frame by a few applications of the principle of the force-triangle. In a trussed beam (*e.g.* a bridge, roof, or bicycle frame) the external forces are the loads carried by the structure, whose magnitude and lines of action are generally known, and the reactions at the supports. If there are two supports the reactions can be determined by the methods of section 17, so that they shall be in equilibrium with the loads.

FIG. 76.

To find the stresses on the individual members of the frame we begin by choosing a pin at which two bars meet and one external load acts ; the magnitude and direction of the latter, and the direction of the forces exerted by the bars on the pin, being known, the force-triangle for the pin can be drawn. Thus, beginning at the pin A_1, on which three forces (the external force F_1, and the thrusts of the bars $A_1 A_2$ and $A_1 A_3$) act, the force-triangle can be at once drawn. Before proceeding with this drawing it will be convenient to use the following notation : Let the spaces into which the bars divide the frame be denoted by a, b, . . . , and the spaces between the external forces F_1, F_2, . . .

by k, l, . . . , then the bar $A_1 A_2$ which divides the spaces a and k will be denoted by $a k$, the stress on this bar will also be denoted by $a k$. The force-triangle for the pin A_1, at which point the spaces a, o and k meet, is $a o k$ (fig. 76). Proceeding now to the pin A_2, at which four forces act, the external force F_2 and that exerted by the bar $a k$ are known, and the direction of the forces exerted by the bars $a b$ and $b l$ are known. Two sides, $a k$ and $k l$, of the force-polygon for the pin A_2 are already drawn, the polygon is completed by drawing $a b$ and $l b$ (fig. 76) respectively, parallel to the bars $a b$ and $l b$ (fig. 75). Proceeding now to the pin A_3, only two forces are as yet unknown, and of the force-polygon two sides, $b l$ and $l m$, are already drawn. The remaining sides, $b c$ and $m c$, are drawn parallel to the corresponding bars (fig. 75). At the pin A_5, four of the forces acting are already known, and the corresponding sides, $n o$, $o a$, $a b$, and $b c$, of the force-polygon are already drawn. The side $n c$ of the force-diagram must therefore be parallel to the corresponding bar of the frame-diagram, and a check on the accuracy of the drawing is obtained.

With the above notation, the letters $A_1 A_2$ . . . and $F_1 F_2$ . . . may be suppressed.

Figure 75 is called the *frame-diagram* and figure 76 the *stress-diagram*, or *force-diagram*. In the force-diagram, the polygon of external forces is drawn in thick lines, and the direction of each force is indicated by an arrow. From these arrows it will be easy to determine whether the stress on any member of the frame is tensile or compressive.

The total force on any member of a framed structure being obtained, its sectional area can be obtained at once by formula (1).

84. **Thin Tubes subjected to Internal Pressure.**— An important case of simple tension is that of a hollow cylinder subjected to fluid pressure ; *e.g.* the internal shell of a steam boiler, or the pneumatic tyre of a cycle wheel. In long cylindrical boilers the flat ends have to be made rigid in order to preserve their form under internal pressure, while the cylindrical shell is in stable equilibrium under the action of the internal pressure. A pneumatic tyre of circular section is also of stable form under internal pressure ; a deformation by external pressure at any point will

be removed as soon as the external pressure at the point be removed.

Let p be the internal pressure in lbs. per sq. in., d the diameter and t the thickness of the tube (fig. 77).   Consider a section by a

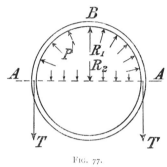

FIG. 77.

plane, A A, passing through the axis of the tube.   The upper half, $A B A$, is under the action of the internal pressure p, distributed over its inner surface, and the forces T due to the pull of the lower part of the tube ; therefore 2 $T =$ the resultant of pressure p on the half tube. This resultant can be easily found by the following artifice : Consider a stiff flat plate joined

at A A to the half tube, so as to form a D tube.   If this tube be subjected to internal pressure, p, and to no external forces, it must remain at rest ; if otherwise, we would obtain perpetual motion.   Therefore, the resultant pressure R_1 on the curved part must be equal and opposite to the resultant pressure R_2 on the flat portion of the tube.   If we consider a portion of the tube 1 in. long in the direction of the axis,

$$R_2 = p\,d,$$

and therefore

$$2\;T = p\,d \quad . \quad . \quad . \quad . \quad . \quad . \quad (7)$$

But if f be the intensity of the tension on the sides of the tube, $T = f\,t$

$$\therefore\; 2\,f\,t = p\,d, \text{ or } f = \frac{p\,d}{2\,t} \quad . \quad . \quad . \quad (8)$$

Example. — A pneumatic tyre $1\frac{3}{4}$ in. inside diameter, outer cover $\frac{1}{16}$ in. thick, is subjected to an air pressure of 30 lbs. per square inch.   The average tensile stress on the outer cover is

$$f = \frac{30 \times 1.75}{2 \times \frac{1}{16}} = 420 \text{ lbs. per sq. in.}$$

85. **Introductory.**—We have in chapter x. considered the stresses on a bar acted on by forces parallel to its axis. We now proceed to consider the stresses on a bar due to forces the lines of action of which pass through the axis, but do not coincide with it. Each force may be resolved into two components, respectively parallel to, and at right angles to, the axis. The components parallel to the axis may be treated as in the previous chapter. Of the transverse forces, the simplest case is that in which they all lie in the same plane, a beam supporting vertical loads being a typical example. Such a beam must be acted on by at least three forces, the *load* and the two *reactions* at the supports.

86. **Shearing-force on a Beam.**—If a bar in equilibrium be acted on by three parallel forces at right angles to its axis (fig. 78), every section by a plane parallel to the direction of the forces will be subjected to a bending stress.

Consider the body divided into two portions by a plane at X. Under the action of the force R_1 the part A will tend to move upwards relative to the part B. The part A therefore acts on the part B with a force R_1' equal and parallel to R_1, and the part B reacts on the part A with an equal opposite force R_1''. The two forces R_1' and R_1'' at X constitute a shearing at the section. It will easily be seen that the shearing-force will be the same for all sections of the beam between the points of application of the forces R_1 and W, and that the shearing-force on the section X_1 will be the algebraic sum of the forces to the left-hand side, or to the right-hand side, of the section. This is true for a beam acted on by any number of parallel forces.

In particular, if a beam be supported at its ends (fig. 78) and loaded with a weight, W, the reactions R_1 and R_2 at the supports will, by section 49, be equal to

$$\frac{b\,W}{a+b}, \quad \frac{a\,W}{a+b} \qquad . \qquad . \qquad . \qquad . \qquad . \qquad . \qquad (1)$$

where a and b are the segments in which the length of the beam is divided at the point of application of the load. The shearing-force on the part A will be equal to R_1, and the shearing-force on the part B will be equal to $R_1 - W = -R_2$.

Shearing-force Diagram.—The value of the shearing-force at any section of a beam is very conveniently represented by drawing an ordinate of length equal to the shearing-force at the corresponding section, any convenient scale being chosen. The shaded figure (fig. 79) is the shearing-force diagram for a beam supported at the ends and loaded with a single weight.

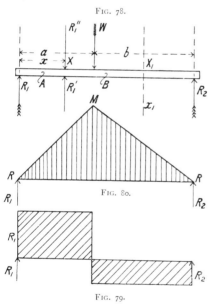

FIG. 78.

FIG. 80.

FIG. 79.

The shearing-force at the section X (fig. 78) is of such a nature that the part on the left-hand side tends to slide *upwards* relative to the part on the right-hand side of the section. The shearing-force at X_1 is of such a nature that the part on the left tends to slide *downwards* relative to the part at the right of the section. Thus shearing-forces may be opposite in sign; if that at X be called positive, that at X_1 will be negative. The diagram (fig. 79) is drawn in accordance with this convention.

87. **Bending-moment.**—If a bar of length, l, be fixed horizontally into a wall (fig. 83), and be loaded at the other end with a

weight W, the said weight will tend to make bar turn at its support, the tendency being measured by the moment Wl of the force. This tendency is resisted by the reaction of the wall on the beam. The section of the beam at the support is said to be subjected to a *bending-moment* of magnitude Wl.

From this definition a weight of 50 lbs. at a distance of 20 inches will produce the same bending-moment as a weight of 100 lbs. at a distance of 10 inches; the bending-moment being 50×20, or $100 \times 10 = 1000$ inch-lbs.

Returning to the discussion of figure 78, it will be seen that the part A is acted on by two equal, parallel, but opposite forces, R_1 and R_1'', constituting a couple of moment $R_1 x$ tending to turn the part A. But the part A is actually at rest; it must, therefore be acted on by an equal and opposite couple. The only other forces acting on A are those exerted by the part B at the section X. The upper part of portion B (fig. 81, which is part of figure 78 enlarged) acts on the

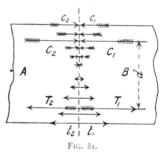

FIG. 81.

portion A with a number of forces, $c_1 c_1$, diminishing in intensity from the top towards the middle of the beam; the resultant of these may be represented by C_1. The lower part of B acts on A with the forces $t_1 t_1$, whose resultant may be represented by T_1. Since the part A is in equilibrium, the resultant of all the horizontal forces acting on it must be zero; therefore T_1 and C_1 are equal in magnitude, and constitute a couple which must be equal to $R_1 x$. If d be the distance between T_1 and C_1, we must therefore have

$$T_1 d = R_1 x.$$

The part A acts on the part B with forces c_2 at the top, and forces t_2 at the bottom of the beam; the resultants being indicated by C_2 and T_2 respectively. The two sets of forces c_1 and c_2 constitute a set of compressive stresses on the upper portion of the beam at X, and the two sets of forces t_1 and t_2 constitute a set of tensile stresses on the lower portion of the beam. The moment of the couple $R_1 x$ is called the bending-moment at the section X; while the moment of the couple $T_1 d$ is called the moment of resistance of the section.

The existence of the shearing-force and bending-moment at any section of a beam can be experimentally demonstrated by

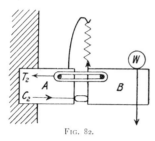

FIG. 82.

actually cutting the beam, and replacing by suitably disposed fastenings the molecular forces removed by the cutting. Figure 82 shows diagrammatically a cantilever treated in this manner. The shearing-force at the section is replaced by the upward pull W of a spiral spring, and the couple acting on the part B formed by the load W, and the

pull of the spring is balanced by the equal and opposite couple formed by the pull T_2 of the fastening bands at the top and the thrust C_2 of the short strut at the bottom of the section.

Bending-moment Diagram. — The bending-moment at any section of a beam can be conveniently represented by a diagram, the ordinate being set up equal in length to the bending-moment at the corresponding section.

Since the bending-moment at the section X (fig. 78) is the product of the force R_1 into the distance x of the section from its point of application, the further the section X be taken from the end of the beam the greater will be the bending-moment. In the case of a beam supported at the ends and loaded at an intermediate point with a weight W, the bending-moment M on the section over which W acts will be given by—

$$M = R_1 a = \frac{a b}{a + b} W \quad . \quad . \quad . \quad . \quad (2)$$

and the bending-moment on any section between R_1 and W will be represented by the ordinate of the shaded area in figure 80.

The bending-moment at the section X_1 (fig. 78) is the sum of the moments of the forces R_1 and W about X_1; or is equal to the moment of the force R_2 about X_1.

88. **Simple Examples of Beams.**—A few of the most commonly occurring examples of beams may be discussed here. Figure 83 shows a cantilever of length, l, supporting a weight, W, at its end. The bending-moment at a section very close to the support is Wl, that at a section distant x from the outer end of the cantilever is Wx. The bending-moment diagram is, therefore, a straight line, the maximum ordinate, Wl, being at the support, that at the end zero. The shearing-force is equal to W for all sections; the shearing-force diagram is, therefore, a straight line parallel to the axis.

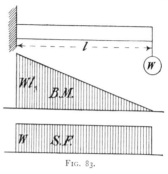

FIG. 83.

Figure 84 shows a cantilever loaded uniformly, the total weight being W. The resultant weight acts at the middle of the cantilever distant $\frac{l}{2}$ from the support, the bending-moment at the support is, therefore, $\frac{Wl}{2}$. At any section distant x from the end of the cantilever, we find the bending-moment as follows: Consider the portion of the cantilever lying to the right of the section, the resultant of the load resting on it is wx, w being the weight per unit of

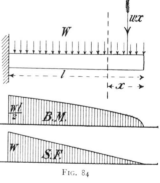

FIG. 84.

length, and acts at a distance $\frac{x}{2}$ from the section. The bending-moment on the section is therefore

$$M = \frac{wx^2}{2} = \frac{Wx^2}{2l} \quad \cdot \quad \cdot \quad \cdot \quad \cdot \quad \cdot \quad \cdot \quad (3)$$

Plotting these values for different values of x, the bending-moment curve is a parabola.

The shearing-force on the section distant x from the end is $w\,x = \dfrac{W\,x}{l}$. Plotting these values for different values of x, we get the shearing-force curve a straight line, having the ordinate W at the support and zero ordinate at the end.

Figure 85 shows a beam of span, l, supporting a load, W, at the middle. The reactions at the support are evidently each equal to

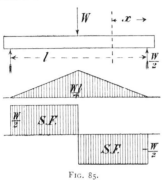

$\dfrac{W}{2}$, the bending-moment at any section distant x from the end is therefore $\dfrac{W x}{2}$, x being less than $\dfrac{l}{2}$. At the middle of the beam the bending-moment is a maximum, and equal to

$$\frac{W}{2} \cdot \frac{l}{2} = \frac{W l}{4} \quad . \quad . \quad (4)$$

FIG. 85.

The bending-moment curve is a triangle, the maximum ordinate being in the middle. The shearing-force is constant and equal to

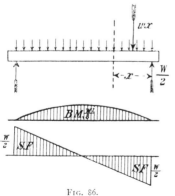

$\dfrac{W}{2}$ from one end up to the middle of the beam, then changes sign and becomes $-\dfrac{W}{2}$ over the other half.

Figure 86 shows a beam supporting a load, W, uniformly distributed. The reaction at each support is evidently $\dfrac{W}{2}$; the bending-moment at a section distant x from the end is the sum of the moments due to

FIG. 86.

the reaction $\dfrac{W}{2}$, and of the resultant load $w\,x$ acting on the right-hand side of the action at a distance $\dfrac{x}{2}$ from the section,

$$\therefore M = \frac{W}{2}x - w\,x \cdot \frac{x}{2} = \frac{W}{2}\left(x - \frac{x^2}{l}\right) \quad . \quad . \quad (5)$$

If x be made equal to $\dfrac{l}{2}$, the above formula gives the bending-moment at the middle of the beam, $M_0 = \dfrac{W}{2}\left(\dfrac{l}{2} - \dfrac{l}{4}\right) = \dfrac{Wl}{8}$. The bending-moment curve is a parabola with its maximum ordinate $\dfrac{Wl}{8}$ at the middle of the beam.

89. **Beam supporting a Number of Loads at Different Points.**—The loads and their positions along the beam being given, the reaction R_1 at one support can be found by taking moments about the other support ; the bending-moment at any section can then be calculated by adding algebraically the moments of all the forces on either one side or other of that section. The reactions R_1 and R_2 at the supports can also be found by the method of sections 47 and 48. Since in this case the forces are all parallel, the construction is simplified ; the force-polygon becomes a straight line, and the corners of the link-polygon lie on the vertical lines of action of the loads and reactions.

Figure 87 shows a beam supporting a number of weights, W_1, W_2, W_3, W_4, and figure 88 the force-polygon a, b, c, d, e. The construction of figure 41 becomes as follows : From any point p_1 on the line of action of W_1 draw a straight line b parallel to the line $O\,b$ (fig. 88). From p_2, where this line cuts the line of action of W_2, draw a straight line, c, parallel to the line $O\,c$; continuing this process until the point p_4 on the line of action W_4 is reached.

Through p_1 and p_4 draw $p_1\,p_r$ and $p_4\,p_r$ parallel to $O\,a$ and $O\,e$ respectively, intersecting each other at p_r and the lines of action of R_1 and R_2 at r_1 and r_2 respectively. The resultant of the loads W_1, W_2, W_3 and W_4 passes through p_r. Through O draw $O\,r$ parallel to $r_1\,r_2$; then the reactions R_1 and R_2 are equal to $r\,a$ and $e\,r$ respectively.

Link-polygon as Bending-moment Diagram.—If the pole O be chosen at random, the closing line $r_1\,r_2$ of the link-polygon will not, in general, be parallel to the axis of the beam. Let a new pole, O^1, be taken by drawing $O\,O^1$ parallel to, and $r\,O^1$ at right angles to, the lines of action of the loads W_1, W_2 . . ., and let a

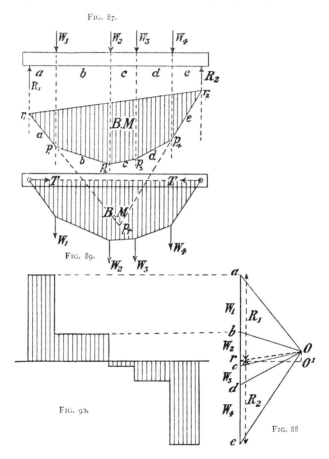

Fig. 87.

Fig. 89.

Fig. 90.

Fig. 88

new link-polygon (fig. 89) be drawn. If a thin wire be made to the same outline as this same polygon and be attached to the beam, and the loads W_1, W_2 . . . attached at the angles, it is evident that the compound structure formed by the bar and wire is subjected to the same bending stresses as the beam (fig. 87). In both cases the dispositions of the loads and reactions are identical ; but in the compound structure the bar is subjected to a thrust, T, represented in the force-diagram (fig. 88) by $O^1 r$. Considering the corner of the wire at which W_1 acts, the tensions on the two portions of the wire, and the force W_1 are in equilibrium, and are represented by the force-triangle $O^1 a b$ (fig. 88).; similarly for the other portions of the wire. It will be noticed that at each part of the wire the horizontal component of the pull is equal to $O^1 r$; that is, equal to T. Taking any vertical section of the compound structure (fig. 89) the mutual actions consist of a thrust, T, on the bar, an equal horizontal pull, T, on the wire, and the vertical component of the pull on the wire. The two former constitute the bending-couple at the section, the latter the shearing-force. The bending-moment on any section of the beam is therefore equal to $T h$, h being the ordinate of the link-polygon ; the link-polygon can therefore be used as a bending-moment diagram.

The shearing-force on any section of the beam (fig. 87) is equal to the vertical component of the pull on the wire (fig. 89), which is equal to the vertical component of the corresponding line from the pole O^1 (fig. 88). A shearing-force diagram (fig. 90) can therefore be constructed by projecting over a base line from r, and straight lines from a, b . . . (fig. 88) to the corresponding divisions of the beam.

Example.—Calculate, and draw, a bending-moment diagram for the frame of a tandem bicycle carrying two riders, each 150 lbs. weight (30 lbs. of which is assumed to be applied at the crank-axle) ; the wheel-base being 64 inches long, the rear crank-axle being 19 inches in front of the rear wheel centre, the crank-axles 22 inches apart, and the saddles 10 inches behind their respective crank-axles.

The figures of illustrations are given in chapter xxiii., page 327.

To calculate the reactions on the wheel spindles, take moments about the centre of the rear wheel—

$$(120 \times 9) + (30 \times 19) + (120 \times 31) + (30 \times 41) - (R \times 64) = 0$$

from which, $R_1 = 103 \cdot 1$ lbs.,

and $R_2 = 196 \cdot 9$ lbs.

The greatest bending-moment, which occurs on the vertical section passing through the front seat, is

$$M = (103 \cdot 1 \times 33) - (30 \times 10) = 3,102 \text{ inch-lbs.}$$

The frame, or beam (fig. 321) is drawn $\frac{1}{32}$nd full size ; the scale of the force-diagram (fig. 323) is 1 inch to 400 lbs., and the pole distance O^1 corresponds to 125 lbs. ; 1 inch ordinate of the bending-moment diagram (fig. 324) therefore represents 32 in. × 125 lbs., *i.e.* 4,000 inch-lbs.

The results got by the graphical and arithmetical methods must agree ; thus a check on the accuracy of the work is obtained.

90. **Nature of Bending Stresses.**—We must now consider more minutely the nature of the stresses t and c (fig. 81) on any section subject to bending.

Let a beam be acted on by two equal and opposite couples at its ends ; it will be bent into a form, shown greatly exaggerated in figure 91. It can be easily seen that the bending-moment on the middle portion of the beam will be of the same value throughout, and if the section is uniform, the amount of bending will be the same at all sections ; that is, the beam, originally straight, will be bent into a circular arc.

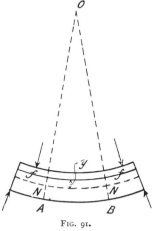

FIG. 91.

Consider the portion of the beam included between two parallel sections A and B. After bending, these sections are inclined, and if produced, will meet at the centre of curvature of the beam. The top fibres of the beam will be shortened and the lower fibres lengthened, while those at some intermediate layer, NN, will be unaltered in length. The surface in which the centres of the fibres NN lie is called the *neutral surface* of the beam, while its line of intersection with a transverse plane is called the *neutral axis* of the section. Now, suppose that the fibres could be laid out flat and of exactly the same length as they are after bending. If the left-hand ends all lay in the plane $A\ A$ (fig. 92) at right angles to $N N$, the other ends must evidently lie in a plane $B^1\ B^1$; $B\ B$ representing the plane in which the ends of the unstretched fibres would lie. The distance, parallel to $N N$, included between the

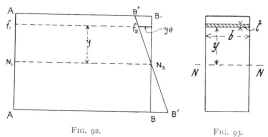

FIG. 92.   FIG. 93.

lines $B\ B$ and $B^1\ B^1$ gives the amount of the contraction or elongation of the corresponding fibres. The elongation or contraction of any fibre is thus seen to be proportional to its distance from $N N$. Now the stress on a bar or fibre is proportional to the extension produced ; therefore the stress on the fibres of a beam varies as the distance from the neutral axis.

Let O be the centre, and R the radius of curvature of $N N$ (fig. 91), y the distance of any fibre f above the neutral axis, θ the angle $N O N$ subtended at the centre O by the portion of the fibre considered. The radius of curvature of the fibre f is $(R - y)$, the length of the arc $f_1\ f_2$ (fig. 92) is therefore $(R - y)\ \theta$; and the length of the arc $N_1\ N_2$ is $R\ \theta$. A fibre at the neutral axis is unaltered in length by bending, so the length $N_1\ N_2$ is the same as in the straight position. The length of the fibre $f_1\ f_2$ was originally equal to that of $N_1\ N_2$; the decrease in its length is therefore

$$R\ \theta - (R - y)\ \theta = y\ \theta ;$$

its compression per unit of length is therefore

$$\frac{y\ \theta}{R\ \theta} = \frac{y}{R}.$$

By section 81, the stress producing this compression is

$$p = \frac{E\ y}{R} \quad . \quad . \quad . \quad . \quad . \quad . \quad (6)$$

That is, the intensity of stress on any fibre of a beam subject to bending is proportional to its distance from the neutral axis, and inversely proportional to the radius of curvature of the neutral axis. If a fibre below the neutral axis be taken, y will be negative, the fibre will be stretched, and the stress on it will be tensile.

Since the material near the neutral axis is subjected to a low stress, it adds very little to the strength of the beam, while it adds to the weight. It is therefore economical to place the material as far as possible from the centre of the section. The framework of the earliest bicycles was made of solid bars ; but a great saving of weight, without sacrificing strength, was effected by using hollow tubes. The same principle is carried out to a fuller extent in a well-designed Safety frame ; the top- and bottom-tubes together forming a beam, in which practically all the material is at a great distance from the neutral axis. If the frame be badly designed, however, the top- and bottom-tubes may form merely two more or less independent beams, instead of one very deep beam.

91. **Position of Neutral Axis.**—Consider the equilibrium of the portion of the beam to the left hand of section A (fig. 91). There are no external horizontal forces acting on this portion, and therefore the resultant of the horizontal forces due to the internal reaction of the particles at the section A must be zero.

Let figure 93 be the transverse section at A (fig. 91), $N N$ being the neutral axis. The part of the section above $N N$ is subjected to compression, that below $N N$ to tension ; the resultant compressive force must therefore be equal to the resultant force of tension. Consider a strip of the section of breadth b, and thickness t, at a distance y from the neutral axis ; the area of this strip is $b\,t$, the stress per square inch is $\dfrac{E\,y}{R}$; the total force on it is therefore

$$\frac{E}{R} \cdot b\,t\,y.$$

The total force on the whole section will be the sum of the forces on all such strips ; compression being considered positive and tension negative. $\dfrac{E}{R}$ is the same for all the strips, therefore the resultant force on the section may be written

$$\dfrac{E}{R} \Sigma \, b \, t \, y,$$

$\Sigma \, b \, t \, y$ indicating the sum of all the products $b \, t \, y$. Since the resultant force on the section is zero, we must have

$$\Sigma \, b \, t \, y = 0 \quad . \quad . \quad . \quad . \quad . \quad (7)$$

Referring to section 50, it will be seen that this condition is equivalent to saying that the neutral axis must pass through the mass-centre of the section.

92. **Moment of Inertia of an Area.**—In figure 93, $b \, t$ is the area of a narrow strip parallel to, and distant y from, the axis $N \, N$; $b \, t \, y^2$ is therefore the product of a small element of area into the square of its distance from the axis. The sum of such products for all the elementary strips into which the given area can be divided is called the *moment of inertia of the area*, and, as shall be shown in the next section, is of fundamental importance in the theory of bending.

The calculation of moments of inertia for areas of given shape is beyond the scope of an elementary work like the present ; a few of the most important results will be given for convenience of reference.

Let I denote the moment of inertia about an axis passing through the mass centre. Then, for a square of side h,

$$I = \dfrac{1}{12} \, h^4 \quad . \quad . \quad . \quad . \quad . \quad (8)$$

For a circle of diameter d,

$$I = \dfrac{\pi}{64} \, d^4 \quad . \quad . \quad . \quad . \quad . \quad (9)$$

For a rectangular section of breadth b and depth h (perpendicular to the neutral axis),

$$I = \dfrac{1}{12} \, b \, h^2 \quad . \quad . \quad . \quad . \quad . \quad (10)$$

For an elliptical section of breadth b and depth h,

$$I = \frac{\pi}{64} b h^3 \quad . \quad . \quad . \quad . \quad . \quad (11)$$

For a hollow circular section of outside and inside diameters, d_1 and d_2 respectively,

$$I = \frac{\pi}{64} (d_1^4 - d_2^4) \quad . \quad . \quad . \quad . \quad (12)$$

Let A be the area, the moment of inertia of which is being considered. Then for a rectangular section $A = b h$, and (10) may be written

$$I = \frac{1}{12} A h^2 \quad . \quad . \quad . \quad . \quad . \quad (13)$$

For a circle $A = \frac{\pi}{4} d^2$, and (9) may be written

$$I = \frac{1}{16} A d^2 \quad . \quad . \quad . \quad . \quad . \quad (14)$$

Similarly, for an ellipse of breadth b and depth h, $A = \frac{\pi}{4} b d$; (11) may therefore be written

$$I = \frac{1}{16} A h^2 \quad . \quad . \quad . \quad . \quad . \quad (15)$$

That is, for each of the three sections considered, the moment of inertia is equal to the product of the area, and the square of the depth at right angles to the axis of inertia, multiplied by a constant factor, which depends on the *shape* of the section. It can be shown that this is true for sections of all shapes, the value of the constant factor being different for different shapes of section, but the same for large or small sections of the same shape.

Moment of Inertia of an Area about Parallel Axes.—The moment of inertia of an area is least about an axis passing through the centre of area.

Let I_0 be the moment of inertia of an area A about any axis through the centre of area. Then it can be easily shown that the moment of inertia about a parallel axis distant y_0 from the centre of area is $I_0 + A y_0^2$.

Moment of Inertia of an Area about different Axes passing through the centre of figure.—The moment of inertia of an area about different axes passing through the centre of figure are in general different, but however complex be the outline of the area, an ellipse can be drawn with its centre coinciding with the centre of the area, such that the moment of inertia relative to any axis drawn through the centre varies inversely as the square of the corresponding radius-vector of the ellipse. This ellipse is called the *ellipse of inertia*, or the *momental ellipse*, of the area. The axes corresponding to the major and minor axes of the ellipse are called the principal axes of the figure.

The momental ellipse for a rectangle, if drawn to a suitable scale, touches its sides. Similarly, for a triangle it can be shown that the ellipse touching the three sides at their middle points can be taken as the momental ellipse.

If the major and minor axes of the momental ellipse are equal, the ellipse becomes a circle, and the moments of inertia about all axes through the centre are equal. For example, since from symmetry the momental ellipse for a square is a circle, the moment of inertia of a square is the same for all axes passing through its centre.

93. **Moment of Bending Resistance.**—The moment about the neutral axis of all the forces p on the fibres of the cross section is called the *moment of resistance to bending* of the section, and is of course equal to the bending-moment on the section due to the external forces.

The moment of the force on the strip $b\ t$ (fig. 93) is

$$\frac{E}{R} b\ t\ y \times y$$

and the moment of all the forces on all the strips is

$$M = \frac{E}{R} \Sigma\ b\ t\ y^2 \ . \ \ . \ \ . \ \ . \ \ . \ \ . \ (16)$$

which may be written

$$M = \frac{E}{R} I \ . \ \ . \ \ . \ \ . \ \ . \ \ . \ \ . \ (17)$$

Substituting the value of $\dfrac{E}{R}$ from (6) in (17) it may be written

$$\frac{M}{I} = \frac{p}{y} \ \ . \ \ . \ \ . \ \ . \ \ . \ \ . \ \ . \ (18)$$

(17) and (18) may be conveniently written together thus :

$$\frac{M}{I} = \frac{E}{R} = \frac{p}{y} \quad . \quad . \quad . \quad . \quad . \quad . \quad (19)$$

94. **Modulus of Bending Resistance of a Section.**—The greatest stress on a section occurs, as has already been shown, on the fibre furthest away from the neutral axis. Let f be this stress, then, denoting the corresponding of y by y_1, (18) may be written

$$M = \frac{I}{y_1} f \quad . \quad . \quad . \quad . \quad . \quad . \quad (20)$$

The quantity $\dfrac{I}{y_1}$, which is a geometrical quantity depending on the area and shape of the section, and not in any way on the material, is called the *modulus* of bending resistance of the section, and will be denoted by the letter Z.   (20) may then be written

$$M = Z f \quad . \quad . \quad . \quad . \quad . \quad . \quad (21)$$

From (21) it is evident that the modulus cf a section bears the same relation to the bending-moment on it, as the area of a section bears to the total direct tension or compression on it. The total pull on a bar is equal to the product of its area into the tensile strength per square inch. The bending-moment on any section of a beam is equal to the modulus of the section multiplied by the greatest stress on the section.

For a rectangular section

$$Z = \frac{b\,h^2}{6} = \tfrac{1}{6}\,A\,h \, . \quad . \quad . \quad . \quad (22)$$

For a circular section,

$$Z = \frac{\pi}{32}\,d^3 = \tfrac{1}{8}\,A\,d \quad . \quad . \quad . \quad . \quad (23)$$

or approximately,

$$Z = \frac{d^3}{10} \quad . \quad . \quad . \quad . \quad . \quad . \quad (24)$$

For a hollow circular section,

$$Z = \frac{\pi}{32}\frac{(d_1^{\,4} - d_2^{\,4})}{d_1} \quad . \quad . \quad . \quad . \quad (25)$$

Table III. gives the sectional areas and moduli for round bars.

From (20) and (23) it is evident that the bending-moment a round bar can resist, *i.e.* its transverse strength, is proportional to the cube of its diameter.

TABLE III.—SECTIONAL AREAS AND MODULI OF BENDING RESISTANCE OF ROUND BARS.

Diameter	Sectional area	Z	Diameter	Sectional area	Z
Inches	Sq. in.	In.³	Inches	Sq. in.	In.³
$\frac{1}{16}$	·0031	·000024	$\frac{13}{16}$	·5185	·0526
$\frac{1}{8}$	·0123	·000192	$\frac{7}{8}$	·6013	·0658
$\frac{3}{16}$	·0276	·000647	$\frac{15}{16}$	·6903	·0809
$\frac{1}{4}$	·0491	·001534	I	·7854	·0982
$\frac{5}{16}$	·0767	·00300	$1\frac{1}{8}$	·9940	·1398
$\frac{3}{8}$	·1104	·00517	$1\frac{1}{4}$	1·2272	·1917
$\frac{7}{16}$	·1503	·00822	$1\frac{3}{8}$	1·4849	·2552
$\frac{1}{2}$	·1964	·01227	$1\frac{1}{2}$	1·7671	·3313
$\frac{9}{16}$	·2485	·0175	$1\frac{5}{8}$	2·0739	·4211
$\frac{5}{8}$	·3068	·0240	$1\frac{3}{4}$	2·4053	·5261
$\frac{11}{16}$	·3712	·0319	$1\frac{7}{8}$	2·7611	·6471
$\frac{3}{4}$	·4418	·0414	2	3·1416	·7854

95. **Beams of Uniform Strength.**—The bending-moment on a beam generally varies from section to section along the axis ; consequently, if of uniform section throughout it will be weakest where the bending-moment is greatest. A *beam of uniform strength* is one in which the section varies with the bending-moment in such a manner that the tendency to break is the same at all sections. This means that *f*, the maximum stress on the section, has the same value throughout, and therefore that *M* is proportional to Z.

For a thin hollow tube of constant external diameter throughout its length, Z is approximately proportional to the thickness ; therefore for a tubular beam in which the bending-moment varies continuously the thickness should also vary continuously, if the beam is required to be of uniform strength. For example, the bending-moment on the handle-bar of a bicycle, due to the pull of the rider, increases from zero at the end to its maximum value at the handle-pillar. If the external diameter of the handle-bar be the same throughout, the lightest possible bar would vary in thickness from the middle to the ends. This ideal handle-bar cannot be conveniently made, but an approximation thereto is sometimes made by inserting a liner at the middle, where the bending-moment is greatest ; there will in this case be three weak sections, the middle section and those just beyond the ends of the liners.

96. **Modulus of Circular Tubes.**—On account of the extensive use of tubes in bicycle making, it will be desirable to give some

additional formula relating to the moment of inertia and the modulus of a tubular section.

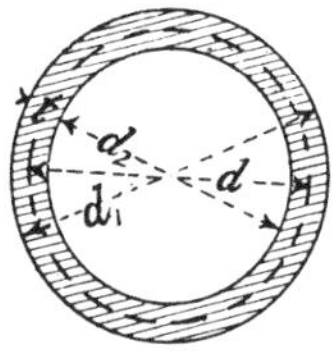

Let d_1, d, and d_2 (fig. 94) be the outside, mean, and inside diameters respectively, t the thickness, and A the area of the transverse section of the tube. From (12) for this section

FIG. 94.

$$I = \frac{\pi}{64} (d_1{}^4 - d_2{}^4) = \frac{\pi}{64} (d_1 - d_2)(d_1 + d_2)(d_1{}^2 + d_2{}^2) \quad . \quad (26)$$

Now, $d_1 - d_2 = 2\,t$, $d_1 + d_2 = 2\,d$, $d_1 = d + t$, $d_2 = d - t$.
Therefore,
$$(d_1{}^2 + d_2{}^2) = (d + t)^2 + (d - t^2) = 2(d^2 + t^2).$$

Substituting in (26)

$$I = \frac{\pi}{64} \cdot 2\,t \cdot 2\,d \cdot 2(d^2 + t^2)$$

But $\pi\,d\,t = A$, therefore,

$$I = \frac{A}{16} (d_1{}^2 + d_2{}^2) = \frac{A}{8} (d^2 + t^2) . \quad . \quad . \quad (27)$$

Now,

$$Z = \frac{I}{\frac{d_1}{2}} = \frac{A}{8} \frac{(d_1{}^2 + d_2{}^2)}{d_1} = \frac{A}{8} \frac{(2\,d_1{}^2 - 4\,d_1\,t + 4\,t^2)}{d_1}$$

$$= \frac{A}{4} \left(d_1 - 2\,t + \frac{2\,t^2}{d_1} \right) = \frac{A}{4} \left(d_2 + \frac{2\,t^2}{d_1} \right) . \quad . \quad . \quad (28)$$

If the tube be *thin*, t^2 will be small in comparison with d^2, and

$\dfrac{2\,t^2}{d_1}$ will be small in comparison with d_2. Equations (27) and (28) may then be written

$$I = \frac{\pi}{8}\, d^3\, t = \frac{A\,d^2}{8} \text{ approximately} \quad . \quad . \quad . \quad (29)$$

$$Z = \frac{\pi}{4}\, d^2\, t = \frac{A\,d_2}{4} \text{ approximately}. \quad . \quad . \quad . \quad (30)$$

The error introduced by using the approximate formula (30) for Z is on the safe side, and is very small for the ordinary tube sections used in cycle construction. Thus for a tube 1 inch diameter, 16 W.G., the exact value of Z is ·04140, that given by (30) is ·04102, the error being less than 1 per cent. in this case. If, however, d or d_1, be used instead of d_2 in formula (30) the error will be on the wrong side.

Table IV. gives the sectional areas, weights per foot run, and moduli of bending resistance for the ordinary sections of steel tubes used in cycle construction, the moduli having been calculated from the exact formula (28).

From (30) the transverse strength of a tube is proportional to its sectional area and to its internal diameter. If the internal diameter be kept constant, the transverse strength is proportional to the thickness. If the sectional area be kept constant, the transverse strength is proportional to the internal diameter. If the thickness be kept constant the strength is approximately proportional to the square of the diameter.

97. **Oval Tubes.**—We have already seen that the moment of inertia of an ellipse with major and minor

axes h and b respectively is $\dfrac{\pi}{64}\, b\, h^3$.

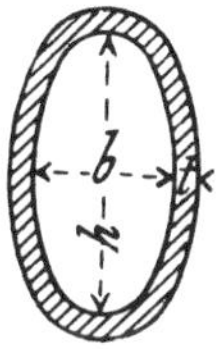

FIG. 95.

Let a second ellipse (fig. 95) be drawn outside the first and concentric with it, having its semi-axes the length t greater. The axes of the second ellipse will be $b + 2\,t$ and $h + 2\,t$ respectively, and its moment of inertia will be

$$\frac{\pi}{64}(b + 2\,t)(h + 2\,t)^3 =$$

$$\frac{\pi}{64}\Big\{ b\, h^3 + (2\, h^3 + 6\, b\, h^2)\, t + (12\, h^2 + 12\, b\, h)\, t^2$$

$$+ (24\, h + 8\, b)\, t + 16\, t^4 \Big\}$$

TAB.

SECTIONAL AREAS, WEIGHTS PER FOOT RUN, A

Outside diameter of tube		3″/8			1″/2			5″/8		
Imperial standard wire gauge	Thickness Inches	W lbs. per foot length	A sq. in.	Z in³	W lbs. per foot length	A sq. in.	Z in³	W lbs. per foot run	A sq. in.	Z in
No. 10	·128	·34	·0993	·0051	·52	·1496	·0116	·69	·1998	·0:
11	·116	·33	·0944	·0051	·48	·1399	·0113	·64	·1855	·0:
12	·104	·31	·0885	·0049	·45	·1294	·0108	·59	·1702	·0:
13	·092	·28	·0818	·0048	·41	·1180	·0103	·53	·1541	·0.
14	·080	·26	·0742	·0046	·36	·1056	·0096	·47	·1370	·0.
15	·072	·24	·0685	·0044	·33	·0968	·0091	·43	·1251	·0.
16	·064	·22	·0625	·0042	·30	·0877	·0085	·39	·1128	·0
17	·056	·19	·0561	·0039	·27	·0781	·0078	·35	·1001	·0
18	·048	·17	·0493	·0036	·24	·0682	·0070	·30	·0870	·0.
19	·040	·15	·0421	·0032	·20	·0578	·0062	·25	·0736	·0:
20	·036	·13	·0383	·0030	·18	·0525	·0057	·23	·0666	·0
21	·032	·12	·0345	·0027	·16	·0470	·0052	·21	·0596	·0
22	·028	·11	·0305	·0025	·14	·0415	·0046	·18	·0525	·0
23	·024	·09	·0265	·0022	·12	·0359	·0041	·16	·0453	·0
24	·022	·08	·0244	·0020	·11	·0330	·0038	·14	·0417	·0
25	·020	·08	·0223	·0019	·10	·0302	·0035	·13	·0380	·0
26	·018	·07	·0202	·0017	·09	·0273	·0032	·12	·0343	!0
28	·0148	·06	·0167	·0015	·08	·0226	·0027	·10	·0284	·0
30	·0124	·05	·0141	·0012	·07	·0190	·0023	·08	·0239	·0
32	·0108	·04	·0124	·0011	·06	·0166	·0020	·07	·0208	·0

Outside diameter of tube		$1\frac{1}{4}″$			$1\frac{3}{8}″$			$1\frac{1}{2}″$		
No. 10	·128	1·56	·4511	·1151	1·73	·5013	·1433	1·91	·5516	·I
11	·116	1·43	·4132	·1074	1·59	·4587	·1334	1·74	·5043	·I
12	·104	1·29	·3744	·0992	1·43	·4152	·1228	1·58	·4560	·I
13	·092	1·16	·3347	·0903	1·28	·3708	·1115	1·41	·4069	·I
14	·080	1·02	·2940	·0809	1·12	·3255	·0996	1·23	·3569	·I
15	·072	·92	·2664	·0742	1·02	·2947	·0912	1·12	·3230	·I
16	·064	·82	·2384	·0673	·91	·2635	·0825	1·00	·2887	·0
17	·056	·73	·2101	·0600	·80	·2320	·0735	·88	·2540	·0
18	·048	·63	·1813	·0525	·69	·2001	·0642	·76	·2190	·0:
19	·040	·53	·1521	·0446	·58	·1679	·0544	·63	·1836	·0
20	·036	·47	·1373	·0405	·52	·1514	·0494	·57	·1656	·0:
21	·032	·42	·1224	·0364	·47	·1350	·0443	·51	·1476	·0
22	·028	·37	·1075	·0321	·41	·1185	·0391	·45	·1295	·0
23	·024	·32	·0924	·0278	·35	·1019	·0338	·38	·1113	·0
24	·022	·29	·0849	·0256	·32	·0935	·0311	·35	·1022	·0

V

MODULI OF BENDING RESISTANCE OF STEEL TUBES

3/4"			7/8"			1"			1 1/8"		
W lbs. per foot run	A sq. in.	Z in.³	W lbs. per foot run	A sq. in.	Z in.³	W lbs. per foot run	A sq. in.	Z in.³	W lbs. per foot run	A sq. in.	Z in.³
·86	·2501	·0336	1·04	·3003	·0493	1·21	·3506	·0681	1·38	·4009	·0900
·80	·2310	·0320	·96	·2766	·0466	1·11	·3222	·0640	1·27	·3677	·0843
·73	·2111	·0301	·87	·2519	·0436	1·01	·2927	·0595	1·15	·3335	·0781
·66	·1902	·0280	·78	·2263	·0402	·91	·2624	·0546	1·03	·2981	·0714
·58	·1684	·0256	·69	·1998	·0365	·80	·2312	·0493	·91	·2626	·0641
·53	·1534	·0238	·63	·1816	·0337	·73	·2099	·0455	·82	·2382	·0589
·48	·1379	·0218	·56	·1631	·0308	·65	·1882	·0414	·74	·2133	·0535
·42	·1221	·0197	·50	·1441	·0277	·57	·1661	·0371	·65	·1881	·0479
·37	·1059	·0175	·43	·1247	·0244	·50	·1436	·0326	·56	·1624	·0419
·31	·0893	·0150	·36	·1050	·0210	·42	·1207	·0279	·47	·1364	·0357
·28	·0807	·0137	·33	·0949	·0191	·38	·1090	·0253	·43	·1232	·0325
·25	·0722	·0124	·29	·0847	·0172	·34	·0973	·0228	·38	·1099	·0292
22	·0635	·0110	·26	·0745	·0152	·30	·0855	·0202	·33	·0965	·0258
·19	·0547	·0096	·22	·0642	·0133	·25	·0736	·0175	·29	·0830	·0223
·17	·0503	·0089	·20	·0590	·0122	·23	·0676	·0161	·26	·0762	·0206
·16	·0459	·0082	·19	·0537	·0112	·21	·0616	·0148	·24	·0694	·0188
·14	·0414	·0074	·17	·0485	·0102	·19	·0555	·0134	·22	·0626	·0170
·12	·0342	·0062	·14	·0400	·0085	·16	·0458	·0111	·18	·0516	·0140
·10	·0287	·0052	·12	·0336	·0071	·13	·0385	·0094	·15	·0433	·0119
.09	·0251	·0046	·10	·0293	·0063	·12	·0336	·0082	·13	·0378	·0104

1 5/8"			1 3/4"			1 7/8"			2"		
2·08	·6019	·2090	2·25	·6521	·2467	2·43.	·7024	·2874	2·60	·7526	·3312
1·90	·5499	·1938	2·06	·5954	·2283	2·21	·6409	·2657	2·37	·6865	·3058
1·72	·4969	·1777	1·86	·5377	·2090	2·00	·5785	·2428	2·14	·6194	·2792
1·53	·4431	·1608	1·66	·4792	·1888	1·78	·5153	·2191	1·91	·5515	·2516
1·34	·3883	·1430	1·45	·4197	·1676	1·56	·4511	·1942	1·67	·4826	·2228
1·21	·3513	·1306	1·31	·3795	·1530	1·41	·4078	·1771	1·51	·4361	·2029
1·08	·3138	·1179	1·17	·3390	·1379	1·26	·3641	·1594	1·34	·3892	·1825
·95	·2760	·1047	1·03	·2980	·1223	1·11	·3200	·1413	1·18	·3420	·1617
·82	·2378	·0911	·9	·2566	·1063	·95	·2755	·1227	1·02	·2944	·1402
·69	·1992	·0770	·74	·2150	·0898	·80	2307	·1036	·85	·2465	·1184
·62	·1797	·0698	·67	·1938	·0814	·72	·2080	·0938	·77	·2221	·1071
·55	·1601	·0625	·60	·1727	·0728	·64	·1853	·0839	·68	·1978	·0958
·49	·1405	·0551	·53	·1515	·0642	·56	·1625	·0739	·60	·1735	·0843
·42	·1207	·0476	·45	·1301	·0554	·48	·1396	·0638	·51	·1490	·0727
·38	·1108	·0438	·41	·1194	·0510	·44	·1281	·0586	·47	·1367	·0669

Therefore the moment of inertia of the area included between the ellipses is

$$I = \frac{\pi}{64} \left\{ (2\,h^3 + 6\,b\,h^2)\,t + (12\,h^2 + 12\,b\,h)\,t^2 \right.$$
$$\left. + (24\,h + 8\,b)\,t + 16\,t^4 \right\} \quad . \quad . \quad . \quad . \quad . \quad (31)$$

If t is small in comparison with b and h, the second, third, and fourth terms in the expression for I are smaller and smaller compared with the first, and may be neglected. Therefore, the moment of inertia of the figure is approximately

$$I = \frac{\pi}{32}\,h^2\,t\,(h + 3\,b) \quad . \quad . \quad . \quad . \quad . \quad . \quad (32)$$

The modulus of bending resistance is approximately

$$Z = \frac{\pi}{16}\,h\,t\,(h + 3\,b) \quad . \quad . \quad . \quad . \quad . \quad . \quad (33)$$

When a tube of circular section is flattened to form an oval tube, its thickness will be nearly uniform throughout, but in the oval tube section, above discussed, the thickness is not constant throughout, but is a little less than t except at the ends of the major and minor axes. The strength of an oval tube of uniform thickness will therefore be under-estimated if the formula (33) be used, so that the error is on the safe side.

The area of the ellipse is $\frac{\pi}{4}\,a\,b$; and in the same way as above it can be shown that the area included between the two ellipses is

$$A = \frac{\pi}{2}\,(b + h)\,t \quad . \quad . \quad . \quad . \quad . \quad . \quad . \quad (34)$$

Therefore, (32) and (33) may be respectively written,

$$I = \frac{(3\,b + h)}{16\,(b + h)}\,A\,h^2 \quad . \quad . \quad . \quad . \quad . \quad . \quad (35)$$

$$Z = \frac{(3\,b + h)}{8\,(b + h)}\,A\,h \quad . \quad . \quad . \quad . \quad . \quad . \quad (36)$$

An oval tube of uniform thickness will be stronger than indicated by formula (36). This is clearly shown by figure 96, which represents a quarter section ; the modulus given in (36) is that of the tube whose inner surface is represented by the dotted line. The inner continuous line represents a tube of the same sectional area *A*, and of uniform thickness. It has the area *b* in excess of the dotted tube, and the areas *a* and *c* deficient, $a + c = b$. It is evident from the figure that the moment of inertia of *b* about the minor axis is greater than that of *a* and *c*; and therefore the tube of uniform thickness is slightly stronger than the section above discussed.

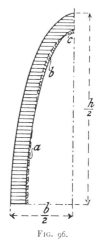

FIG. 96.

98. **D Tubes.**—Tubes of **D** section have been recently introduced for the lower back fork of a bicycle; it will be instructive to investigate their bending resistances here. We will assume that the outline of the **D** tube is made up of a semicircle and its diameter (fig. 97). Let *r* be the radius and *h* the diameter of the semicircle, *t* the thickness, which we will consider very small in comparison with *r*, and *A* the sectional area of the **D** tube. First, consider the moment of inertia about the axis *a a* at right angles to the flat side of the tube. The moment of inertia of the rectangle of depth *h* and width *t* is $\frac{1}{12} h^3 t$, the moment of inertia of the semicircle is $\frac{\pi}{16} h^3 t$, therefore the moment of inertia of the **D** tube about the axis *a a* is

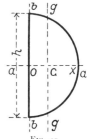

FIG. 97.

$$I = \left(\frac{1}{12} + \frac{\pi}{16}\right) h^3 t \quad . \quad . \quad . \quad (37)$$

The modulus of the section, about the same axis, is

$$Z = \left(\frac{1}{6} + \frac{\pi}{8}\right) h^2 t \quad . \quad . \quad . \quad . \quad . \quad . \quad . \quad (38)$$

Consider now the moment of inertia about the axis *b b*, coinciding with the flat side of the **D** tube. The moment of inertia of the flat side is $\frac{1}{12} h\, t^3$, that of the semicircle is $\frac{\pi}{16} h^3\, t$; therefore the moment of inertia of the section of the **D** tube is

$$\frac{1}{12} h\, t^3 + \frac{\pi}{16} h^3\, t.$$

If t be small in comparison with h, the first term in this expression may be neglected in comparison with the second, and therefore,

$$I = \frac{\pi}{16} h^3\, t, = \frac{\pi}{2} r^3\, t, \text{ approximately} \quad . \quad . \quad . \quad (39)$$

But the I just found is not about an axis through the centre of figure ; this we now proceed to find. Let G be the centre of figure ; the distance $O\,G$ can be found as follows : The moment of the semicircle about the axis $b\,b$ is $2\, r^2\, t$ (see sec. 50), that of the straight side about the same axis is zero, the total moment of the **D** tube about the axis $b\,b$ is therefore $2\, r^2\, t$. But the total moment is also equal to the total area multiplied by the distance $O\,G$; therefore $2\, r^2\, t = (2\,r + \pi\,r)\, t \times OG$,

$$\text{And } O\,G = \frac{2\, r^2}{(2 + \pi)\, r} = \frac{2}{2 + \pi} r = .389\, r. \quad . \quad . \quad . \quad (40)$$

Let I_0 be the moment of inertia about an axis $g\,g$ passing through G parallel to $b\,b$; then by section 92

$$I = I_0 + (2 + \pi)\, r\, t \cdot \frac{4}{(2 + \pi)^2}\, r^2$$

$$\text{Therefore, } I_0 = \frac{\pi}{2}\, r^3\, t - \frac{4\, r^3\, t}{(2 + \pi)} = \left(\frac{\pi}{2} - \frac{4}{2 + \pi} \right) r^3\, t \quad . \quad . \quad (41)$$

But $A = (2 + \pi)\, r\, t$; therefore we may write

$$I_0 = \left\{ \frac{\pi}{2\,(2 + \pi)} - \frac{4}{(2 + \pi)^2} \right\} A\, r^2 \quad . \quad . \quad . \quad (42)$$

Z, the modulus of bending resistance about the axis $g\,g$ is equal to $\frac{I_0}{G\,X}$, X being the extremity of the radius through $O\,G$. Now,

$$G\,X = O\,X - O\,G = r - \frac{2}{2 + \pi}\, r = \frac{\pi}{2 + \pi}\, r.$$

$$\therefore Z = \left\{ \frac{1}{2} - \frac{4}{\pi\,(2 + \pi)} \right\} A\, r = .2524\, A\, r \quad . \quad (43)$$

Let d be the diameter of a round tube equal in perimeter to the **D** tube. Then $\pi d = (2 + \pi)r$,

$$\therefore r = \frac{\pi}{2 + \pi}d = \cdot 6110\, d.$$

Substituting this value of r in (43), we get,

$$Z = \left\{ \frac{\pi}{2(2 + \pi)} - \frac{4}{(2 + \pi)^2} \right\} A\, d = \cdot 1542\, A\, d \quad (44)$$

The Z of the original round tube is approximately $\frac{1}{4} A\, d$, so the strengths of a round tube and the **D** tube into which it can be pressed are in the ratio of $\cdot 2500$ to $\cdot 1542$, *i.e.* 1000 to 617.

But since a **D** tube is used when the space $O X$ is limited, it would seem fairer to compare it with a round tube of equal weight and of diameter $O X$. The Z of a round tube of diameter $O X$ is $\cdot 25\, A\, r$. Comparing this value with that in (43), it is seen that the strength of the **D** tube is slightly greater than that of a round tube of equal weight, and of diameter equal to the smallest diameter of the **D** tube, the ratio being $\cdot 252$ to $\cdot 2500$, *a difference of less than one per cent. in favour of the* **D** *tube.*

FIG. 98.

99. Square and Rectangular Tubes.—Consider the I of a square tube of section $A B C D$ (fig. 98), about an axis $a\, a$ parallel to the side $A B$. The I of each of the sides $B C$ and $D A$ is $\frac{h^3 t}{12}$, that of each of the sides $A B$ and $C D$ is $ht \cdot \frac{h^2}{4}$; therefore, for the whole section

$$I = 2 \cdot \frac{h^3 t}{12} + 2 \cdot h t \cdot \frac{h^2}{4} = \frac{2}{3} h^3 t \cdot \quad . \quad . \quad . \quad . \quad (45)$$

The total sectional area is $4\, h\, t$, therefore

$$I = \frac{1}{6} A\, h^2 \cdot \quad . \quad . \quad . \quad . \quad . \quad . \quad . \quad . \quad (46)$$

also, $\quad Z = \dfrac{I}{\frac{h}{2}} = \frac{1}{3} A\, h \quad . \quad . \quad . \quad . \quad . \quad . \quad . \quad . \quad (47)$

Let d be the diameter of a round tube of the same perimeter as the square tube ; then $4\,h = \pi\,d$

$$\therefore h = \frac{\pi}{4}\,d = \cdot7854\,d,\text{ and}$$

$$Z = \frac{\pi}{12}\,A\,d = \cdot2618\,A\,d \quad . \quad . \quad . \quad . \quad (48)$$

hence, comparing with (30), the moduli of bending resistance of the square tube and of the original round tube are in the ratio of $\frac{\pi}{12}$ to $\frac{1}{4}$, or of π to 3, *i.e.* 1047 to 1000, in favour of the square tube. Compared with a round tube of equal sectional area, but of the same diameter as the side of the square tube, the ratio is $\frac{A\,d}{3}$ to $\frac{A\,d}{4}$, *i.e.* $\frac{1}{3}$ to $\frac{1}{4}$, or 133·3 to 100 ; *i.e. the square tube is* 33·3 *per cent. stronger than the round tube of equal area and diameter.*

Rectangular Tubes.—If a round tube be drawn into a rectangular tube of the same thickness, perimeter, and sectional area, it can be shown that the strength of the latter will be greatest when its depth h is three times its width b.

For any rectangular section, approximately

$$I = 2\,b\,t\left(\frac{h}{2}\right)^2 + 2\ \frac{t\,h^3}{12} = \frac{t\,h^2}{2}\left(b + \frac{h}{3}\right) \quad . \quad . \quad (49)$$

$$Z = t\,h\left(b + \frac{h}{3}\right) \quad . \quad . \quad . \quad . \quad . \quad . \quad . \quad . \quad (50)$$

For the strongest rectangular tube, (49) becomes

$$I = 9\,t\,b^3 = \frac{1}{8}\,A\,h^2 \quad . \quad . \quad \quad . \quad . \quad . \quad . \quad (51)$$

and,$\qquad Z = \frac{1}{4}\,A\,h \quad . \quad . \quad . \quad . \quad . \quad . \quad . \quad . \quad . \quad (52)$

Comparing (33) and (50), it is seen that a thin rectangular tube is stronger than an elliptical tube of the same depth, width, and thickness in the ratio $16 : 3\,\pi$. Now the ratio of the perimeters, and therefore the weights, is never greater than $4 : \pi$; this being the value when the ellipse and rectangle become a circle and square respectively. Weight for weight, then, the rectangular has at least $\frac{4}{3}$ times the strength of the elliptical tube.

That the rectangular is stronger than the elliptical tube of equal depth, width, and sectional area, can be easily shown from first principles, as follows : Figure 99 shows quadrants of rectangular and elliptical tubes of equal sectional area. Since the perimeter of the ellipse is less than that of the rectangle, its thickness is greater. Let a portion a of the ellipse be marked off equal in width to the corresponding part of the rectangle, so that the moments of inertia about the axis $O\,X$ are equal. The part b is common to both ellipse and rectangle, and there remain only the parts c. That belonging to the rectangle is at a much greater distance from the axis $O\,X$ than that belonging to the ellipse ; its moment of inertia is therefore greater, and the rectangular is stronger than the elliptical tube to resist bending.

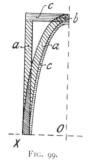

FIG. 99.

100. **Compression.**—The laws relating to simple compressive stress are exactly the same as those of simple tension, the formula (1), (2), (3), and (4), of chapter x. will apply, p being in this case the compressive stress, e the compression per unit of length, and E the modulus of elasticity for compression. For a homogeneous material with perfect elasticity, as above defined, E would be the same for tension and compression.

On a bar which is short in comparison to its diameter, if the compressive stress be increased above the elastic limit of compression, the bar gives way ultimately by lateral yielding. If the material be hard, the bar may actually split up into several pieces. If of a soft, ductile material it will bulge gradually in the middle while being shortened in length.

101. **Compression or Tension combined with Bending.**—If a bar be simultaneously subjected to bending, and a pull or thrust parallel to its axis, the maximum stress on the section is the sum of the separate stresses due to the separate straining actions. If the bar be subjected to a pull P, and a bending-moment M, A being the area and Z the modulus of the section, the maximum tensile stress is

$$f = \frac{P}{A} + \frac{M}{Z} \quad . \quad . \quad . \quad . \quad . \quad . \quad . \quad (1)$$

and the minimum tensile stress is

$$f^1 = \frac{P}{A} - \frac{M}{Z} \quad . \quad . \quad . \quad . \quad . \quad . \quad . \quad (2)$$

For circular tubes of small thickness, substituting the value of Z from (30), section 96,

$$f = \frac{P}{A} + \frac{4\,M}{A\,d} \quad . \quad . \quad . \quad . \quad . \quad . \quad (3)$$

The bending-moment may be produced by applying the pull P at a distance x from the neutral axis of the section (fig. 100). In this case $M = Px$ and (3) may be written

$$f = \frac{P}{A} + \frac{4\,Px}{A\,d} \qquad . \quad . \quad . \quad . \quad . \quad . \quad (4)$$

If the bar be subjected to a compression P and a bending-moment M, equations (1), (3), and (4) give the maximum *compressive* stress on the section, equation (2) the minimum compressive stress.

102. **Columns.**—If a long bar be subjected to tension, any slight deviation from straightness (fig. 100) will, under the action of the forces, tend to get less. If, on the other hand, the bar be subjected to compression, the deviation from straightness will tend to get greater, and the bar will give way by bending (fig. 101).

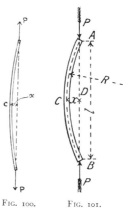

The stresses on a straight short column supporting a load, placed eccentrically, are given by formulæ (1) and (2).

FIG. 100.   FIG. 101.

Example.—A bicycle tube 1 in. diameter, 16 W.G., is subjected to a compressive force, the axis of which is $\frac{1}{4}$ in. from the axis of the tube. Find the breaking load, the breaking stress of the material being 30 tons per sq. in. From Table IV., $A = .1882$ sq. in., $Z = \cdot 0414$ in.³, also $M = \frac{1}{4}\,P$ inch-lbs. $f = 30 \times 2240$ lbs. per sq. in. ; substituting in (1)

$$30 \times 2240 = \frac{P}{\cdot 1882} + \frac{P}{4 \times \cdot 0414},$$

from which, $P = 5921$ lbs.

If the load were placed exactly co-axial with the tube, it would reach the value given by,

$$\frac{P}{\cdot 1882} = 30 \times 2240$$

i.e., $P = 12650$ lbs.

103. **Limiting Load on Long Columns.**—If, under the action of the load, the deviation x becomes greater, the bending-moment also becomes greater without any addition being made to the load ; thus the deviation once started, may rapidly increase until fracture of the column takes place.

Let the section of the column be such that, under the action of the load, its neutral axis bends into a circular arc $A\,C\,B$ (fig. 101) of radius R. Let $A\,D\,B$ be the chord, $C\,D$ the greatest deviation, and $C\,E$ a diameter of the circle. Then, by the well-known proposition in elementary geometry,

$$C\,D \times D\,E = A\,D \times D\,B.$$

i.e. neglecting the difference between $C\,E$ and $D\,E$,

$$2\,R\,x = \frac{l^2}{4} \text{ approximately.}$$

But $R = \dfrac{E\,I}{M}$, from (17), chap. xi., and $M = P\,x$. Substituting,

$$P = \frac{8\,E\,I}{l^2} \quad . \quad . \quad . \quad . \quad . \quad . \quad . \quad . \quad (5)$$

If the load be less than that given by (5), no deviation will take place.

If the column be of constant section throughout its length its neutral axis bends into a curve of sines, and it can be shown that the limiting load is

$$P = \frac{\pi^2\,E\,I}{l^2} \quad . \quad . \quad . \quad . \quad . \quad . \quad . \quad (6)$$

If the middle section of the column be prevented from deviating laterally, it will bend into the form shown in figure 102. In this case the length of the segment of the curve corresponding to figure 101 is half the total length, and the corresponding load will be

$$P = \frac{4\,\pi^2\,E\,I}{l^2}. \quad . \quad . \quad . \quad . \quad . \quad . \quad (7)$$

FIG. 102.

Again, if the ends of the column be held in such a manner that the directions of the axis at the end are always the same, it will give way by bending as shown in figure 103.   The segment

$b\,d$ in this case is of the same shape as the curve in figure 101, while the portions $a\,b$ and $e\,d$ are of the same form as $b\,c$ and $d\,c$. In this case, therefore, the length of the segment $b\,d$ is $\dfrac{l}{2}$, and the corresponding limiting load is given by the formula

$$P = \frac{4\,\pi^2\,E\,I}{l^2} \quad . \quad . \quad . \quad . \quad . \quad . \quad . \quad (8)$$

If the column be fixed at one end (fig. 104), held laterally but free to turn at the other,

$$P = \frac{2\,\pi^2\,E\,I}{l^2}. \quad . \quad . \quad . \quad . \quad . \quad . \quad (9)$$

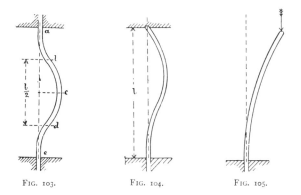

FIG. 103.   FIG. 104.   FIG. 105.

If the column be fixed at one end and quite free at the other end (fig. 105),

$$W = \frac{\pi^2\,E\,I}{4\,l^2} \quad . \quad . \quad . \quad . \quad . \quad . \quad . \quad (10)$$

These are known as Euler's formulæ, and are only applicable to bars or columns in which the length l is great as compared with the least transverse dimension. l is the length before bending ; though in the figures, in which the bending is greatly exaggerated, it is marked as *after* bending.

104. **Gordon's Formula for Columns.**—The pieces of tube used in bicycle building are too long to have the simple compression formula applied to them, and too short for the application

of Euler's formula. A great many experiments on columns, principally cast iron, have been made by Hodgkinson, and Gordon has suggested an empirical formula which agrees very closely with his experiments. For thin tubes, Gordon's formula becomes

$$\frac{W}{A} = \frac{f}{1 + \dfrac{8\,l^2}{c\,d^2}} \quad \cdots \quad \cdots \quad (11)$$

f and c being constants depending on the material.

Actual experiments on the compressive strengths of weldless steel tubes are wanting, but taking $f = 30$ tons per sq. in., and $c = 32,000$, Gordon's formula becomes

$$\frac{W}{A} = \frac{67200}{1 + \dfrac{l^2}{32000\,d^2}} \quad \cdots \quad \cdots \quad (12)$$

Example.—A tube is 1 in. diameter, No. 16 W.G., 20 in. long ; required the crushing load by Gordon's formula.

From Table IV., page 113, $A = \cdot 1882$;

$$\therefore \frac{W}{\cdot 1882} = \frac{67200}{1 + \dfrac{400}{32000 \times 1}} = \frac{67200}{1 \cdot 0125}$$

from which,

$$W = 12490 \text{ lbs.,}$$

slightly less than for a short length of the same tube (sec. 102).

105. **Shearing.**— Let $A\,B\,C\,D$ (fig. 106) be a small square prism of unit width perpendicular to the paper, subjected to

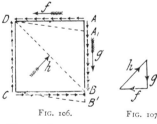

FIG. 106. FIG. 107.

shearing stress on the planes $A\,B$ and $C\,D$. If the planes $A\,B$ and $C\,D$ be very close to each other, the shearing stress will be the same on both. If q be the shearing stress per unit of area, the downward force acting at $A\,B$ and the upward force at $C\,D$ will each be $q \times \overline{A\,B}$. But since the portion $A\,B\,C\,D$ is at rest, the couple formed by the forces at $A\,B$ and $C\,D$ must be balanced by an equal and opposite couple, formed by forces acting at $A\ D$ and $B\ C$, since no force acts normally at the surfaces $A\ B$ and $C\ D$. Thus the shearing stress on the sides $A\ D$ and $B\ C$ is equal to that on $A\ B$ and $D\ C$; or the shearing stress on a plane is always accompanied by an equal shearing stress on a plane at right angles to the former, and to the direction of the shearing stress on the former plane.

Transverse Elasticity.—Under the action of the shearing forces the square $A\ B\ C\ D$ (fig. 106) will be distorted into a rhombus, $A^1\ B^1\ C\ D$, the angle of distortion $A\ D\ A^1$ being proportional to the shearing stress. Let ϕ be this angle and q the shearing stress producing it ; then

$$q = C\,\phi \quad . \quad . \quad . \quad . \quad . \quad . \quad . \quad (13)$$

C being the *modulus of transverse elasticity*, or the *coefficient of rigidity* of the material.

Shearing Stress equivalent to Simultaneous Tension and Compressive Stresses.—Draw a diagonal $B\ D$ (fig. 106) ; the triangular prism $A\ D\ B$ is in equilibrium under the action of the three forces, f, g, and h, acting on its sides, which can therefore be represented by the sides of a triangle (fig. 107). f and g being equal, the force h is evidently at right angles to the side $B\ D$. The triangles $A\ B\ D$ and $f\ g\ h$ are similar ; that is, the forces f, g and h are proportional to the lengths of the sides on which they act ; the stress per unit area must therefore be the same for the three sides $A\ B$, $B\ D$, and $D\ A$. Thus, the stress on the plane $B\ D$ is a compressive stress of the same intensity as the shearing stress on the planes $A\ B$ and $A\ D$.

In the same way it may be shown that a *tensile* stress of equal magnitude exists on the plane $A\ C$. Thus, in any body a pair of shearing stresses on two planes at right angles are equivalent to a pair of compressive and tensile stresses respectively on two planes mutually at right angles, and inclined 45° to the former planes.

106. **Torsion.**—If a long bar be subjected to two equal and opposite couples acting at its ends, the axes of the couples being parallel to the axis of the bar (fig. 108), it is said to be subjected to *torsion*. The moment of the couple applied is called the *twisting-moment* on the bar. If one end be rigidly fixed, the

other end will, under the action of the twisting-moment, be displaced through a small angle, and any straight line on the surface of the bar originally parallel to the axis will be twisted into a spiral curve *a a*. If the twisting-moment be increased indefinitely, the bar will ultimately break, the total angle of twist before breaking depending on the nature of the material.

Let figure 109 be the longitudinal elevation of a thin tube of mean radius *r* and thickness *t*, subjected to a twisting-moment

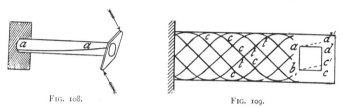

FIG. 108.   FIG. 109.

T foot-lbs.   A square, *a b c d*, drawn on the surface of the tube becomes distorted while in a strained condition into the rhombus *a b c′ d′*.   Thus, every transverse section of the tube is subjected to a shearing stress.   If the tube be of uniform diameter and thickness, this shearing stress, *q*, will be the same throughout, provided the thickness is very small in comparison with the diameter.

The sectional area of the tube is $2\pi t r$; and since *q* is the shear on unit area, the total shear on the section is $2\pi q t r$. The shearing-force on each element of the section acts at a distance *r* from the centre of the tube; the moment of the total shearing-force is therefore $2\pi q t r^2$.   This must be equal to the twisting-moment T_1 applied to the end; therefore

$$T_1 = 2\pi q t r^2 \quad . \quad . \quad . \quad . \quad . \quad . \quad (14)$$

Thus the twisting-moment which can be transmitted by a thin tube of circular section is proportional to the square of its radius or diameter and to its thickness.

107. **Torsion of a Solid Bar.**—In a solid cylinder of radius r_1, imagine the square *a b c d* (fig. 109) drawn on a concentric cylindrical surface of radius *r*; it is easily seen that the angle of distortion of the fibres, ϕ, or *d a d′*, is proportional to *r*.   If ϕ_1 be

the angle of distortion for a square drawn on the surface of the cylindrical rod, q and q_1 the shearing stresses at radii r and r_1 respectively, then evidently

$$\phi = \frac{\phi_1}{r_1}\, r\,;$$

and therefore

$$q = \frac{q_1}{r_1}\, r.$$

If now the solid rod be considered to be divided into a number of thin concentric tubes, all of the same thickness, t, the area of the tube of radius r is $2\pi t r$, the total shear on this tube is

$$\frac{2\pi q_1}{r_1}\, t\, r^2,$$

and the twisting-moment resisted is

$$\frac{2\pi q_1}{r_1}\, t\, r^3.$$

The sum of the moments of all the concentric tubes into which the rod is divided is easily found, by one of the simplest examples in the integral calculus, to be

$$T = \frac{\pi q_1}{2}\, r_1^{\,3},$$

or,

$$T = \frac{\pi}{16} d^3 q_1 = \frac{d^3 q}{5} \text{ approx. } \quad . \quad (15)$$

108. **Torsion of Thick Tubes.**—If r_1 and r_2 be the external and internal radii of a hollow tube, the sum of the twisting-moments (45) of the very thin concentric tubes into which it may be divided —and, therefore, the twisting-moment such a tube can resist—is

$$\frac{\pi (r_1^{\,4} - r_2^{\,4})\, q_1}{2\, r_1}$$

or

$$T = \frac{\pi}{16}\, \frac{(d_1^{\,4} - d_2^{\,4})\, q_1}{d_1} \quad . \quad . \quad . \quad . \quad (16)$$

The quantity $\dfrac{\pi}{16} \dfrac{(d_1^{\,4} - d_2^{\,4})}{d_1}$ depends simply on the dimensions

of the section of the tube, and may be called the *modulus of
resistance to torsion* ; it may be denoted by the symbol Z_T.   Then

$$T = Z_\mathrm{T} \cdot q_1.$$

Comparing Z_T with Z, chapter xi., it will be seen that the
modulus of resistance of a circular tube or solid bar to torsion is
twice its modulus of resistance to bending.   The strength of any
tube to resist bending can therefore be obtained by multiplying
the modulus from Table IV., page 112, by twice the maximum
shear q_1.

109. **Lines of Direct Tension and Compression on a Bar
subject to Torsion.**—From what has been said in section 105,
there will be a compressive stress on the plane *a c*, and a tensile
stress on the plane *b d* (fig. 109).   This holds for every point on
the surface of the tube.   Now if the tube be split up into a
number of narrow strips by the spiral lines *t t*, inclined 45° to the
axis (fig. 10), the tensile stresses can be transmitted just as before.
The spiral lines *t t* are said to be *tension lines*, and the spiral lines
c c at right angles *compression lines*.   If the twisting-moment be in
the opposite direction, however, the pressure and tension spiral
lines will be interchanged, and the split tube will not be able to
transmit the twisting-moment.

110. **Compound Stress.**—If the straining actions on any part
of a structure be all parallel to one plane, the stress on any plane
section, at right angles to the plane of the straining actions, can
be resolved into a normal stress, tension or compression—and a
tangential stress, shearing.   It can be shown that any system of
stress in two dimensions is equivalent to a pair of normal stresses

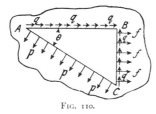

FIG. 110.

on two planes mutually at right
angles, and that the stress on one
of these planes is greater than, that
on the other plane less than, on
any other plane section of the material.   On any other plane the
stress will have a tangential component.

An important case of compound stress is that of a shaft subjected to bending and torsion ; a section at right angles to the

axis of the shaft is subjected to a normal stress, f, and simultaneously to a torsional shearing stress, q. Consider a small portion of a material (fig. 110) subjected to stresses parallel to the plane of the paper. Let $A\ B\ C$ be a small prism, of unit depth at right angles to the paper, the face $B\ C$ being subjected to a normal stress, f, and a tangential stress, q. From section 106 we know that an equal shearing stress, q, must exist on the face $A\ B$. Let us find the magnitude of the stress p on the face $A\ C$, on which the stress shall be wholly normal.

Considering the equilibrium of the prism $A\ B\ C$, and resolving the forces on the three faces parallel to the side $A\ B$, we have

$$p \,.\, A\ C \,.\, sin\ \theta - q \,.\, A\ B - f \,.\, B\ C = 0$$

or

$$(p - f)\ tan\ \theta = q \quad . \quad . \quad . \quad . \quad . \quad (17)$$

Similarly resolving the forces parallel to $B\ C$, we get,

$$p \,.\, A\ C \,.\, cos\ \theta - q \,.\, B\ C = 0$$

or

$$p = q\ tan\ \theta \,.\, \quad . \quad . \quad . \quad . \quad . \quad . \quad . \quad (18)$$

Multiplying (17) and (18) together, we get

$$p\ (p - f) = q^2$$

from which

$$p = \tfrac{1}{2} \{f \pm \sqrt{f^2 + 4\ q^2}\} \quad . \quad . \quad . \quad . \quad . \quad (19)$$

the two values of p in (19) are the maximum and minimum normal stresses on the material. That is, the tension f and the shear q, on the face $B\ C$, produce on some plane $A\ C$ the maximum tensile stress $\tfrac{1}{2}\{f + \sqrt{f^2 + 4\ q^2}\}$, and on another plane the minimum tensile stress $\tfrac{1}{2}\{f - \sqrt{f^2 + 4\ q^2}\}$; the latter plane being at right angles to the former.

If the stresses on two planes at right angles be wholly normal and of equal intensity, it can easily be shown that the stress on any other plane is wholly normal and of the same intensity. If the normal stress be compression, the whole system of stress is of the nature of fluid pressure. If there be a tensile stress on one plane and an equal compressive stress on the plane at right angles, it has already been shown that this is equivalent to shearing stresses of the same intensity on two planes at angles of 45° with the planes of the normal stresses. This pair of shearing stresses tends to distort the body but not to alter its volume, whereas fluid pressure tends to alter the volume but not the shape of the body.

Any set of stresses in two dimensions can be expressed as the sum of a fluid stress and a shearing stress. Let two planes, A and B, at right angles be subjected to normal tensile stresses of intensity, p and q, respectively. Then this state of stress is equivalent to the sum of two states of stress, the first being a tensile stress $\frac{p+q}{2}$ on both planes A and B, the second a tensile stress $\frac{p-q}{2}$ on A and an equal compressive stress on the plane B. For $p = \frac{p+q}{2} + \frac{p-q}{2}$, and $q = \frac{p+q}{2} - \frac{p-q}{2}$. This principle will be made use of when discussing the outer cover of a pneumatic tyre.

111. **Bending and Twisting of a Shaft.**—In a circular shaft of diameter, d, subjected to a bending-moment, M, and a twisting-moment, T, the normal stress due to the bending-moment is

$$f = \frac{M}{\frac{\pi}{32} d^3},$$

and the shearing-stress due to the twisting-moment is

$$q = \frac{T}{\frac{\pi}{16} d^3}.$$

Substituting these values in (19),

$$p = \frac{1}{\frac{\pi}{16} d^3} \left\{ M + \sqrt{M^2 + T^2} \right\}.$$

if the shaft be subjected to a twisting-moment, T_e, which would produce the same stress, p,

$$p = \frac{T_e}{\frac{\pi}{16} d^3},$$

and T_e is said to be the twisting-moment equivalent to the given bending-moment and twisting-moment acting simultaneously. Comparing the two expressions for p, we get

$$T_e = M + \sqrt{M^2 + T^2} \quad . \quad . \quad . \quad . \quad (20)$$

Similarly, the equivalent bending-moment is

$$M_e = \tfrac{1}{2} T_e = \tfrac{1}{2} \left\{ M + \sqrt{M^2 + T^2} \right\} \quad . \quad . \quad (21)$$

112. **Stress, Breaking and Working.**—Each part of a machine must be capable of resisting the greatest straining actions that may come on it. This condition fixes, as a rule, the smallest possible section of the part below which it is not permissible to go. In ordinary machines, where mere mass is sometimes requisite, the section actually used may often with advantage be considerably greater than the minimum; but in bicycles, since 'lightness' is always sought after, though it should always be secondary to 'strength,' the actual section used must not be very much greater than the minimum consistent with safety. The magnitude of the stress on any piece depends on the general configuration of the machine and of the arrangement of the external forces acting on it. The strength of the various parts depends on the physical qualities of the materials of which they are made, as well as on their section; this we will now proceed to discuss.

Breaking Stress.—If a load be applied at the end of a bar and be gradually increased, the bar will ultimately break under it. If the bar be of unit section—one square inch—the load on it at the instant of breaking is called the *breaking tensile strength* of the material. A great number of experiments have been made from time to time on the strength of materials, and the values of the breaking tensile strength for all materials used in construction are fairly accurately known.

Factor of Safety.—One method of designing parts of a machine or structure is to fix arbitrarily on a *working stress* which shall not be exceeded. This working stress is got by dividing the breaking stress of the material, as determined by experiment, by an arbitrary number called a *factor of safety*. This factor of safety varies with the nature of the material used, and with the conditions to which the structure is subjected. Professor Unwin, in 'Elements of Machine Design,' gives a table of factors of safety, the factor varying from 3 for wrought iron and steel supporting a dead load, to 30 for brickwork and masonry subjected to a varying load. The factor of safety should be large for a material that can be easily broken by impact, and may be low for a material that undergoes considerable deformation before fracture actually takes place.

113. **Elastic Limit.**—We have already seen (sec. 81) that the application of a load to a bar of what might be popularly called a rigid material produces an elongation, and that this elongation is proportional to the load applied up to a certain limit. If not loaded beyond this limit, on removing the load the bar returns to its original length, and no permanent alteration has been made. If, however, the load applied be greater than the above limit, the elongation produced by it becomes greater proportionally, and on the load being removed the bar is found to be permanently increased in length. The stress beyond which the elongation is no longer proportional to the load, is called the *elastic limit*.

Since the elongation is in most metals proportional to the load applied up to this point, it has also been called the proportional limit (German, ' Proportionalitätsgrenze '). In a few metals— cast iron, brass—there is no well-defined proportional limit.

The total elongation of a bar loaded up to a stress just inside the elastic limit is a very small fraction of its original length. On increasing the load beyond the elastic limit and up to the breaking point, the elongation before fracture occurs, in the case of most materials, is a very much greater fraction of the original length.

Table V. gives the breaking and elastic strengths and coefficients of elasticity of most of the materials used in cycle making ; the figures are taken from Professor Unwin's 'Elements of Machine Design.'

114. **Stress-strain Diagram.**—The relation between the elongation and the load producing it can be conveniently exhibited in the form of a diagram. Let the stress be represented by an ordinate $O y$ drawn vertically (not shown on the diagram), and

TABLE V.—ULTIMATE AND ELASTIC STRENGTHS OF MATERIALS, AND COEFFICIENTS OF ELASTICITY IN LBS. PER SQUARE INCH.

Material	Breaking Strength — Tension	Breaking Strength — Pressure	Elastic Strength — Tension	Elastic Strength — Pressure	Coefficient of Elasticity — Direct, E	Coefficient of Elasticity — Transverse, C
Cast iron	30,500	130,000	10,500	21,000	23,000,000	7,600,000
	17,500	95,000			17,000,000	6,300,000
	10,800	50,000			14,000,000	5,000,000
Wrought-iron bars	67,000	50,000	30,000	30,000	31,000,000	10,500,000
	57,600				29,000,000	
	33,500				27,000,000	
Iron boiler-plates	47,000		24,000	24,000	26,000,000	14,000,000
Steel plates, ¼ per cent. carbon	65,000		42,000	38,000	31,000,000	13,000,000
,, ,, ½	78,000		47,000	49,000	31,000,000	
,, ,, 1	110,000		67,000	71,000	31,000,000	
Cast steel, untempered	150,000		80,000	80,000	30,000,000	12,000,000
	120,000					
	84,000					
,, tempered	31,000		190,000	190,000	36,000,000	14,000,000
Copper rolled plates	45,000		5,600	4,000	15,000,000	5,600,000
,, annealed wire	58,000				16,000,000	
,, hard drawn wire { From	17,500				17,000,000	
To	29,000					
Brass	52,000				13,500,000	
	27,000					
Gun metal or bronze	23,000		6,200		13,500,000	
Delta metal, cast	36,000		17,000		12,000,000	
,, rolled	74,000		51,000		13,000,000	
Phosphor bronze	58,000		19,700		14,000,000	5,250,000

the corresponding extension be a line $y\,p$ drawn horizontally from y. The locus of the point p will be the stress-strain curve of the material. Stress-strain curves for a number of different materials subjected to tension are shown in figure 111.

It has been proposed to represent the comparative values of materials for constructive purposes by figures derived from their stress-strain curves. The work done in breaking a test piece, reckoned per cubic inch of volume, may be used. This is proportional to the area included between the base and the stress-strain curve. Tetmajer's 'value-figure' for a material is the product of the maximum stress and the elongation per unit length. It is the area of the rectangle formed by drawing from the final point of the stress-strain curve lines parallel to the axes. Of the materials represented in figure 111, 'Delta' metal and aluminium bronze have the highest 'value-figures.'

115. **Mild Steel.**— Figure 111 shows the stress-strain curve for mild steel, such as the material from which weldless steel tubes are made. The straight portion $O\,a$ represents the action within the elastic limit. If the load be increased beyond that represented by a, the extension takes place at a more rapid rate, as shown by the slightly curved portion $a\,b$. At a point, b, somewhat above the elastic limit, a, a sudden lengthening of the bar takes place without any increase of load, this being represented by the portion $b\,c$ of the curve. The stress at which this occurs is called the *yield-point* of the material. On further increasing the load, extension again takes place, at first comparatively slowly, but afterwards more rapidly, until the maximum stress at the point d is reached. Under this stress the bar elongates until it breaks. If, however, the stress be partially removed after the maximum stress, d, is reached, as can be done in a testing machine, the curve falls gradually, as at $d\,e$, then more rapidly until fracture occurs at f. The elongation represented by the curve up to E takes place uniformly over the whole length of the bar, that represented by $e\,f$ only on a small portion in the neighbourhood of the fracture.

In wrought iron, the yield-point is not so distinctly marked as in mild steel ; the stresses at the elastic limit and at breaking are less, the elongation before fracture is also less. The specific gravity of wrought iron and mild steel is, on an average, 7·7.

116. **Tool Steel.**—For a tool steel of good quality, containing about one per cent. carbon, the maximum stress may be much higher ; the stress-strain curve takes the form shown in figure 111,

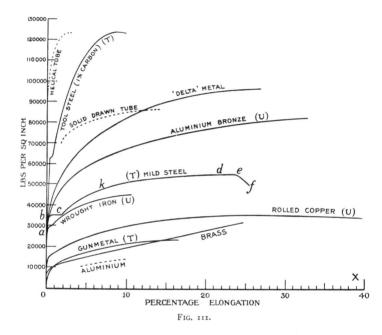

FIG. 111.

the extension being smaller, though the tenacity is very much greater, than that of mild steel.

117. **Cast Iron** has no well-defined elastic limit ; in fact, the stress-strain curve is not straight for any part of its length, so that for cast iron the term 'elastic limit,' though often used, has no definite meaning.

118. **Pure Copper** varies greatly in tensile strength, according to the mechanical treatment to which it has been subjected. Rolling and wire-drawing both increase its tenacity. The stress-strain curve for rolled copper (fig. 111) is from Professor Unwin's 'The Testing of Materials of Construction.'

119. **The Alloys of Copper** with other metals form a most

important series. Their mechanical properties are most fully discussed in Professor Thurston's 'Brasses, Bronzes, and other Alloys.'

Brass contains 66–70 per cent. copper, and 34–30 per cent. zinc ; sometimes a little lead. The stress-strain diagram (fig. 111) shows that the stress at the elastic limit is very low in comparison with the ultimate breaking stress.

Gun-metal is an alloy of copper and tin. The stress-strain diagram (fig. 111) is from a metal containing 98 per cent. copper, 2 per cent. tin.

Ternary alloys of copper, zinc, and tin have been exhaustively investigated by Professor Thurston. He finds the best proportion, when toughness as well as tenacity is important, is copper 55, tin 0·5, zinc 44·5.

Aluminium Bronze.—Copper and aluminium form a most useful series of alloys. The stress-strain curve (fig. 111) is from an alloy containing about 10 per cent. aluminium ; it shows clearly the great strength and ductility of the material.

Alloys containing a much larger proportion of aluminium are valuable where lightness is the first consideration, but since they possess little strength and ductility, they can only be sparingly used in structural work.

Delta metal is a copper-zinc-iron alloy, which can be cast and worked hot or cold. It possesses great strength and ductility, as is shown by the stress-diagram (fig. 111) from a bar ·79 sq. in. sectional area, tested by Mr. A. S. E. Ackermann at the Central Technical College.

120. **Aluminium.**—A specimen of squirted aluminium, containing 98 per cent. of the pure metal, was tested at the Central Technical College by Mr. Ackermann ; the tenacity was 6·32 tons per sq. in. ; the elongation in 8″ was 1·12″, of which ·53″ was in the immediate neighbourhood of the fracture ; the general elongation may, therefore, be taken as 10 per cent. For comparison this result is plotted in figure 111.

Pure aluminium has not sufficient strength and toughness to be of much value as a structural material, though its lightness as compared with other metals is a desirable quality. Some alloys, containing a small percentage of aluminium, possess great strength, but they are, of course, heavy. It remains to be seen whether a strong alloy, containing a large percentage of aluminium, and therefore light, can be discovered. Such an alloy may possibly be of value in cycle making.

The specific gravity of sheet aluminium is 2·67, of mild steel 7·7.

121. **Wood** is not so homogeneous as most metals ; it is, as a rule, much stronger along than across the grain. The fact that wood joints are generally of low efficiency is against its use in tension members of a frame. For compression members, where there is no loss of strength at the joints, it may be used with advantage in some cases, its compressive strength (see Table VI.) being not much inferior, weight for weight, to that of the metals. In beams of short span subjected to bending, it is, in some important cases, immensely superior, weight for weight, to metal. The strength of a rectangular beam is proportional to its width, the *square* of its depth, and the strength of the material from which it is made (sec. 94), *i.e.* proportional to $b z^2 f$. If beams of equal weight be made from wood and steel, the width b being the same in both, the depth d of the wood beam will be greater than that of the steel beam ; and the product $z^2 f$ will be much greater for the wood than the steel beam.

The rim of a bicycle wheel is subjected to compression and bending (sec. 255). Since its width must be made to suit the tyre, a wood rim will be much stronger than a solid steel rim of the same weight ; or, for equal strengths, the wood rim will be the lighter. A *hollow* steel rim will possibly be stronger than a wood rim of equal weight.

Table VI., taken from Laslett's ' Timber and Timber Trees,' gives the weights and strengths of a few woods.

122. **Raising of the Elastic Limit.**—Let a bar be subjected to a stress—represented by the point k (fig. 111)—considerably above its elastic limit. If the load be removed and the bar be again tested, it will be found that it is elastic up to a stress as high as that indicated by k. Thus the elastic limit in tension of a material like mild steel can be raised by simply applying an initial stress a little above the limit required.

An important application of this principle occurs in the case

TABLE VI.

SPECIFIC GRAVITY AND STRENGTH OF WOODS.

Name of wood	Specific gravity, water being taken 1·000	Transverse load on pieces $2'' \times 2'' \times 72''$	Tensile stress on pieces $2'' \times 2'' \times 30''$	Vertical stress on pieces $2'' \times 2'' \times 2''$
		lbs.	lbs. per sq. in.	lbs. per sq. in.
Ash, English . . .	·736	862	3,780	6,963
,, American . .	·480	638	5,495	5,494
Elm, English . . .	·558	393	5,460	5,785
,, Canadian . .	·748	920	9,182	7,418
Fir, Dantzic . . .	·582	877	3,231	7,104
,, Spruce, Canada .	·484	670	3,934	4,852
Kauri, New Zealand .	·530	816	4,543	5,880
Larch, Russian . .	·646	626	4,203	5,985
Oak, English . . .	·735	776	7,571	7,640
,, French . .	·976	878	8,102	7,942
,, White, American .	·983	804	7,021	6,964
Pine, Yellow . . .	·554	505	2,027	4,172
,, Pitch, American .	·659	1,049	4,666	6,462

of Southard's twisted cranks. Here the cranks are given a considerable initial twist in the direction in which they are strained while driving ahead; their strength is considerably increased thereby. A twist (sec. 109) is equivalent to a direct pull along certain fibres, and a direct compression along other fibres at right angles. The initial twist in Southard's crank is, therefore, equivalent to raising the elastic limit of tension of the fibres under tensile stress, and the elastic limit of compression of the fibres under compressive stress.

123. **Complete Stress-strain Diagram.**—The complete stress-strain diagram of a material should extend below the axis $O X$; in other words, it should give the contractions of the bar under compressive stresses, as well as elongations under tensile stresses. Figure 112 represents such a curve, the point a denoting the elastic limit in tension, and b the elastic limit in compression. If the bar has had its elastic limit in tension raised artificially to the point k (fig. 111), it is found experimentally that the elastic limit in compression has been lowered, and thus the new stress-strain curve would be somewhat as represented in figure 113.

These considerations, when applied to the case of Southard's cranks, detract from the value of the initial twist. The line *t t*

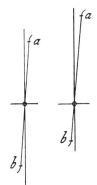

FIG. 112.   FIG. 113.

(fig. 109), which is the tension line when the rider is pedalling ahead, has had its elastic limit in tension artificially raised, and its elastic limit in compression artificially lowered by the initial twist. When back-pedalling, *t t* becomes the compression line. A twisted crank is therefore weaker for back-pedalling than an untwisted crank of the same material.

124. **Work done in Breaking a Bar.**—A material that gives very little extension before breaking is said to be wanting in *toughness*, and is not so suitable for structural purposes as a material with a larger extension. The total elongation of a material is usually expressed as a percentage of its original length. If the actual instead of the percentage elongations be set off horizontally (fig. 111), the area included between the stress-strain curve, its end ordinate, and the axis *O X*, represents the work done in breaking the bar. A bicycle is a structure subjected not to steadily applied forces but to impact. The relative value of a hard and a tough material for resisting such straining actions may be illustrated by an example.

Example.—Take a material like hardened steel, elastic up to its breaking-point, so that its stress-strain diagram is as shown at figure 114. Let its breaking-stress be 60 tons per square inch, and $E = 12,000$ tons per square inch. Then the extension at breaking-point is

$$E = \frac{60}{12000} = \cdot 005.$$

If the original length of the bar be 10 inches, the total elongation *O x* (fig. 114) will be ·05 inches, and the work done will be the area of the triangle *O a x*,

$$= \tfrac{1}{2} \times 60 \times \cdot 05 = 1 \cdot 5 \text{ inch-tons.}$$

Take now a material like mild steel, and consider that its stress-strain curve is quite straight up to the yield-point *b* (fig. 115).

Let the yield-point occur at 15 tons per square inch ; then, taking E, as before, 12,000 tons per square inch, and the original length of the bar 10 inches, $O\,x$ will be ·0125 inches. The work done in stretching the bar up to the yield-point will be

$$\tfrac{1}{2} \times 60 \times \text{·0125} = \text{0·375 inch-tons.}$$

Consider both bars to be acted on by a force of impact equivalent to a weight of 10 tons falling through a height of $\tfrac{1}{5}$ inch. The work stored up in this falling weight will be

$$10 \times \tfrac{1}{5} = 2 \text{ inch-tons.}$$

This must be taken up by the bar. But the work done in breaking the hard steel bar of high tenacity is only 1·5 inch-tons ; it would therefore be broken by such a live load. The mild-steel bar would be stretched an additional length, $x\,x_1$, until the total area, $O\,b\,b^1\,x_1$, was equal to 2 inch-tons. The area, $b\,b_1\,x_1\,x$, is therefore

$$2 - \text{0·375} = \text{1·625 inch-tons.}$$

The distance $x\,x_1$ will be

$$\frac{\text{1·625}}{15} = \text{·108 inch.}$$

Thus the only effect of the impulsive load on the mild steel bar is to stretch it a small distance, though the same load is sufficient to break the bar of much higher tensile strength but with little or no elongation before fracture.

The above examples show that the elongation before fracture of a material is almost as important as its breaking strength in determining its value as a material for bicycle building.

125. **Mechanical Treatment of Metals.**—The tenacity of a metal is almost invariably increased by rolling, or by drawing through dies. A metal to be drawn into wire or tube must be strong and ductile. The finest wire is made from a metal in which the ratio of the elastic to the ultimate strength is low. A metal with very high tenacity has not generally the ductility necessary for drawing into tubes or wire. The Premier Cycle Company,

Fig. 114.   Fig. 115.

instead of using drawn tubes, which must be made from a steel having a comparatively low tenacity, build up their tubes from flat sheets bent into spirals, each turn of the spiral overlapping the adjacent one, so that there are two thicknesses of plate at every part of the tube (fig. 116). A steel of much higher tenacity can

FIG. 116.

be used for this process than could be successfully drawn into tubes. These 'helical' tubes, therefore, have greater tenacity but less ductility than weldless steel tubes, as is shown by the comparative tests of helical and solid-drawn tubes 1 inch external diameter, recorded in Table VII. For comparison with other materials, the results of these tests are plotted on figure 111 ; the final points of the stress-strain diagrams being the only ones obtainable from the data, the curves are drawn dotted.

TABLE VII.

TENSILE STRENGTH OF HELICAL AND SOLID-DRAWN TUBES.

Description	Sectional area	Ultimate stress	Extension in 10 inches	Appearance of fracture
	Sq. in.	lbs. per sq. in.		
Helical 14A .	0·105	117,000	3·1	12 per cent. silky 88 per cent. granular
,, 20A .	0·107	122,000	1·5	Granular
,, 20C .	0·134	130,000	3·4	Granular
Solid-drawn C_1 .	0·106	80,000	18·7	Silky
,, H_1 .	0·106	94,000	8·0	Silky

126. **Repeated Stresses.**—If a bar be subjected to a steady load just below its breaking load, it will support it for an indefinite period provided the load remains constant, neither being increased or diminished. If the load is variable, however, the condition is quite different. Wöhler has shown that if the load vary from a maximum T_1 to a minimum T_2, fracture will occur when T_1 is less than the statical breaking load T, after a certain number of alterations from T_1 to T_2. The number of repetitions of the load

before fracture takes place depends not only on T_1 but on the difference $T_1 - T_2$, between the maximum and minimum loads. With a great range of stress the number of repetitions before fracture is less than with a smaller range.

A steel axle tested by Bauschinger, which had a statical tensile strength of 40 tons per square inch, stood at least two or three million changes of load before breaking, with the following ranges of stress :

Minimum stress tons per sq. in.	Maximum stress tons per sq. in.	Range of stress tons per sq. in.
− 10·5	+ 10·5	21·0
0	19·7	19·7
20	32·1	12·1
40	40	0

A fuller discussion of this subject is given in Professor Unwin's ' Machine Design ' and ' The Testing of Materials of Construction.'

The running parts of a bicycle—the wheels, chain, pedal-pins, cranks, and crank-axle—are subjected, during riding, to varying stresses. The range of stress on the spokes is probably small, so that a high maximum stress may be used without running any risk of fracture after the machine has been in use a considerable time. The stress on a link or rivet of the chain varies from zero, when on the slack side, to the maximum on the tight side. The double change of stress on the pedal-pins, cranks, and crank-axle takes place once during each revolution of the latter. A distance of 5,000 miles ridden on a bicycle geared to 60″ corresponds to 1,500,000 double changes of stress on the cranks and axle. If these be made light (see chapter xxx.), no surprise need be expressed if fracture occurs at any time, after having run satisfactorily for one or two years.

PART II
CYCLES IN GENERAL

PART II

CYCLES IN GENERAL

CHAPTER XIV

DEVELOPMENT OF CYCLES : THE BICYCLE.

127. **Introductory.**—Wheeled vehicles drawn by horses have probably been used by all civilised nations. The *chariot* of the ancients was two-wheeled, the wheels revolving upon the axle. Coming down to later times, the *coach*, a covered vehicle for passengers, appears to have been first made in the thirteenth century, the earliest record relating to the entry of Charles of Anjou and his queen into Naples in a small *carretta*. The first coaches in England are said to have been made by Walter Rippon for the Earl of Rutland in 1555, and for Queen Elizabeth in 1564. The weight of these early coaches was probably so great that for centuries it seemed utterly impracticable to make a vehicle that could be propelled by the rider. With the growth of the mechanical arts, at the beginning of this century, more attention was given to the subject. Starting from the four-wheeled vehicles drawn by a horse, the most obvious step towards getting a pedomotive vehicle was to make one of the axles cranked, and let the rider drive it either direct or by a system of levers, the wheels being rigidly fastened to the ends of the axle. Such a cycle is illustrated in figures 117, 118. If this cycle had to travel in straight lines or curves of large radius, as on a railway, it might have been, apart from its weight, fairly satisfactory. A grave mechanical defect was that in moving round a sharp curve one or both driving-wheels slipped, as well as rolled, on the ground, with a corresponding waste of energy in friction.

The first attempts at overcoming this difficulty consisted in fastening only one wheel rigidly to the driving-axle, the other running freely. This gave, however, a machine which did not always respond to the steering gear as the rider wished ; in fact, while a driving effort was being exerted, the machine tended to turn to the side opposite to the driving-wheel (see chap. xviii.). The introduction of the differential driving-axle, which allows both

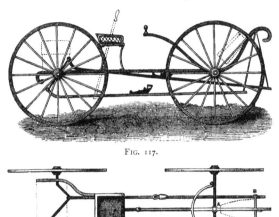

FIG. 117.

FIG. 118.

wheels to be driven at different speeds, overcame this difficulty completely without introducing any new ones.

The weight of the four-wheeler, and even of the three-wheeler, was, however, so great that it was not in this direction that cycles were at first developed. A wooden frame for supporting two wheels was, of course, much lighter than one for three wheels ; for this reason principally, bicycles were brought to a fair degree of perfection before tricycles. The use of steel tubes for the various parts of the frame made it possible to combine the strength and lightness necessary for a practicable cycle, and laid on a sure basis the foundations of the cycle-making industry.

Without attempting to give an exhaustive history of the development of bicycles and tricycles, a short account of the various types that have from time to time obtained public favour may be given here.

128. **The Dandy-horse.**—Figure 119 may be taken as the first velocipede man-motor carriage. This was patented in France in 1818 by Baron von Drais. In 'Ackermann's Magazine,' 1819, an account of this pedestrian hobby-horse is given. " The principle

FIG. 119.

of the invention consists in the simple idea of a seat upon two wheels propelled by the two feet acting on the ground. The riding seat or saddle is fixed upon a perch on two short wheels running after each other. To preserve the balance a small board covered and stuffed is placed before, on which the arms are laid, and in front of which is a little guiding pole, which is held in the hand to direct the route. The swiftness with which a person well practised can travel is almost beyond belief, 8, 9, and even 10 miles may, it is asserted, be passed over within the hour on good level ground."

129. **Early Bicycles.**—Messrs. Macredy and Stoney, in 'The Art and Pastime of Cycling,' write : "To Scotland, it appears, belongs the honour of having first affixed cranks to the bicycle ;

Fig. 120.

and, still stranger to relate, it was not to the 'hobby-horse,' but to a low-wheeled rear-driver machine, the exact prototype of the present-day Safety. The honour of being the inventor has now been fixed on Kirkpatrick M'Millan, of Courthill, Dumfries-

Fig. 121.

shire, though prior to 1892 Gavin Dalzell of Lesmahagow was the reputed inventor. It seems, however, that Dalzell only copied and probably improved on a machine which he saw with M'Millan.

M'Millan first adapted crank-driving to the 'hobby-horse' about the year 1840, and it was not earlier than 1846 that Dalzell built a replica of M'Millan's machine, a woodcut of which we reproduce (fig. 120). M'Millan is said to have frequently ridden from Court-hill to Dumfries, some fourteen miles, to market on his machine, keeping pace with farmers in gigs." Figure 121 illustrates the 'French' bicycle or Bone-shaker,' which was in popular favour during the sixties. The improvement on the Dandy-horse consisted principally in the addition of cranks to the front wheel, so that the rider was supported entirely by the machine.

In 'Velocipedes, Bicycles, and Tricycles,' published by George Routledge & Sons in 1869, descriptions and illustrations of the bicycles, tricycles, and four-wheelers then in use are given. The concluding paragraph of this little book may be quoted : " Ere I say farewell, let me caution velocipedists against expecting too much from any description of velocipede. They do not give power, they only utilise it ; there must be an expenditure of power to produce speed. One is inclined to agree with the temperate remarks of Mr. Lander, C.E., of Liverpool, rather than with the extravagant enthusiasm of American or French riders. As a means of healthful exercise it is worthy of attention. Certainly not more than forty miles in a day of eight hours can be done with ease ; Mr. Lander thinks only thirty. If this is correct, it does not beat walking, though velocipedists affirm that double the distance can be done with ease. Much will and must depend on the skill of the rider, the state of the roads, and the country to be travelled."

130. **The Ordinary.**—What has since been called the 'Ordinary' bicycle came into use early in the seventies. Figure 122 illustrates one made by Messrs. Humber & Co., Limited. The great advance on the bicycle illustrated in figure 121 consisted mainly in the use of indiarubber tyres, thus diminishing vibration and jar, and consequently diminishing the power necessary to propel the machine. As a direct consequence of this, a larger driving-wheel could be driven with the same ease as the comparatively small driving-wheel of the French bicycle. The design of the 'Ordinary' is simplicity itself, and it still remains the embodiment of grace and elegance in cycle construction, though superseded by its more speedy rival, the rear-driving Safety. The

motive power of the rider is applied direct to the driving-wheel ; wheel, cranks and pedal-pins forming one rigid body. In this respect it has the advantage over bicycles of later design, with gearing of some kind or other between the pedals and driving-wheel.

In the 'Ordinary' the mass-centre of the rider was nearly directly over the centre of the wheel, so that any sudden obstruction to the motion of the machine frequently had the effect of sending

FIG. 122.

the rider over the handle-bar. This element of insecurity soon led to the introduction of other patterns of bicycles.

131. **The 'Xtraordinary'** (fig. 123), made by Messrs. Singer & Co., was one of the first Safety bicycles. The crank-pin was jointed to a lever, one end of which vibrated in a circular arc (being suspended by a short link from near the top of the fork), the other end was extended downwards and backwards, and supported the pedal. A smaller wheel could thus be used, and the saddle placed further back than was possible in the 'Ordinary.'

FIG. 123.

132. The Facile.—In the 'Facile' bicycle a smaller driving-wheel was used, and the mass-centre of the rider brought further behind the centre of the driving-wheel.   This was accomplished

FIG. 124.

by driving the crank by means of a short coupling-rod from a point about the middle of a vibrating lever; the end of this vibrating lever forming the pedal. The fork of the front wheel was continued downwards and forwards to provide a fulcrum for the lever. The motion of the pedal relative to the machine was thus one of up-and-down oscillation in a circular arc, and was quite different from that of the uniform circular motion in the 'Ordinary.' From the position of the mass-centre of the rider relative to the centre of the driving-wheel, it is evident that this bicycle possessed a much greater margin of safety than the 'Ordinary.' Also, from the fact that the machine and rider offered a less surface to wind resistance, the machine was easier to propel under certain circumstances than the 'Ordinary.' In 1883, Mr. J. H. Adams rode $242\frac{1}{2}$ miles on the road within twenty-four hours ; this was at that time the best authentic performance on record.

133. **Kangaroo.**—Figure 125 illustrates the 'Kangaroo' type of front wheel crank-driven Safety introduced by Messrs. Hillman,

FIG. 125.

Herbert, and Cooper, 1884. A smaller driving-wheel is used than in the 'Ordinary'; the crank-axle is placed beneath and a little behind the centre of the driving-wheel. The crank-axle is divided into two parts, since its axis passes through the driving-wheel ; the front-wheel fork is continued downwards to support the crank-axle bearings ; the motion of each portion of the crank-axle is transmitted by chain-gearing to the driving-wheel.   In a 100-mile road race on September 27, 1884, organised by the makers of the machine, the distance was covered by Mr. G. Smith in 7 hours 7 minutes and 11 seconds, the fastest time on record for any cycle then on the road.

A geared dwarf bicycle is superior to an ' Ordinary ' in two important respects, which more than compensate for the friction of the extra mechanism.   Firstly, the rider being placed lower, the total surface exposed by the machine and rider is much less, the air resistance is therefore less, this advantage being greatest at high speeds.   Secondly, since the speeds of the driving-wheel and crank-axle may be arranged in any desired ratio, the speed of pedalling and length of crank can be chosen to suit the convenience of the rider, irrespective of size of driving-wheel ; while in an ' Ordinary ' the length of crank is less, and the speed of pedalling greater, than the best possible values.

As regards safety, the ' Kangaroo ' is a little better than the ' Ordinary,' but not so good as the ' Rover ' or ' Humber ' Safety. Two serious defects, which ultimately made it yield in popular favour to the rear-driving Safety, existed.   A narrow tread must be kept between the pedals, and the consequent narrow width of bearing of the crank-axle gives a bad design mechanically. Again, the chains, however carefully adjusted initially, will, after a time, get a trifle slack.   In pressing the pedal downwards the front side of the chain is tight, but when the pedal is ascending, since it cannot be lifted direct by the rider, it is pulled up by the chain, the rear side of which gets tightened.   This reversal, taking place twice every revolution, throws a serious jar on the gear. This defect cannot, as in the ' Humber ' with only one driving-chain, be overcome by skilful pedalling.

134. **The Rear-driving Safety** was invented by Mr. H. J. Lawson in 1879, but it was a few years later before it was in great demand.   The ' Rover ' Safety (fig. 126), made by Messrs. Starley and Sutton in 1885, was the first rear-driving bicycle that attained popular favour.   The cranks and pedals are placed on a separate axle, the motion of which is transmitted by a single driving-chain to the driving-wheel.   This type is absolutely safe as regards headers

over the handle-bar. Compared with the 'Kangaroo' gearing, the single driving-chain is a great improvement, as its driving side

FIG. 126

may be kept tight continuously. The steering-head of the front wheel was vertical, and an intermediate handle-pillar was used, with coupling-rods to the front fork. In a later design (fig. 127)

FIG. 127.

the front fork was sloped, and the steering made direct ; this machine thus formed the prototype of the modern rear-driving bicycle.

Figure 128 is an illustration of the 'Humber' Safety dwarf-roadster, made in 1885. In this all the arrangements of the 'Ordinary' may be said to be reversed ; the proverbial Irishman's description of it being " The big wheel is the smallest, and the hind wheel is in front." The driving-wheel is changed from front to back, the small wheel is placed in front, and the mass-centre of the rider is brought nearer the centre of the rear wheel.

The 'Humber' Safety of 1885 is essentially the same machine as that in greatest demand at the present day. The improvements

FIG. 128.

effected since 1885, though undoubtedly of very great practical importance, are merely improvements in details. Change in the relative size of the front and back wheels, different design of frame, and last, but not least, the introduction of pneumatic tyres, account for the different appearances of the earliest and latest Safeties.

Rear-driving Safeties were made by all the makers, the difference in bicycles by different makers being merely in detail. About this time (1886) the number of Safety bicycles made per annum began to increase very rapidly, while a few years later the number of 'Ordinaries' began to diminish.

135. **Geared Facile.**—The 'Facile' bicycle, with its small driving-wheel and direct link-driving from the pedal lever, necessitated

very fast pedal action on the part of the rider. The 'Geared Facile' (fig. 124) enabled the pedalling to be reduced to any desired speed. The connecting link in the 'Geared Facile' did not work directly on the driving-wheel, but the crank shaft ran loose co-axially with the driving-wheel, a sun-and-planet gear being inserted between the crank and the wheel. Figure 129

FIG. 129.

shows a 'Geared Facile' rear-driving bicycle, the usual sun-and-planet gear being modified to suit the altered conditions.

136. **Diamond-frame Rear-driving Safety.**—From the date of its introduction, the rear-driving Safety advanced steadily in popular favour until, in 1887, it was the bicycle in most general demand. In the preface to 'Bicycles and Tricycles of the Year 1888,' Mr. H. H. Griffin says : "We made careful inquiries of all those in a position to know as to the proportion of Dwarf Safeties and Ordinary bicycles, and were not a little surprised to hear that, taking the average through the trade, at least six Dwarf Safeties are made to one Ordinary." Up to the year 1890 the greatest possible variety existed in the frames of the rear-driving Safety, but they all agreed in having the distance between the rear and front wheels reduced to a minimum. The crank-bracket was placed just sufficiently in front of the driving-

wheel to have the necessary clearance, the steering-wheel sufficiently far in front to allow it in steering to swing clear of the
pedals and the rider's foot.   The down-tube, from the saddle to

Fig. 130.

the crank-bracket, was usually curved, both in the diamond-frame
and the cross-frame, or omitted altogether, as in the open-frame.
Up till 1890 the nearest approach to the now universally adopted

Fig. 131.

frame was that made by Humber & Co. (fig. 130).   During these
years the diamond-frame was being more and more generally
adopted, and after Messrs. Humber introduced their rear-driving

Safety, with long wheel-base and diamond-frame (fig. 131), it became almost universal. By having several inches clearance between the crank-bracket and the driving-wheel, it was possible to use a straight tube from the saddle to the crank-bracket, while the long wheel-base rendered the steering more reliable. In the chapter on 'Frames' the reasons for the survival of the diamond-frame and the practical extinction of all others will be given.

137. **Rational Ordinary.**—The admirers of the 'Ordinary' bicycle were loth to let their favourite machine fall into disuse, and attempts were made to make it safer and more comfortable, by placing the saddle further behind the driving-wheel centre, by sloping the front fork, and by making the rear wheel larger than was usual in the 'Ordinary.' Such a machine was called a '*Rational* Ordinary.'

138. **Geared Ordinary** and **Front-driving Safety.**—In 1891, the Crypto Cycle Company—with whom Messrs. Ellis & Co., the

FIG. 132.

makers of the 'Facile' and 'Geared Facile' had amalgamated—brought out a *Geared Ordinary*. This bicycle was in external appearance just like a 'Rational'; but the cranks, instead of being rigidly connected to the driving-wheel, drove the latter by means of an epicyclic gear (see sec. 306) concealed in the hub. The number of revolutions of the driving-wheel could thus be made greater than those of the crank ; in fact, the machine could be geared up, just like a rear-driving Safety. The size of the driving-wheel being reduced, a front-driving Safety was obtained. Figure 132 shows the 'Bantam,' the latest development of the front-driver in this direction, with the front wheel 24 inches in diameter, and geared to 66 inches. The resemblance, in general arrangement at least, to the French bicycle (fig. 121) will be apparent, though as regards efficiency of action the two machines are as wide apart as the poles. Figure 243 shows

the 'Bantamette,' in which the frame is so arranged that the bicycle
may be ridden by a lady.

139. **The Giraffe** and **Rover Cob.**—The 'Ordinary' had undoubtedly many good points which are missing in the modern
Safety, among which may be mentioned greater lateral stability
and steadiness in steering due to the high mass-centre. The
'Giraffe' (fig. 133), by the New Howe Machine Company, is a
high-framed Safety, the saddle being raised as high as in the

FIG. 133.

'Ordinary.' In the introduction to Leechman's 'Safety Cycling,'
Mr. Henry Sturmey gives an enthusiastic account of the 'Giraffe,
and a comparison with the low-framed Safety.

The 'Rover Cob' (fig. 134), made by Messrs. J. K. Starley & Co.,
is at the opposite extreme, the frame being made so low that the
pedals will just clear the ground when rounding a corner at slow
speed. It is intended for those who may have fear of falling ;
the mounting can be done by simply pushing off from the ground.

140. **Pneumatic Tyres.**—Whether judged by speed perform-
ances on the road or racing track, or from additional comfort
and ease of propulsion to the tourist, the greatest advance in
cycle construction due to a single invention must be credited to

Mr. James Dunlop, the inventor of the pneumatic tyre. A patent for a pneumatic tyre had been taken out by Thompson in 1848, but there is no record that he made a commercial success of his invention. In 1890, Mr. James Dunlop, of Dublin, made a pneumatic tyre for his son, and the results obtained by its use being so astounding, arrangements were very soon made for their manufacture. While in 1889 a pneumatic tyre was unheard of, at the Stanley Bicycle Club Show, November–December, 1891, from an analysis [1] of the machines exhibited, it appears that 40 per cent. of the tyres exhibited were *pneumatic*,

FIG. 134.

$32\frac{1}{2}$ per cent. *cushion*, $16\frac{1}{2}$ per cent. *solid*, 10 per cent. *inflated*, and the remainder, about 1 per cent., were classed as nondescript. In the above classification, under pneumatic tyres are included only those with a separate inner tube, the inflated being really single-tube pneumatic tyres. Cushion tyres were made and used as a kind of compromise between solids and pneumatics. The proportion of pneumatic tyres to the total has grown greater year by year, until now there is hardly a cycle made, for use in Britain at least, with any other than pneumatic tyres.

141. **Gear-cases.**—The most troublesome portion of a modern rear-driving bicycle is undoubtedly the chain and the accompanying gear. The chain, however well made originally, is found to stretch slightly under the heavy stresses to which it is subjected

[1] *The Cyclist's Annual and Year-book for* 1892.

in ordinary working.   If the distance between the centres of the two chain-wheels—on the crank-axle and driving-wheel hub respectively—over which the chain passes is unalterable, the chain will ultimately get so slack that there will be a great risk of it over-riding the teeth of the wheels, to the danger of the rider.   All chain-driven cycles are consequently provided with some means of tightening the chain, *i.e.* of increasing the distance between the centres of the two chain-wheels.   Again, in an exposed chain, it is practically impossible to lubricate perfectly the rubbing parts, very little of the oil applied to the outside surface finding

FIG. 135.

its way in between the rivet-pins and the blocks of the chain. Dust and grit from the road soon adhere to the chain and chain-wheel, so that the frictional resistance of the chain as it is wound on and off the chain-wheel is rapidly increased.

These considerations led Mr. Harrison Carter to introduce the *gear-case*, the function of which is to exclude dust and mud, and provide an oil-bath in which the lowest portion of the chain may dip.   The reduction of frictional resistance is perhaps one of the least of the advantages pertaining to the use of the gear-case ; one great advantage is that less trouble is given to the rider, and chain adjustments need not be made so frequently.   In fact, some makers claim that with an oil-tight gear-case the chain does not stretch perceptibly, and no chain adjustments are necessary. The author is not aware, however, that any maker has ventured to place on the market a bicycle with gear-case but no chain adjustment.

142. **Tandem Bicycles.**—When the success of the bicycle for a single rider was assured, attempts were soon made to make a bicycle for two riders. Figure 135 shows the ' Rucker ' Tandem bicycle, made in 1884, one of the first successful tandem bicycles. This consists practically of two ' Ordinary ' driving-wheels and forks connected together by a straight tubular backbone. At the front

FIG. 136.

end of this backbone there is an ' Ordinary ' steering centre ; at the other end it is connected to the head of the rear-wheel fork by a frame which permits it to twist sideways. Figure 136 shows a later andem bicycle, also made by Mr. Rucker—probably the first practicable machine of this type. It is practically a tandem ' Kangaroo.' In a paper on ' Construction of Cycles,' read before the Institution of Mechanical Engineers in 1885, Mr. R. E Phillips says, " This tandem bicycle . . . eclipses the earlier, and bids fair to prove the fastest cycle yet produced. The weight is only 55 lbs., and it is, therefore, the lightest machine yet made to carry two riders "

Figure 137 shows a front-driving chain-driven Safety Tandem, made by Hillman, Herbert, and Cooper, 1887. Both riders drive the front wheel, and both wheels are moved in steering.

FIG. 137.

The 'Invincible' Tandem Safety (fig. 138), and the 'Ivel' Tandem Safety (fig. 139), which was made convertible so that it

FIG. 138.

could be used as a single Safety, were among the first approximations to the present popular type of Tandem Safety, both riders.

being placed between the wheels, and both driving the rear wheel. It will be noticed that the front crank-axle is connected by chain gearing to the rear crank-axle, the two axles rotating at the same speed ; the second chain passes over the larger wheel on the rear crank-axle and the chain-wheel of the driving-axle.

FIG. 139

Both riders have control of the steering, a light rod connecting the front fork to the rear steering-pillar. The long wheel-base of these bicycles adds to the steadiness of the steering at high speeds, since (see fig. 202), for the same deviation of the handle-bars, a machine with long wheel-base will move in a curve of larger radius than one with a shorter wheel-base. The distance between the wheel centres being much greater than in the single machine, the frame is subjected to very much greater straining actions, and imperfect design will be much more serious than in the single machine.

FIG. 140.

Figure 140 is an example of the present popular type of Tandem bicycle made by Messrs. Thomson and James. The machine is kinematically the same as that of figure 138, the particular difference being in the rear frame, which is of the diamond type, completely triangulated.

CHAPTER XV

143. **Early Tricycles.**—No sooner was a practicable bicycle made than attention was turned to the three-wheeler as being the safer of the two machines, and offering some advantages, such as the possibility of sitting while the machine is at rest. It was very early found that the greater safety of the three-wheeler was more apparent than real. 'Velox,' writing in 1869, says, "Strange as it may appear to the un-initiated, the tricycle is far more likely to upset the tyro than the bicycle."

Figure 141 (from 'Velox's' book) represents a simple form of tricycle made in the sixties by Mr. Lisle, of Wolverhampton, known as the 'German' tricycle. It was, in fact, a converted 'Bone-shaker' bicycle, with the rear wheel removed and replaced by a pair of wheels running free on an axle two feet long. The motive power

FIG. 141

was applied by pedals and cranks attached to the axle of the front wheel. A number of tricycles were made on the same general principle ; but the weight of the rider being applied vertically over a point near the front corner of the wheel-base triangle, the margin of lateral stability was small. Mr. Lisle also made a ladies' double-driving tricycle (fig. 142), in which the power was applied by treadles and levers acting on cranks on the axle of the rear wheels. Nothing is said about the axle of the rear wheels being divided,

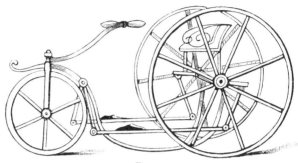

FIG. 142.

so it is probable that in turning round a corner the rear wheels skidded, just as is the case with railway rolling stock.

In the 'Dublin' tricycle (fig. 143) the driving-wheel was behind, and two steering-wheels placed in front ; the margin of stability in case of a stoppage was much greater than in the 'German' tricycle (fig. 141). Another point of difference consisted in the application of the lever gearing ; the pedals were fixed on oscillating levers, the motions of which were communicated by crank and connecting-rods to the driving-wheel.

FIG. 143.

The 'Coventry' bicycle was at first made with lever gearing, but chain gearing was very soon afterwards applied to it. The 'Coventry Rotary' (fig. 144) was the most successful of the early single-driving tricycles. It may be interesting to note that this type has been revived recently, the Princess of Wales having selected a tricycle of this type.

FIG. 144.

If the mass-centre be vertically over the centre of the wheel-base triangle, the pressure on each wheel will be one-third of the

FIG. 145.

total weight. Under certain circumstances this pressure is insufficient for adhesion for driving, hence arose the necessity for double-driving tricycles. In the ' Devon ' tricycle, made in 1878 (fig. 145), which is fitted with chain gearing, the cog-wheels co-axial with the driving-wheels are fitted loose on their axles, and each cog-wheel drives its axle by means of a ratchet and pawl. In rounding a corner, the inside wheel is driven by the chain, while the outside wheel overruns its cog-wheel, the pawls of the ratchet-wheel being arranged so as to permit of this.

In the 'Club' tricycle (fig. 146), made by the Coventry Machinists Company in 1879, one of the wheels was thrown

FIG. 146.

automatically out of gear when turning to one side or the other. Later, the same company used a clutch gear, somewhat similar in principle to the ratchet gear, but which had the advantage that the clutch could come into action at any point of the revolution, instead of only at as many points as there were teeth in the ratchet-wheel. The tricycle illustrated in figure 146 has only two tracks, which, in the early days of tricycles, was supposed to be of some advantage, in so far that it was easier to pick out two good portions along a bad piece of road than three.

A number of single and side-driving, rear-steering tricycles (fig. 147) were made about the years 1879 and 1880, but on account of their imperfect steering they were sometimes found extremely dangerous, and their manufacture was soon abandoned in favour of double-driving rear-steerers, of which the 'Cheyles-more' (fig. 148), made by the Coventry Machinists Company, was

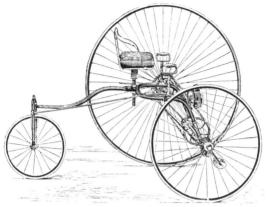

Fig. 147.

one of the most successful.   Tradesmen's carrier tricycles are still made of this type.

144. **Tricycles with Differential Gear.**—The front-steering, double-driving tricycle with loop frame, as in figure 145, next

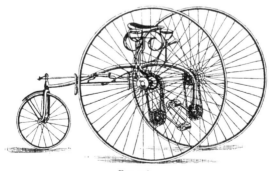

Fig. 148.

became more and more popular.   The invention by Mr. Starley of the 'Differential' tricycle axle or *balance-gear* marks a great step in the development of the three-wheeler.   This gear, or its

equivalent, has been ever since used for double-drivers, clutch and
ratchet gears having been abandoned.

As improvements in detail were slowly introduced, the lever
gear fell into disuse (which is easily accounted for by the fact that
with it gearing either up or down is impossible), and chain gearing
became universal.   With chain gear, and the possibility of gearing
up, the driving-wheels were made gradually smaller and smaller,
with a consequent reduction in the weight of the machine.

The 'Humber' tricycle met with great success on the racing
path, but, on account of its tendency to swerve on passing over a

FIG. 149.

stone, its success as a roadster was not so marked.   When used as
a tandem (fig. 149), with one rider seated on the front-frame supporting the driving-axle, the tendency to swerve was reduced and
the safety increased (see sec. 183).   In a later type this difficulty
was overcome by converting the machine into a rear-steerer, the
steering-pillar being connected by light levers and rods to the
steering-wheel.

The loop-frame tricycle was gradually superseded by one with
a central frame, in which the steering-wheel was actuated direct
by the handle-bar, the result being the 'Cripper' tricycle (fig.
150).   In this, as made by Messrs. Humber & Co., the chain lies

in the same plane as the backbone ; the crank-bracket being
suspended from the backbone and the gear being exactly central.

FIG. 150.

The axle is supported by four bearings, though the axle-bridge,
with four bearings, had already been used in the ' Humber '
tricycles.

FIG. 151.

Among the successful tricycles of this period may be mentioned the ' Quadrant,' in which the steering-wheel was not

mounted in a fork, but the ends of the spindle ran on guides in the frame (see fig. 254), and the 'Rudge Royal Crescent' (fig. 151), in which the fork of the steering-wheel was horizontal, and the steering-axis intersected the ground some considerable distance between the point of contact of the steering-wheel.

Up to the year 1886 the 'Ordinary' bicycle had a very great influence on tricycle design, the driving-wheels of tricycles being usually made very large (in fact, sometimes they were geared *down* instead of up) and the steering-wheel small. The weight of two large wheels was a serious drawback, while the excessive vibration from the small steering-wheel was a source of great

Fig. 152.

discomfort to the rider. The distance between the wheel centres was usually made as small as possible, the idea being that the tricycle should occupy little space. Common measurements for , Cripper' tricycles at this time were : Driving-wheels, 40 in. diam. ; steering-wheel, 18 in. diam. ; distance between driving- and steering-wheel centres, 32 in. ; driving-wheel tracks, 32 in. apart. Weight : Racers, 40 lbs. ; roadsters, 70–80 lbs.

The size of the driving-wheel has been gradually diminished, that of the steering-wheel increased, until now (1896) 28 in. may be taken as the average value for the diameter of each of the three wheels. The wheel centres have been put further apart,

42–45 in. being now the usual distance, the comfort of the rider and the steadiness of steering being both increased thereby.

FIG. 153.

The design of frame has also been greatly improved, so that the weight of a roadster has been reduced to 40–45 lbs. without in any way sacrificing strength.

FIG. 154.

Figure 152 shows a tricycle by the Premier Cycle Company, Ltd., embodying these improvements. The frame and chain gearing is almost identical with that of the bicycle ; the balance-gear and axle-bridge, with its four bearings, being added.

Figure 153 shows a tricycle by Messrs. Starley Bros., in which the bridge is a tube surrounding, and concentric with, the axle,

FIG. 155.

and the gear is exactly central ; so that the frame is considerably simplified, and the appearance of the machine vastly improved.

FIG. 156.

This may be taken as the highest point reached in the development of the ' Cripper ' type of tricycle.

145. Modern Single-driving Tricycles.—Several successful single-driving rear-driver tricycles have been made, among them being the 'Facile Rear-Driver' (fig. 154) and the 'Phantom'

FIG. 157.

(fig. 155). In these the two idle (or non-driving) wheels run freely on an axle supported by the front frame. These tricycles

FIG. 158.

are subject to the same faults of swerving as the 'Humber'
tricycle.

FIG. 159.

An important improvement is effected by mounting each
wheel on a short axle, which can turn about a vertical steering-

FIG. 160.

head placed as close as possible to the wheel, as in the 'Olympia'
(fig. 160), one of the most successful of modern tricycles.

146. **Tandem Tricycles.**—Tricycles for two riders were soon
brought to a relatively high state of perfection, and were almost,

if not quite, as popular as tricycles for single riders.   Among the
earliest may be mentioned the ' Rudge Coventry Rotary ' (fig. 156),

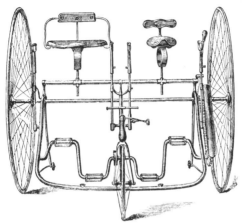

FIG. 161.

the 'Humber' (fig. 149), the ' Invincible ' rear-steerer (fig. 157), and
the 'Centaur' (fig. 158). Later, the 'Cripper' (fig. 159) and the ' Royal

FIG. 162.

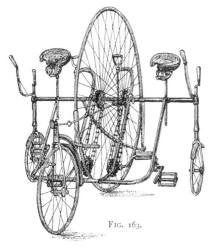

FIG. 163.

FIG. 164.

Crescent' (fig. 151) were made as tandems. In all these tandems, with the exception of the 'Coventry Rotary,' one of the riders overhangs the wheel-base, so that the load on the steering-wheel is actually less than when a single rider used the machine. The 'Coventry Rotary' is a single-driver, the others are double-drivers.

The most successful modern tandem tricycle is the 'Olympia' (fig. 160), a single-driver.

147. **Sociables**, or tricycles for two riders sitting side by side, were at one time comparatively popular. Figure 161 shows one with a loop frame made by Messrs. Rudge & Co., which, by the removal of certain parts, could be converted into a single tricycle ; figure 162, a 'Sociable' formed by adding another driving-wheel, crank-axle, and seat to the 'Coventry Rotary' (fig. 144).

In the 'One-track Sociable' (fig. 163),

made in 1886 by Mr. J. S. Warman, the weight of the rider rested mainly on the two central wheels, the small side wheels merely preventing the machine overturning when starting and stopping. It was, in fact, a sociable bicycle with two side safety-wheels added.

In the 'Nottingham Sociable' tricycle (fig. 164), made by the Nottingham Cycle Co. in 1889, each rider sat directly over the

FIG. 165.

rear portion of a 'Safety' bicycle, and the heads of the two frames were united by a trussed bridge to a central steering-head.

148. **Convertible Tricycles.**—A great many machines for two riders were at one time made by adding a piece to a tricycle so as to form a four-wheeler. Of these *convertible tricycles*, as they were called, the 'Royal Mail' two-track machine (fig. 165) and the 'Coventry Rotary Sociable' (fig. 162) may be noticed.

Figure 166 shows the 'Regent' tandem tricycle, formed by

Fig. 166.

Fig. 167

coupling the front wheel and backbone of a 'Kangaroo' bicycle to the rear portion of a 'Cripper' tricycle ; affording an example of a *treble-driving* cycle.

Figure 167 shows a four-wheeler formed by coupling together the driving portions of a 'Humber' and a 'Cruiser' tricycle, affording an example of a *quadruple-driving* cycle, all four wheels being used as drivers.

149. **Quadricycles.**—With the exception of the convertible tricycles above referred to, comparatively few four-wheeled cycles have been made. In 1869 'Velox' wrote : " No description of velocipedes would be perfect without some allusion to the favourite

Fig. 168.

our-wheeler of the past generation of mechanics." Figures 117 and 118 show one of the best as manufactured by Mr. Andrews, of Dublin. The frame was made of the best inch square iron 7 feet long between perpendiculars, and was nominally rigid, so that in passing over uneven ground either the frame was severely strained or only three wheels touched the ground. The two driving-wheels were fixed at the ends of a double cranked-axle driven by lever gear, the path of each pedal being an oval curve with its longer axis horizontal. While moving in a circle, the driving-wheels skidded as well as rolled, since the outer had to move over a greater distance than the inner.

Bicycles and tricycles have almost monopolised the attention of cycle makers, and no practicable quadricycle was made until Messrs. Rudge & Co. produced their 'Triplet' quadricycle (fig. 168) in 1888.   The front-frame supporting the two side steering-wheels can swing transversely to the rear-frame, so that the four wheels always touch the ground, however uneven, without straining the frame.   The same design was applied to a quadricycle for a single rider.

CHAPTER XVI

CLASSIFICATION OF CYCLES

150. **Stable and Unstable Equilibrium.**—Cycles may be divided into two great classes, according as the static equilibrium during the riding is stable or unstable. The former class may be

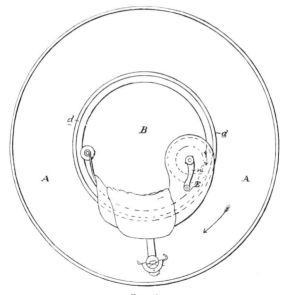

Fig. 169.

further separated into three divisions : (*a*) Tricycles, in which the frame, supported as it is at three points, is a statically determinate structure ; (*b*) Multicycles, having four or more wheels, the frame generally having a hinge or universal joint, so that the wheels may adjust themselves to any inequalities of the ground. If the frame be absolutely rigid it will be a statically indeterminate structure. (*c*) Dicycles of the 'Otto' type, with two wheels, in which the mass-centre of the machine and rider is lower than the axle. No machine of this class has ever been made, to the author's knowledge.

Cycles with unstable equilibrium may be divided into three classes, according to the direction in which the unstable equilibrium exists : *Monocycles*, having only one wheel ; *Bicycles*, having two wheels forming one track ; and

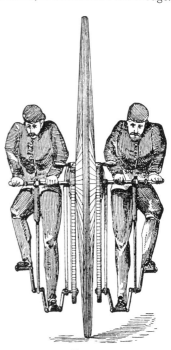

FIG. 170.   FIG. 171.

Dicycles, having two wheels mounted on a common axis. In all monocycles the transverse equilibrium is unstable ; they may be subdivided into two sub-classes, according as the longitudinal equilibrium is stable or unstable. An example of the former sub-class is shown in figure 169, in which the frame, carrying seat, pedal-axle, and handle, runs on an inner annular wheel, *d*, on the driving-wheel *A* ; the central opening, *B*, being large enough for the body of the rider, while his legs hang on each side of the main wheel. An example of the latter is shown in figure 170, and a sociable monocycle of the former class for two riders in figure 171.

In bicycles, the transverse equilibrium is unstable and the longitudinal equilibrium stable. In dicycles, the transverse equilibrium is stable. They may be subdivided into two sub-classes, according as their longitudinal equilibrium is stable or unstable.

The 'Otto' di-cycle (fig. 172) is the only example of the former sub-class, while none of the latter class, as already remarked, have attained any commercial import-ance. A dicycle of the latter type would be made with very large driving-wheels, and the mass-centre of machine and rider lower than the axis of the driving-wheel.

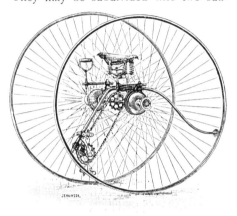

FIG. 172.

151. **Method of Steering.**—Proceeding to the further division and classification of bicycles, the first subdivision that suggests itself takes account of the method of steering; a bicycle being said to be a *front-* or *rear*-steerer, according as the steering-wheel is in front or behind, while among tricycles there are also *side*-steerers. A few bicycles have been made with *double-steering*.

The complete frame of the machine is usually divided into two parts, called respectively the *front-frame* and the *rear-frame*, united at the steering centre; though sometimes that part to which the saddle is fixed is called the 'frame,' to the exclusion of the other portion carrying the steering-wheel. It should be pointed out that the steering portion will sometimes be the larger and heavier of the two, the 'Humber' tricycle (fig. 149) affording an example of this. In the 'Chapman Automatic-Steering'

Safety (fig. 173) the saddle is not fixed direct to the rear-frame, but moves with the steering fork. The complete frame is in this case divided into three parts, which can move relative to each

FIG. 173.

other, on which are fixed the driving-gear, the steering-wheel, and the saddle respectively.

Examples of double-steering are afforded by the ' Adjustable' Safety (fig. 174), made by Mr. J. Hawkins in 1884, and by the

FIG. 174.

' Premier' Tandem Safety (fig. 137), in each of which the forks of both wheels move relative to the backbone.

There have been very few rear-steering bicycles made, though their only evident disadvantage is, that in turning aside to avoid an obstacle, the rear-wheel *may* foul, though the front-wheel has already cleared. Nearly all successful types of bicycles have been front-steerers.

152. **Bicycles, Front-drivers.**—Bicycles may be divided into front-drivers and rear-drivers, according to which wheel is used for driving. The 'Rucker' Tandem (fig. 135) is an example of a bicycle in which both wheels are used as drivers ; but generally only one wheel is used for driving. Each of these divisions may again be subdivided into *ungeared* and *geared*.

Among ungeared front-drivers we have the 'Bone-shaker,' the Ordinary,' the 'Rational,' the 'Facile,' the 'Xtraordinary,' and the 'Claviger' (fig. 504). In this classification we regard as ungeared those machines in which one revolution of the driving-wheel is made for each complete cycle of the pedal's motion. Thus, any bicycle with only lever gearing will be classed as un-geared, since with such mechanism it is, in general, impossible to gear up or gear down.

Geared bicycles may be subdivided into toothed-wheel geared, chain geared, and clutch geared. Among wheel geared front-drivers we have the 'Geared Ordinary,' 'Front-Driver,' the 'Bantam,' the 'Geared Facile,' the 'Sun-and-Planet' bicycle (fig. 479), and the 'Premier' Tandem Dwarf Safety (fig. 137). Among chain geared safeties we have the 'Kangaroo,' with two driving chains, one on each side of the driving-wheel, the 'Adjustable' Safety Road-

FIG. 175.

ster (fig. 174), and the 'Shellard Dwarf' Safety Roadster (fig. 175).

A combination of toothed-wheel and chain gear was used in the 'Marriott and Cooper' Front-Driver.

Clutch geared bicycles have never been very successful, the

Brixton Merlin Safety (fig. 176) being about the only example of this type. In the Merlin gear, a drum rotates on the axle at each side of the wheel, round which is coiled a leather strap, the other

FIG. 176.

end being fastened to the pedal lever. When the pedal is pushed outwards by the rider the drum is locked by a clutch to the axle, and the effort is transmitted to the wheel. On the upstroke a spring raises the pedal lever. With this gear any length of stroke may be taken, but the imperfect action of the clutch is such that the great advantages due to the possibility of varying the length of stroke are more than neutralised.

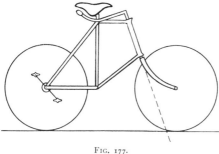

FIG. 177.

Figure 177 shows a possible front-driving rear-steering geared bicycle, the front hub having a 'Crypto' or equivalent gear.

153. **Bicycles, Rear-drivers.**—Among ungeared rear-drivers

may be mentioned the Rear-driving 'Facile' and the American 'Star' (fig. 178).

Among toothed-wheel geared rear-drivers we have the 'Burton,' the Geared 'Facile' Rear-driver (fig. 129), the 'Claviger' Geared (fig. 507), the 'Fernhead' Chainless Safety, driven by bevel-gearing. Of chain geared rear-drivers, the present popular Safety

FIG. 178.

of the 'Humber' or 'Rover' type is the most important representative.

In the 'Boudard-geared' Safety a combination of toothed-wheels and chain gear is used, while the same may be said of the two-speed gears that are applied to the ordinary type of chain-driven safety.

This classification is represented diagrammatically on page 194. From this diagram it will be seen that no successful type of rear-steering bicycle has been evolved. Experimenters might with advantage direct their energies to this comparatively untrodden domain.

154. **Tricycles, Side-steering**.—The classification of tricycles may go on on similar lines to that of bicycles. There would be three

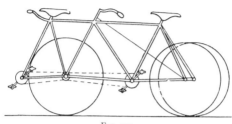

types — front-steering, side-steering, and rear-steering. Of side-steering tricycles there are two subdivisions : the 'Rudge Coventry Rotary' (fig. 156) being a side-driver, while the 'Dublin'

FIG. 179.

(fig. 143) and the 'Olympia' (fig. 160) are back-drivers. No side-steering, front-driving tricycle has been made, to our know-

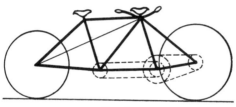

ledge ; though we can see nothing at present to prevent tandem tricycles of this type (figs. 179–181) from being successful roadsters. That shown in figure

FIG. 180.

179 could be ridden by a lady in ordinary costume on the front seat ; it would, perhaps, be slightly deficient in lateral

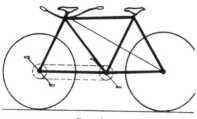

stability, as the mass-centre would be near the forward corner of the wheel-base triangle. That shown in figure 180 would be superior in this respect, while the weight on the driving-wheel would still probably

FIG. 181.

be sufficient for all ordinary requirements. A type intermediate (fig. 181) might be made with a 'Crypto' gear on the front wheel hub, the two crank-axles being connected by a chain ; the frame would be simpler than in figures 179 and 180.

Tricycles are either single-driving or double-driving, according as there are one or two driving-wheels. The only treble-driving tricycle which has been yet put on the market is the tandem made by Messrs. Trigwell and Co., by coupling the front wheel and backbone of a ' Kangaroo ' to the rear portion of a 'Cripper' (fig. 166). The driving-wheels of a double-driving tricycle are invariably mounted on the same axle, and since in going round a corner the wheels, if of equal size, must rotate at different speeds, the driving-axle must be in two parts. In the 'Cheylesmore' tricycle two separate driving chains were used between the crank- and wheel-axles, the cog-wheel on the wheel-axle being held by a clutch when driving in a straight line, while in rounding a corner the wheel which tended to go the faster overran the clutch, and all the driving effort was transmitted through the more slowly moving wheel. Starley's differential gear (see sec. 189), allowing, as it does, both wheels to be drivers under all circumstances, is now universally used for double-driving.

155. **Front-steering Front-driving Tricycles**.—The early ' Bone-shaker ' tricycle (fig. 141) is an ungeared example of this class, while the 'Humber' tricycle (fig. 149) is a geared tricycle of this same class. The ' Humber ' is a double-driver.

Single-driving tricycles of this division may be made by taking a 'Crypto' or 'Kangaroo' bicycle, and having two back wheels at the end of a long axle. They would, however, be deficient in lateral stability, unless used as tandems, on account of the load being applied over a point near the forward apex of the triangular wheel-base.

156. **Front-steering Rear-driving Tricycles**.— Of ungeared cycles, Lisle's early Ladies' tricycle (fig. 142) and the 'Club' (fig. 146) are examples.

The geared tricycles may be subdivided into single-drivers and double-drivers. Of the former class the 'Olympia' (fig. 160), the 'Phantom' (fig. 155), the 'Facile' (fig. 154), the 'Claviger,' and the 'Trent' convertible (fig. 182) are examples.

The double-drivers may be conveniently subdivided into direct-steerers and indirect-steerers. The 'Cripper' (figs. 150, 152, 153), of which probably more examples have been made than all the other types put together, is a direct-steerer, so also is the Merlin (fig. 183). Among indirect-steerers we may mention the 'Devon' tricycle (fig. 145), the 'Club' (fig

FIG. 182.

146). The 'Nottingham Sociable' (fig. 164) formed by conversion of two bicycles, and Singer's Omnicycle with clutch gear (fig. 184), made in 1879, also belong to this division.

FIG. 183.

This classification of tricycles is shown diagrammatically on page 195.

157. **Rear-steering Front-driving Tricycles.**—The 'Veloci-

man,' a hand-tricycle made by Messrs. Singer & Co., of which figure 241 represents an improved design for 1896, is an example of this class.   The 'Cheylesmore' (fig. 148), made by the Coventry Machinists Co., was one of the most successful of the early tricycles.   Several tandem tricycles were made on this type, one of the most popular being tne 'Invincible' (fig. 157), made by the Surrey Machinists Co., Limited.

FIG. 184.

The tandem tricycles in figures 179–181, if made with both rear wheels running freely on the same axle, fixed to a rear frame, would afford examples of single-drivers of this class.

A rear-steering side-driving tricycle was the 'Challenge' (fig. 147), made by Messrs. Singer in 1879.

158. **Quadricycles.**—A great many quadricycles were made at one time by adding a piece to a tricycle, so as to form a machine for two riders (see sec. 148).   The attachment of the extra portion was usually made by means of a universal joint.   The one track Sociable (fig. 163) may really be classified as a four-wheel cycle, though from the lack of the universal joint in the frame it differs essentially from those mentioned above.

Rudge's quadricycle (fig. 168), giving only two tracks and a rectangular wheel base, is a very well designed machine of this type.   The steering gear is similar in principle to that used in the 'Olympia' tricycle.   The front portion of the frame supporting the two side-steering wheels is connected to the rear portion by a horizontal joint at right angles to the driving-axle, so that the four wheels may each touch the ground, however uneven, without straining the frame.   It is made as a single, tandem, and triplet. Its stability is discussed in section 161.

The quadricycle with two tracks has some advantages as compared with the tricycle, and may well repay further consideration

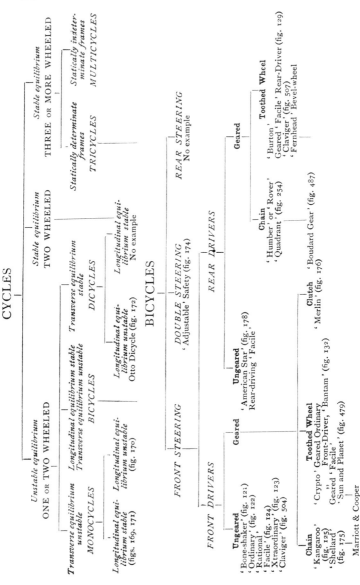

CYCLES

Unstable equilibrium
ONE OR TWO WHEELED

Transverse equilibrium unstable
MONOCYCLES

Longitudinal equilibrium stable (figs. 169, 171)

Longitudinal equilibrium unstable (fig. 170)

Stable equilibrium
THREE OR MORE WHEELED

Statically determinate frames
TRICYCLES

Statically indeterminate frames
MULTICYCLES

Stable equilibrium
TWO WHEELED

Longitudinal equilibrium stable, Transverse equilibrium unstable
BICYCLES

Transverse equilibrium stable
DICYCLES

Longitudinal equilibrium unstable
Otto Dicycle (fig. 172)

Longitudinal equilibrium stable
No example

BICYCLES

FRONT STEERING

DOUBLE STEERING
'Adjustable' Safety (fig. 174)

REAR STEERING
No example

FRONT DRIVERS

REAR DRIVERS

Ungeared
'American Star' (fig. 178)
Rear-driving 'Facile'

Geared

Ungeared
'Bone-shaker' (fig. 121)
'Ordinary' (fig. 122)
'Rational'
'Facile' (fig. 124)
'Xtraordinary' (fig. 123)
'Claviger' (fig. 504)

Geared

Chain
'Kangaroo' (fig. 125)
'Shellard' (fig. 175)
Marriott & Cooper

Toothed Wheel
'Crypto' Geared Ordinary
 ,, Front-Driver, 'Bantam' (fig. 132)
Geared 'Facile'
'Sun and Planet' (fig. 479)

Chain
'Humber' or 'Rover'
'Quadrant' (fig. 254)

Clutch
'Merlin' (fig. 176)

Geared

'Boudard Gear' (fig. 487)

'Burton,' **Toothed Wheel**
Geared 'Facile' Rear-Driver (fig. 129)
'Claviger' (fig. 507)
Fernhead' Bevel-wheel

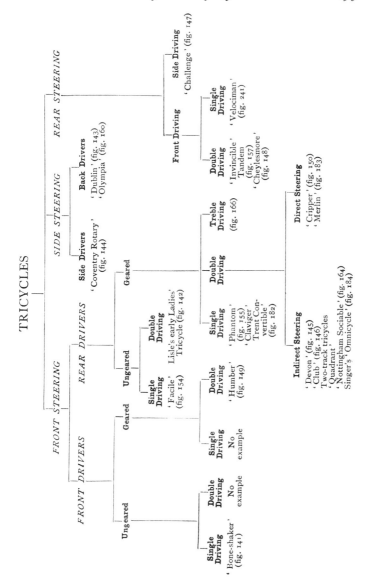

TRICYCLES

FRONT STEERING

FRONT DRIVERS

Ungeared

Single Driving
'Bone-shaker' (fig. 141)

Double Driving
No example

Geared

Single Driving
No example

Double Driving
'Humber' (fig. 149)

REAR DRIVERS

REAR DRIVERS

Ungeared

Single Driving
'Facile' (fig. 154)

Double Driving
Lisle's early Ladies' Tricycle (fig. 142)

Geared

Single Driving
'Phantom' (fig. 155)
'Claviger'
Trent Convertible (fig. 182)

Double Driving

Indirect Steering
'Devon' (fig. 145)
'Club' (fig. 146)
Two-track tricycles
'Quadrant'
'Nottingham Sociable' (fig. 164)
Singer's Omnicycle (fig. 184)

Direct Steering
'Cripper' (fig. 150)
'Merlin' (fig. 183)

Treble Driving
(fig. 166)

SIDE STEERING

Side Drivers
'Coventry Rotary' (fig. 144)

Back Drivers
'Dublin' (fig. 143)
'Olympia' (fig. 160)

REAR STEERING

Front Driving

Double Driving
'Invincible' Tandem (fig. 157)
'Cheylesmore' (fig. 148)

Single Driving
'Velociman' (fig. 241)

Side Driving
'Challenge' (fig. 147)

by cycle makers and designers.   If a satisfactory mode of supporting the frame on the wheel axles by springs could be devised, the
horizontal joint might be omitted, the design of frame simplified,
the stability of the machine increased, and additional comfort
obtained by the rider.   If the two steering-wheels revolved independently on a common axle, as in the 'Phantom' tricycle
(fig. 155), the design of the machine would be further simplified ;
the relation of the wheels to the frame being exactly the same as
is a four-wheeled vehicle drawn by a horse.   This type of quadri-

cycle would, however, possess the same objection-
able features as to swerving as the tricycles shown
in figures 149, 154, and 155.   In a horse vehicle
the front axle is fixed to the shafts to which the
horse is harnessed, so that the axle cannot swerve
when one wheel meets an obstacle without dragging
the horse sideways.   In this respect the horse
performs the same function as the front wheel of
a 'Cripper' tricycle.   A hansom cab is equivalent
to a 'Cripper' tricycle, and a four-wheeler to a

FIG. 185.

pentacycle (fig. 185), in which the rear portion trails after the
front.

159. **Multicycles.** — By stringing together a number of
'Humber' or 'Cripper' frames with their crank-axles and pairs of
driving-wheels, a cycle of 4, 6, 8, or any even number of wheels
may be obtained.   The steering of such a multicycle should be
effected by the front rider, the intersection of the first two axles
determining the radius of curvature of the path.   The following
wheels should be merely trailing wheels, so that they may follow
in the required path.

CHAPTER XVII

160. **Stability of Tricycles.**—If $a\,b\,c$ (fig. 186) be the points of contact of the three wheels of a tricycle with the ground, it will be in equilibrium under the action of the rider's weight, provided the perpendicular from the mass-centre of the rider and machine falls within the triangle $a\,b\,c$. If this perpendicular fall at the point d, the pressures of the wheels on the ground can easily be found by the principle of moments. Let W be the total weight of the rider and machine, w_a, w_b, and w_c the pressures of the wheels at a, b, and c on the ground. Then taking moments about the line $b\,c$, draw perpendiculars $a\,a_1$ and $d\,d_1$ to $b\,c$. We then have

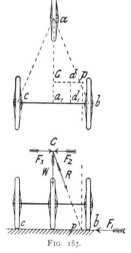

FIG. 186.

$$W \times d\,d_1 = w_a \times a\,a_1$$

or $$w_a = \frac{d\,d_1}{a\,a_1}\,W \quad . \quad . \quad . \quad . \quad (1)$$

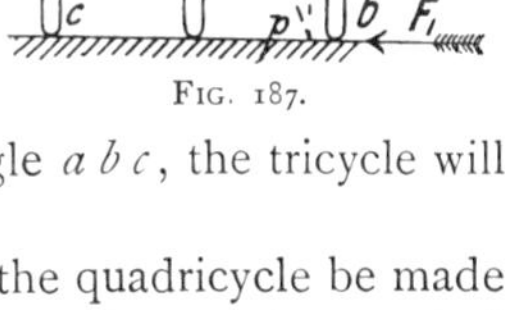

Similar expressions for w_b and w_c can be found.

FIG. 187.

If the point d fall outside the triangle $a\,b\,c$, the tricycle will topple over.

161. **Stability of Quadricycles.**—If the quadricycle be made with the steering-axle capable of turning only round a vertical axis, as in the case of an ordinary four-wheeled carriage drawn by horses, the mass-centre of the machine and rider may lie vertically

above the rectangle $a\,b\,c\,d$ (fig. 188), a, b, c and d being the points of contact of the wheels with the ground. But if one of the axles be hinged to the frame, so as to allow the four wheels

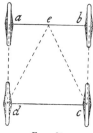

to be always in contact with the ground, however uneven—as in the case of the 'Rudge' quadricycle (fig. 168)—the mass-centre of machine and rider, exclusive of front portion $a\,b$, must lie vertically above the triangle $e\,c\,d$, e being the intersection of the plans of the steering-axle and hinge joint. If the perpendicular from the mass-centre of machine and rider fall between $e\,c$ and $b\,c$, the wheel at d will lift from the ground, and the portion $e\,c\,d$ of the machine will continue

Fig. 188.

to overturn until stopped by coming in contact with the portion $a\,b$.

In a tandem quadricycle formed by attaching a trailing wheel, d (fig. 189), to a 'Cripper' tricycle, $a\,b\,c$, by means of a universal

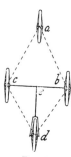

joint at e, the mass-centre of the machine and riders must lie vertically above and inside the quadrilateral $a\,b\,c\,d$. If the joint e be behind the axle, $b\,c$, another condition must be satisfied, viz. the vertical downward pressure at e, due to the weight on the trailing frame, must not be sufficient to tilt the triangle $a\,b\,c$ about the axle $b\,c$. This condition will in general be satisfied if the joint e be not far behind the axle.

162. **Balancing on a Bicycle**.—A bicycle has only two points of contact with the ground, and a perpendicular from the mass-centre of machine

Fig. 189.

and rider must fall on the straight line joining them. If the bicycle and rider be at rest, the position is thus one of unstable equilibrium, and no amount of gymnastic dexterity will enable the position to be maintained for more than a few seconds. If the mass-centre get a small displacement sideways, the displacement will get greater, and the machine and rider will fall sideways. In riding along the road with a fair speed the mass-centre is continually receiving such a displacement. If the rider steer his bicycle in an exact straight line this displacement will get greater, and he and his bicycle will be overturned, as when at rest. But, as every learner knows, when the machine is felt to be falling to the left-hand side, the rider steers to the left—that is, he guides the bicycle in a circular arc, the centre of which is situated at the left-hand side. In popular language, the centrifugal force due to the circular motion of the machine and rider now balances the tendency of the machine to overturn ; in fact, the expert rider automatically steers the bicycle in a circle of such a diameter that the centrifugal force slightly overbalances the tendency to overturn, and the machine again regains its perpendicular position. The rider now steers for a short interval of time exactly in a straight line. But probably the perpendicular position has been slightly overshot, and the machine falls slightly to the right-hand side. The rider now unconsciously steers to the right hand, that is, in a circle having its centre to the right-hand side.

If the track of a bicycle be examined it will be found to be, not a straight line, but a long sinuous curve. With beginners the waviness of the curve will be more marked than with expert riders ; but even with the latter riding their straightest the *sinuosity* is quite apparent. A patent had actually been taken out for a lock to secure the steering-wheel of an ' Ordinary ' bicycle, the purpose being to make it move automatically in a straight line. The above considerations will show, as clearly as the actual trial of his device probably did to the inventor, the absurdity of such a proceeding.

It would be possible to ride a bicycle in a perfectly straight line with the steering-wheel locked, by having a fly-wheel capable of revolving in a vertical plane at right angles to that of the bicycle wheels, and provided with a handle which could be turned by the rider. If the bicycle were falling to the right, the fly-wheel should be driven in the same direction ; the reaction on the rider and frame of bicycle would be a couple tending to neutralise that due to gravity causing the machine to fall.

Lateral Oscillation of a Bicycle.—From the above explanation of the balancing on a bicycle, it will be seen that the machine and rider are continually performing small oscillations sideways—the axis of oscillation being the line of contact with the ground—simultaneously with the forward motion. The bicyclist and his machine may thus be roughly compared to an inverted pendulum. The time of vibration of a simple pendulum is proportional to the square root of its length, a long pendulum vibrating more slowly than a short one. In the same way, the oscillations of a high bicycle are slower than those of a low one ; *i.e.* the time taken for the mass-centre to deviate a certain angle from the vertical is greater the higher the mass-centre ; a rider equally expert on high and low bicycles will thus be able to keep a high bicycle nearer the exact vertical position than he will a low bicycle. In other words, the angle of swing from the vertical is greater in the ' Safety ' than in the ' Ordinary.'

The track of an ' Ordinary ' will therefore be straighter—that is, made up of flatter curves—than that of a ' Safety,' both bicycles being supposed ridden by equally expert riders.

163. **Balancing on the Otto Dicycle.**—In an ' Otto ' dicycle at rest the mass-centre of the frame and rider is, in its normal

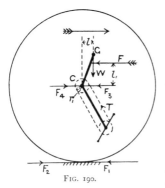

FIG. 190.

position, vertically above the axle of the wheels ; the machine is thus in stable equilibrium laterally and in unstable position longitudinally. In driving along at a uniform speed against a constant wind resistance, F (neglecting at present other resistances), the mass-centre, G, is in its normal position, a short distance, l, in front of the axle (fig. 190). While the rider exerts the driving effort the wheel exerts the force F_1 on the ground, directed backwards, and the reaction of the ground on the wheel is an equal force, F_2, in the direction of motion. The force F_2 is equivalent to an equal force F_4 at the axle and a couple Fr, r being the radius of the driving-wheel. The couple Fr is applied by the pull of the chain to the rigid body formed by the driving-wheel and axle ; therefore, if T be the magnitude of this pull and r_1 the radius of the cog-wheel on the axle, $T r_1 = F r$

Consider now the forces acting on the rigid body formed by the frame and rider : these are, the reaction at the bearing C, the weight W acting downwards, the wind-resistance, F, and the pull of the chain T. Since the frame is in equilibrium, the moment of all these forces about any point must be zero. Taking the moments about C we get

$$W l - F l_1 = T r_1 = F r \quad . \quad . \quad . \quad . \quad (2)$$

Suppose now the mass-centre, G, to fall a little forward of the position of equilibrium, so that the moment of W about C becomes $W l^1$; in order that equilibrium may be established the pull of the chain must have a greater value, T^1, thus $W l^1 - F l_1 = T^1 r_1$. This increased pull on the chain is produced by the rider pressing harder on the pedals ; in other words, by driving harder ahead.

In the same way, should the mass-centre, G, fall a little behind the position of equilibrium, the tendency to fall backward is checked by the rider easing the pressure on the pedals, *i.e.* by slightly back-pedalling.

The frame and rider in an 'Otto' dicycle thus perform oscillations about the axle of the machine ; the length of the inverted pendulum is much less than in the 'Ordinary' or even the 'Safety' bicycle, and the backward or forward oscillation is greater than the lateral oscillation in a bicycle.

164. **Wheel load in Cycles when driving ahead.**—A great deal of misconception exists as to the modification of the wheel loads, due to driving ahead. If the cycle move uniformly, and the several resistances be neglected, the wheel loads will, of course, be the same as if the cycle were at rest, and therefore will depend only on the position of the mass-centre of machine and rider relative to the wheels. If the only resistance considered is the wind pressure F_1 (fig. 191), the load on the front wheel will be decreased, and that on the rear wheel increased, by the amount R, determined by the equation

$$F_1 h_1 = R l ; \quad . \quad . \quad . \quad . \quad . \quad . \quad (3)$$

l being the wheel-base, and h_1 the distance of the centre of wind pressure above the ground. Frictional resistances, including the friction of the bearings and gearing and the rolling friction of the wheels on the ground, make no modification of the distribution of wheel load ; the former, because they are internal forces, and do not in any way affect the external forces, the latter because they act tangentially to the ground, and must be balanced by an equal and opposite reaction of the ground on the driving-wheel.

If the speed of the cycle be increased, the forces due to acceleration can be easily shown as follows : Consider the mass of the machine and rider to be concentrated at the mass-centre G, and that the wheels and frame are weightless ; then, to produce

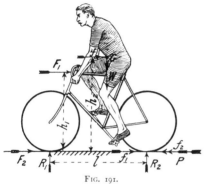

the acceleration, the frame must act on the mass, and the mass react on the frame with an equal but opposite force, f. Introduce at the point of contact of the driving-wheel with the ground two equal and opposite forces, f_1 and f_2 (fig. 191), each equal and parallel to f ; then f is equivalent to the force f_1, and the couple

formed by the equal and opposite forces f and f_2. The force f_1 must be equilibrated by the reaction P of the ground on the driving-wheel, the couple tends to diminish the weight on the front wheel, and increases that on the rear wheel, by an amount, R, given by the equation

$$R\,l = f\,h_2 \quad . \quad . \quad . \quad . \quad . \quad . \quad (4)$$

h_2 being the height of the mass-centre, G, above the ground.

In the most general case, the external forces acting on the system of bodies formed by the machine and rider are shown in figure 191. These are the resistance f, due to the increase of speed, the wind pressure F_1, the resistance of the wheels to rolling, F_2, the reaction of the ground on the driving-wheel, P, the weight, W, of the machine and rider, and the vertical reactions, R_1 and R_2, on the wheels. P, R_1 and R_2 are determined so as to produce equilibrium with the other forces. Pressure exerted on the pedal does not in any way modify the reactions R_1 and R_2, except so far as it affects, or is affected by, the resistances F_1, F_2, and f; *i.e. work spent in overcoming resistances of the mechanism does not in any way affect the wheel loads.*

165. **Stability of Bicycle moving in a Circle.**—Let r be the radius of the circle in which the cycle is moving, W the weight of the rider and machine, and G the position of the mass-centre (fig. 192). We have already seen that a body of mass, W lbs., moving in a circle of radius, r, with speed v, has a radial acceleration, $\dfrac{v^2}{r}$; and must be acted on by a radial force $\dfrac{W v^2}{g r}$ lbs. Now, considering the weight of the rider and bicycle concentrated at G, and that it is transmitted from G to the ground by a weightless frame, the only forces acting on the frame are the weight W, acting vertically downwards at G, and the reaction from the ground, R. The resultant, C, of the two forces, W and R, must therefore be equal to the horizontal radial force

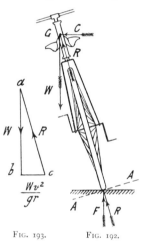

FIG. 193.   FIG. 192.

$$\frac{W v^2}{g r} \qquad . \quad . \quad . \quad . \quad . \quad . \quad . \quad (5)$$

required to give the mass the circular motion, and the line of action of R must therefore pass through G. Draw $a\,b$ equal to W (fig. 193) vertically downwards, and $b\,c$ equal to $\dfrac{W v^2}{g r}$ horizontal. Then the reaction, R, is represented in magnitude and direction by $c\,a$. When the rider is moving steadily in a circle the machine must be inclined at the angle $c\,a\,b$ to the vertical, so that the reaction, R, from the ground may pass through G (see sec. 45).

166. **Friction between Wheel and Ground.**—When there

is no friction between two surfaces in contact the mutual pressure is at right angles to the surfaces. Any component of force parallel to the common surface of contact can only be due to friction. In the case of a bicycle moving in a circle, the centripetal force is supplied by the friction between the wheel and the ground. If the surface of the road be greasy, the friction is insufficient to provide the proper amount of force, and the force of reaction of the ground, F, together with the weight of the machine and rider, W, form a couple (fig. 192) tending to overturn the machine.

Now when a couple acts on a rigid body free to move, the body turns about its mass-centre (see sec. 66). In the case of the bicycle (fig. 192), the mass-centre, G, will have a simultaneous motion downwards, so that the final result will be that the wheel will slip to the right.

Figure 192 also illustrates the forces acting on a bicycle which is being steered in a straight line, and which has already attained a slight inclination to the vertical; the weight, W, of the rider and the reaction of the ground, F, form a couple tending to increase still further the deviation from the vertical.

167. **Banking of Racing Tracks.**—In racing tracks, the surface of the ground at the corners is sloped, as at $A\ A$ (fig. 192), so as to be perpendicular to the average slope of the bicycles going round the corner. From (5) it is evident that this slope depends on the speed of the cyclists and the radius of the track. Table VIII. gives the necessary slopes for different speeds and radii of track.

Example.—Taking a speed of twenty-four miles per hour and the radius of the track 160 feet, $v = \dfrac{24 \times 5280}{3600} = 35\cdot2$ ft. per second, $\dfrac{W v^2}{g r}$ becomes $\dfrac{35\cdot2^2}{32\cdot2 \times 160} W = \cdot24\ W$; that is, $b\ c = \cdot24\ a\ b$ (fig. 193), and therefore the surface of the track must be laid at a slope of 24 vertical to 100 horizontal. If the track be laid at this slope, the wheel of a bicycle moving at a less speed than twenty-four miles an hour will tend to slip downwards towards the inside of the track, that of a bicycle moving at a higher speed will tend to move upwards towards the outside.

TABLE VIII.— BANKING OF RACING TRACKS.

Parts Vertical Rise in 100 *Parts Horizontal.*

Mean radius of track	Speed, miles per hour.				
	20	25	30	35	40
50 ft.	53·4	83·4	120·2	163·7	213·7
100 ft.	26·7	41·7	60·1	81·7	106·8
150 ft.	17·8	27·8	40·1	54·5	71·2
200 ft.	13·3	20·9	30·0	40·9	53·4
250 ft.	10·7	16·7	24·0	32·7	42·7
300 ft.	8·9	13·9	20·0	27·2	35·6

If the width of the track be considerable, the slope should be greater at the inner than at the outer edge, for a given speed. In

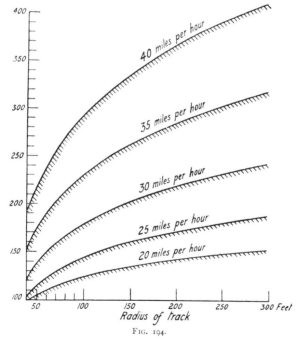

FIG. 194.

this case it can be shown by an easy application of the integral calculus, that if R be the radius at any point of the track and

y the corresponding height above a certain horizontal datum level

$$y = \frac{v^2}{g} \log_e R \quad . \quad . \quad . \quad . \quad . \quad . \quad (6)$$

feet and seconds being the units.

If V be the speed in miles per hour,

$$y = \cdot 15383 \; V^2 \log R^2 \quad . \quad . \quad . \quad . \quad (7)$$

y and R being in feet, and log R being the ordinary tabular logarithm.

Table IX. contains the values of y for different values of R from 40 to 300 feet, and at various speeds from 20 to 40 miles per hour, and figure 194 shows cross sections of tracks for these various speeds.

TABLE IX.—BANKING OF RACING TRACKS.

Elevation above a datum level, in feet.

Radius feet	Speed, miles per hour				
	20	25	30	35	40
40	98·4	153·8	221·5	301·4	393·7
50	104·5	163·3	235·2	320·1	418·1
60	109·4	170·9	246·2	335·0	437·6
70	113·5	177·4	255·4	347·6	454·1
80	117·1	183·0	263·5	358·6	468·4
90	120·2	187·9	270·5	368·2	481·0
100	123·1	192·3	276·9	376·9	492·2
110	125·6	196·3	282·6	384·6	502·4
120	127·9	199·9	287·8	391·8	511·7
130	130·5	203·2	292·6	398·3	520·2
140	132·0	206·3	297·1	404·3	528·1
150	133·9	209·2	301·3	410·0	535·6
175	138·0	215·6	310·5	422·6	552·0
200	141·6	221·2	318·6	433·6	566·3
225	144·7	226·1	325·6	443·2	578·9
250	147·5	230·5	332·0	451·8	590·1
275	150·1	234·5	337·7	459·4	600·3
300	152·4	238·1	342·9	466·7	609·6

Since the circumference of the inner edge of the track is less than that of the outer edge, when record-breaking is attempted, the rider keeps as close as he safely can to the inner edge ; consequently the average speed of riding is greatest at the inner edge. On this account, the convexity of the cross-section is, with advantage, made greater than shown in figure 194.

168. **Gyroscopic Action.**—In the above investigation, it has been assumed that the weight of the wheels is included in that of the rider and machine, and no account has been taken of their gyroscopic action. We have already seen (sec. 70) that if a wheel, of moment of inertia I, have a rotation, ω, about a horizontal axis, and a couple, C, be applied to the axle tending to make it turn in a vertical plane, the axle will actually turn in a horizontal plane with an angular velocity of precession

$$\theta = \frac{C}{I\omega} \quad . \quad . \quad . \quad . \quad . \quad . \quad (8)$$

Thus, in estimating the stability of a wheel rolling along a circular arc, both centrifugal and gyroscopic actions must be considered.

Let R be the radius of the track described by the bicycle, r the outside radius and r_1 the radius of gyration of the wheels, V the speed of the cyclist, and w the weight of the wheels; then

$$\theta = \frac{V}{R}, \quad I = \omega r_1^2, \quad \omega = \frac{V}{r}.$$

Substituting in formula (8) we get

$$C = \frac{w V^2 r_1^2}{R r} \quad . \quad . \quad . \quad . \quad . \quad (9)$$

i.e. the gyroscopic couple required, in addition to the centrifugal couple, is proportional to the square of the speed, inversely proportional to the radius of the track, and approximately proportional to the radius of the cycle wheels.

Example.—If the total weight of the machine and rider be 180 lbs., the weight of the wheels 8 lbs., speed 30 miles per hour $=44$ feet per sec., the radius of the track 100 feet, r the radius of the wheel 14 in. $= \frac{14}{12}$ feet, and $r_1 = 13$ in. $= \frac{13}{12}$ feet, we get $V = 44$ ft. per sec., and

$$C = \frac{8 \times 44^2 \times 13^2}{100 \times 14 \times 12} = 155\cdot8 \text{ foot-poundals}$$

$$= 4\cdot84 \text{ foot-lbs.}$$

i.e. the mass-centre of the machine and rider will have to be $\frac{4\cdot84}{180} = \cdot027$ feet, or $\cdot32$ inches further from the vertical than if the wheels were weightless, and gyroscopic action could be neglected.

From the above example it will be seen that gyroscopic action in bicycles of the usual types is negligible, except at the highest speeds attainable on the racing-path, and on tracks of small radius. If a fly-wheel were mounted on a bicycle and geared higher than the driving-wheel, the gyroscopic action might be, of course, increased. If the fly-wheel were parallel to, and revolved in the same direction as the driving-wheel, the rider, while moving in a circle, would have to lean further over than would be necessary without the fly-wheel. If, on the other hand, the fly-wheel revolved in the opposite direction, the rider would have to lean over a less distance ; in fact, by having the $I\,\omega$ of the fly-wheel large enough it might be possible for a bicyclist to keep his balance while leaning towards the *outside* of the curve being described.

The same gyroscopic action takes place when a tricycle moves in a circle.

169. **Stability of a Tricycle moving in a Circle.**—A tricycle moving round a curve is subjected to the same laws of centrifugal force as a bicycle, the only difference being that the frame of the machine cannot tilt so as to adjust itself into equilibrium with the forces acting.

Let figure 186 be the plan and figure 187 the elevation of a tricycle moving in a circle, the centre of which lies to the left. Let G be the mass-centre of the machine and rider, a, b and c the points of contact of the wheels with the ground. Considering the mass of the machine and rider concentrated at G, a horizontal force, F_2, applied at G is necessary to give the body its circular motion. This force is supplied by the horizontal component of the reaction of the saddle on the rider. There will be an equal horizontal force, F_3, exerted on the frame at G_1 by the rider. This force tends to make the wheels slip sideways on the ground, an equal but oppositely directed force, F_1, will be exerted by the ground on the wheels. The force F_2 gives the body its necessary radial acceleration, while the forces F_1 and F_3 acting on the machine form a couple tending to overturn it. If the resultant R

of the forces F_3 and W cut the ground at a point, p, outside the wheel base, $a\,b\,c$, the machine will overturn. Hence the necessity for tricyclists leaning over towards the inside of a curve when moving round it.

Again, if the force F_1 be greater than $\mu\,W$, the tricycle will slip bodily sideways, μ being the co-efficient of sliding friction between the tyres and the ground. This slipping is often experienced on greasy asphalte or wood paving.

170. **Side-slipping.**—The side-slipping of a bicycle depends on the coefficient of friction between the wheels and the ground, and the angle of inclination of the bicycle to the vertical. The coefficient of friction varies with the condition of the road, being very low when the roads are greasy; when the roads are in this condition the bicyclist, therefore, must ride carefully. The condition of the roads is a matter beyond his control, but the other factor entering into side-slipping is quite within his control. In order to avoid the chance of side-slipping, no sharp turns should be made on greasy roads at high or even moderate speeds. To make such turns, we have seen (sec. 165) that the bicycle must be inclined to the vertical, this slope or inclination increasing with the square of the speed and with the curvature of the path. At even moderate speeds this inclination is so great that on greasy roads there would be every prospect of side-slipping taking place. If a turn of small radius must actually be effected, the speed of the machine must be reduced to a walking pace or even less.

A well-made road is higher at the middle than at the sides. When riding straight near the gutter the angle made by the plane of the bicycle with the normal to the surface of the ground is considerable. If the rider should want to steer his bicycle up into the middle of the road, in heeling over this angle is increased. This may be safely done when the road is dry, but on a wood pavement saturated with water it is quite a dangerous operation. With the road in such a condition the cyclist should ride, if traffic permit, along its crest.

The explanation given above (sec. 162) that in usual riding the lateral swing of a 'Safety' is greater than that of an 'Ordinary,' explains why side-slipping is more often met with in the lower machines. The statement of some makers that their particular arrangement of frame, gear, or tread of pedals, &c., prevents side-slipping is utterly absurd ; the only part of the machine which can have any influence on the matter being the part in contact with the ground—that is, the tyres. Again, the statement of riders that their machines have side-slipped when going straight and steadily cannot be substantiated. A rider may be going along quite carefully, yet if his attention be distracted for a moment, and he give an unconscious pull at the handles, his machine may slip.

Side-slipping with Pneumatic Tyres.—A pneumatic tyre has a much larger surface of contact with the ground than the old solid tyre of much smaller thickness. This fact, which is in its favour as regards ease of riding over soft roads, is a disadvantage as regards side-slipping on greasy surfaces. The narrow tyre on a soft road sinks into it, the bicycle literally ploughing its way along the ground ; and on hard roads the narrow tyre is at least able to force the semi-liquid mud from beneath it sideways, until it gets actual contact with the ground. The pressure per square inch on the larger surface of a pneumatic tyre in contact with the ground being very much smaller, the tyre is unable to force the mud from beneath it ; it has no *actual* contact with the ground, but floats on a very thin layer of mud, just as a well lubricated cylindrical shaft journal does not actually touch the bearing on which it nominally rests, but floats on a thin film of oil between it and the bearing. The coefficient of friction in such a case very small, and a slight deviation of the bicycle from the vertic position—*i.e.* steering in any but a very flat curve—may cau side-slip.

The non-slipping covers, now almost entirely used on roadst pneumatic tyres, are made by providing projections of such sm area that the weight of the machine and rider presses the through the thin layer of mud into actual contact with the groun The coefficient of friction under these circumstances is high , and the risk of side-slip correspondingly reduced.

Apparent Reduction of Coefficient of Friction.—While the driving-wheel rests on a greasy road a comparatively small driving force may cause the wheel to slip circumferentially on the road, instead of rolling on it. This skidding of the wheel, though primarily making no difference in the conditions of stability, in a secondary manner influences side-slipping considerably.

Let a body M (fig. 195) of weight W, resting on a horizontal plane, be acted on by two horizontal forces, a and b, at right angles. Let μ be the coefficient of friction, and let at first only one of the forces, b, be in action. To produce motion in the direction $M\,X$, b must be greater than $\mu\,W$. Now, suppose the body M is being driven, under the action of a force a, in the direction $M\,Y$, in this case a much smaller force, b, will suffice to give the body a component motion in the direction $M\,X$. The actual motion will be in the direction $M\,R$, and since friction always acts in a direction exactly opposite to that of the motion, the resultant force on the body M must be in the direction $M\,R$. Let F be this resultant force ; its components in the directions $M\,X$ and $M\,Y$ must be b and a respectively. Now, if the force a be just greater than $\mu\,W$, it will be sufficient to cause the body to move in the direction $M\,Y$, and any force, b, however small, will give M a component motion in the direction $M\,X$.

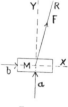

FIG. 195.

A familiar example illustrating the above principle, which has probably been often put into practice by every cyclist, is the adjusting of the handle-pillar in the steering-head. If the handle-pillar fits fairly tightly, as it ought to do, a direct pressure or pull parallel to its axis may be insufficient to produce the required motion, but if it be twisted to and fro—as can easily be done on account of the great leverage given by the handles—while a slight upward or downward pressure is exerted, the required motion is very easily obtained.

In the ' Kangaroo' bicycle the weight on the driving-wheel was less than in either the ' Rover Safety' or in the ' Ordinary.' On greasy roads it was easy to make the driving-wheel skid circumferentially by the exercise of a considerable driving pressure. This circumferential slipping once being established, the very smallest inclination to the vertical would be sufficient to give the wheel a sideway slip, which would, of course, rapidly increase with the vertical inclination of the machine.

171. **Influence of Speed on Side-slipping**.—The above dis-

cussion on side-slipping presumes that the speed of the machine and rider is not very great, so that the momentum of moving parts does not seriously influence the question. If the speed be very great, however, the momentum of the reciprocating parts, due principally to the weight of the rider's legs, pedals, and part of the weight of the crank, may have a decided influence on side slipping.

Let G be the mass-centre of the machine and rider (fig. 196), let the total mass be W lbs., let the linear speed of the pedals

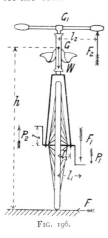

FIG. 196.

relative to the frame of the machine be v, and let w be the mass in lbs. of one of the two bodies to which the vertical components of the pedals' velocity is communicated : w will approximately be made up of the pedal, half the crank, the rider's shoe, foot, and leg from the knee downwards, and about one-third of the leg from the knee to the hip-joint. If the rider's ankle-action be perfect, the mass w may be considerably less, depending on the actual vertical speeds communicated to the various portions of the leg. Let the centre of the mass w be distant l_1 from the central plane of the bicycle. When the pedal is at the top of its path this mass possesses no velocity in a

vertical direction, and therefore no vertical momentum. When the crank is horizontal and going downward, the vertical velocity is at its maximum, and the momentum is $w\,v$. Let t be the time in seconds taken to perform one revolution of the crank, the time taken to impress this momentum is $\dfrac{t}{4}$; and if f^1 be the *average* force in poundals acting during this time to produce the change, we must have (sec. 63) :

$$f^1\,\frac{t}{4} = w\,v.$$

Therefore

$$f^1 = \frac{4\,w\,v}{t}.$$

If f be the average force in lbs., $f^1 = gf$, and the above equation may be written,

$$f = \frac{4\,w\,v}{g\,t} \quad \cdot \quad \cdot \quad \cdot \quad \cdot \quad \cdot \quad \cdot \quad (10)$$

If l be the length of the crank, the length of the path described in one revolution by the pedal-pin is $2\,\pi\,l$, and the time taken to perform one revolution is $\dfrac{2\,\pi\,l}{v}$. Substituting in (10) we get,

$$f = \frac{2\,w\,v^2}{g\,\pi\,l} \quad \cdot \quad \cdot \quad \cdot \quad \cdot \quad \cdot \quad (11)$$

Now leaving out of consideration for an instant the action of any force at the point of contact of the machine with the ground, and considering the machine and rider as forming one system, the above force f is an internal force, and can thus have no action on the mass-centre, G, of the whole system. But two parts of the system have each been impressed with a moment of momentum, $w\,v\,l_1$, about the mass-centre G, the remaining part ($W - 2\,w$) will be impressed with a momentum numerically equal but of opposite sense. Let G_1 be the mass-centre of this remaining part. Then the up-and-down motion of the two pedals being as indicated by the arrows p_1 and p_2, the point G_1 must move to the left with a velocity, v_1, such that

$$2\,w\,v\,l_1 = (W - 2\,w)\,v_1 \times \overline{G G_1}.$$

Thus, if there be absolutely no friction between the wheel and the ground, the point of contact of the wheel must slip sideways to the right.

Let F be the *average* frictional resistance, in lbs., required to prevent this slipping, then

$$F\,h = 2\,f\,l_1,$$

or

$$F = \frac{4\,w\,v^2\,l_1}{g\,\pi\,l\,h} \quad \cdot \quad \cdot \quad \cdot \quad \cdot \quad \cdot \quad (12)$$

If n be the number of turns per second made by the crank, $v = 2\,\pi\,n\,l$, and (12) may be written

$$F = \frac{16\,\pi\,n^2\,l_1\,l\,w}{g\,h} \quad \cdot \quad \cdot \quad \cdot \quad \cdot \quad (13)$$

From (12) and (13) the lateral force F, or what may be called the 'tendency' to side-slip, is proportional to the masses which partake of the vertical motion of the pedals, to the width of the tread, and inversely proportional to the height of the mass-centre from the ground ; from (12) it is proportional to the square of the speed of the pedals, and inversely proportional to the length of the crank ; from (13) it is proportional to the square of the number of revolutions of the crank-axle and to the length of the crank.

The force F changes in direction twice during one revolution of the crank-axle. It is equivalent to an equal force acting at G, and a couple Fh. The force acting at G, changing in direction, will therefore cause the mass-centre of the bicycle and rider to move in a sinuous path, even though the track of the wheel be a perfectly straight line. The less this sinuosity, other things being equal, the better ; *i.e.* in this respect a high bicycle is better than a low one for very high speeds.

It must be carefully noted that in the above investigation the pressure exerted on the pedal by the rider does not come into consideration. When moving at a given speed the tendency to side-slip is therefore quite independent of whether pressure is being exerted on the pedal or not.

172. **Pedal Effort and Side-slip.**—The idea that the pressure on the pedal causes a tendency to side-slip is so general that it may be worth while to study in detail the forces acting on the rider, the wheel and pedals, and the frame of the machine. For simplicity we will consider an 'Ordinary,' in which the rider is vertically over the crank-axle. The investigation will be of the same nature, but a little longer, for a rear-driving 'Safety.' The weight of the machine will be neglected.

Let W be the weight of the rider, F_1 the vertical thrust on the pedal, F_2 the upward pull on the handle-bar, F_3 the vertical pressure on the saddle ; let l_1 and l_2 be the distances of the lines of action of F_1 and F_2 respectively, and l_3 the distance of the crank axle-bearing from the central plane of the machine (fig. 196).

Consider first the forces acting on the rider ; these are, his weight, W, acting downwards at G ; the pull, F_2, of the handle-bar downwards ; the reaction, F_1, from the pedal upwards ; and the reaction, F_3, of the saddle. These forces are all parallel, and since the rider is in equilibrium we must have

$$W - F_1 + F_2 - F_3 = 0 \quad . \quad . \quad . \quad . \quad (14)$$

Also, the moments of these forces about any point is zero ; therefore, taking moments about the mass-centre, G, if the rider has not shifted sideways when exerting the pressure F_1 on the pedals,

$$F_1 l_1 - F_2 l_2 = 0 \quad . \quad . \quad . \quad . \quad . \quad (15)$$

If the rider does not pull at the handles he must either grip tightly on to the saddle, or shift sideways, so that the moment of the force F_1 is balanced.

Consider next the forces acting on the frame, which, for clearness of illustration, is shown isolated (fig. 197) ; these are, the pull, F_2, on the handle-bar upwards ; the pressure, F_3, of the rider on his saddle downwards ; and the upward reaction of the bearings f_1 and f_2. These forces are all parallel, and since they are in equilibrium,

$$F_2 - F_3 + f_1 + f_2 = 0 ;$$

that is,

$$f_1 + f_2 = F_3 - F_2 . \quad . \quad (16)$$

Since the force $(F_3 - F_2)$ has no horizontal component, neither will the force $(f_1 + f_2)$. By taking moments of all the forces about the point of application of f_2, the value of f_1 may be found, and then f_2 can be determined.

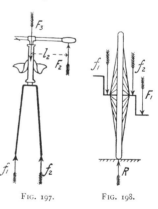

Fig. 197.　　　Fig. 198.

Now, consider the forces acting on the wheel (fig. 198), including cranks and pedal-pin, which together form one rigid body. Besides the forces $F_1, f_1,$ and f_2, there is only the reaction of the ground, R, and since the wheel is in equilibrium vertically,

$$R - F_1 - f_1 - f_2 = 0.$$

Substituting the value of $f_1 + f_2$ from (16) we get

$$R = F_1 - F_2 + F_3 = W \quad . \quad . \quad . \quad . \quad (17)$$

R being vertical, there is no tendency to side-slip.

The above result can be more simply obtained, thus : considering the bicycle and rider as forming one system of bodies, the external forces acting are in equilibrium ; and since these consist only of the weight, W, and the reaction, R, R must be (sec. 71) equal, parallel but opposite to W. W being vertical, R must also be vertical. The force F_1 exerted by the rider on the pedal is an internal force, and has not the slightest influence on the external forces acting on the system.

173. **Headers.**—Taking a 'header' over the handle-bar was quite an every-day occurrence with riders of the 'Ordinary' bicycle. In the 'Ordinary,' the mass-centre of the rider and machine was situated a very short distance behind a vertical through the centre of the front wheel, so that the margin of stability in a forward direction was very small ; any sudden check to the progress of the machine by an obstruction on the road, by the rider applying the brake, or back-pedalling, was in many cases sufficient to send him over the handle-bar. Two classes of headers have to be distinguished : (I) That in which the front wheel may be considered rigidly fixed to the frame ; the header being caused either by the application of the brake to the front wheel, or by back-pedalling in a Front-driver. (II) That in which the front wheel is quite free to revolve in its bearings ; the header being caused by an obstruction on the road, application of the brake to the back wheel, or back-pedalling in a Rear-driver.

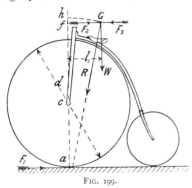

FIG. 199.

(I) Let l_1 (fig. 199) be the distance of the mass-centre, G, from a vertical through the wheel centre, c ; then, in order that the wheel, frame, and rider may turn as one body about the point a as centre, a moment, $W l_1$, must be applied. If d be the diameter of the driving-wheel, μ the coefficient of friction of the brake, and P the pressure of the brake just necessary to lock the frame on the wheel and so cause a header,

$$\frac{\mu P a}{2} = W l_1 \quad . \quad . \quad . \quad . \quad . \quad . \quad (18)$$

If the pressure actually applied to the brake be equal to or greater than P, determined by the above equation, the wheel will be locked to the frame.

Let the circle through G with centre a cut the vertical through c at h. From G draw a horizontal to cut $c h$ in f. In taking a header, the weight of the machine and rider has to be lifted a distance $f h$. If v be the speed of the machine, the kinetic energy stored up in it is $\dfrac{W v^2}{2 g}$, and the work done in lifting it through the height $f h$ is $W \times \overline{f h}$; therefore, if the speed v be greater than that determined by the formula

$$\frac{v^2}{2 g} = \overline{f h} \quad . \quad . \quad . \quad . \quad . \quad . \quad (19)$$

a header will occur if the brake-pressure be applied strongly.

If the check to the speed of a Front-driver be made by back-pedalling, r be the radius of the crank, and P_1 the back-pedalling force applied, we have,

$$P_1 r = W l_1 \quad . \quad . \quad . \quad . \quad . \quad . \quad (20)$$

The action of back-pedalling in a Front-driver is the same as that of applying the brake to the front wheel, as regards the locking of the front wheel to the frame. The speed at which a header will occur if vigorous back-pedalling be applied is in this case also given by equation (19).

Example I.—In a 54-inch 'Ordinary,' the point G (fig. 199) may be 60 inches above the ground and 10 inches behind the wheel-centre c. The height, $f h$, will then be about 1·2 inch $= \frac{1}{10}$ foot. Substituting in (19)

$$\frac{v^2}{2 \times 32\cdot2} = \frac{1}{10}, \text{ from which } v = 2\cdot5 \text{ feet per second,}$$
$$= 1\cdot9 \text{ mile per hour.}$$

Example II.—In a 'Safety' (fig. 200) the height, fh, may be 2 feet.   Substituting in (19),

$$\frac{v^2}{2 \times 32\cdot2} = 2, \text{ from which } v = 11\cdot1 \text{ feet per second,}$$

$$= 7\cdot6 \text{ miles per hour.}$$

The subject may be looked at from another point of view. Let F_1 be the horizontal force of retardation which must be

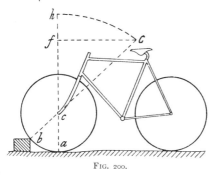

FIG. 200.

supplied by the action of the ground on the wheel. This is transmitted through the wheel, so that an equal force, F_2, acts on the mass at G, and the mass reacts on the frame with an equal and opposite force, F_3.   Then, in order that stability may be maintained, the resultant R of W and F_3 must not cut the ground in advance of the point of contact a.   If R cuts the ground in front of a, the machine will evidently roll over about a as centre.

(II) *Brake on Back Wheel.*—If the brake be applied to the rear, instead of the front wheel, the bicycle is much safer as regards headers.   If the brake, in this case, be applied too suddenly, the retarding force causes an incipient header, the frame turning about the front wheel centre c as axis, and the rear wheel immediately rises slightly from the ground.   The retarding force being thus removed, the development of the header is arrested, the rear wheel again falls to the ground, and the process is repeated, a kind of equilibrium being established.

Headers through Obstructions on the Road.—If the check to the progress of the machine be caused by an obstruction on the road, the only difference from the case treated above is that the front wheel is free to revolve in its bearings ; the header is taken about the point c as a centre, and the resultant R of the weight W and the force F_3 must not pass in front of the wheel centre c.

The direction of the forces between two bodies in contact is (neglecting friction) at right angles to the surface of contact. In a bicycle wheel with no friction at the hub, the direction of the pressure exerted by a stone at the rim must therefore pass through the wheel centre. This condition enables us to determine the size of the largest stone which can be ridden over at high speed without causing a header. Join the mass-centre, G, to the front wheel centre, c (fig. 201), and produce the line to cut the circum-

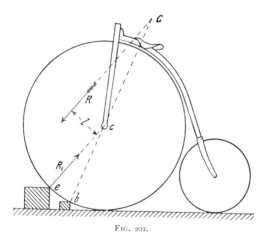

FIG. 201.

ference of the wheel at b. A stone touching the rim at a point higher than b may cause a header at high speed ; a stone touching at a lower point may be ridden over at any speed. Figure 200 is the same diagram for a 'Safety' bicycle, a glance at which shows that with this machine a much larger stone can be safely surmounted than with an 'Ordinary.'

The above discussion presupposes that at the instant the front wheel strikes the stone no driving force is being exerted. If the rider is driving the front wheel forward at the instant, a larger obstacle may be safely surmounted. Let e (fig. 201) be the point of contact of a large stone ; the reaction R_1 is in the direction $e\,c$. The resultant force R on the mass at G must be equal and parallel but opposite to R_1. The forces R and R_1 form a couple $R\,l$, tending to turn the frame and rider about the centre c, l being the length of the perpendicular from G on $e\,c$ produced. If the rider apply to the front wheel a turning moment in the forward direction equal to or greater than $R\,l$, there will be a couple of equal magnitude acting on the frame tending to turn it in the opposite direction, which will neutralise the couple $R\,l$. The final result is that the wheel safely surmounts the obstacle, turning about e as centre.

CHAPTER XVIII

174. **Steering in General.**—When a bicycle moves in a straight line, the axes of its wheels are parallel to each other. The steering is effected by changing the direction of one of the wheel spindles relatively to the other. In order to effect this change of direction, the frame carrying the wheels is made in two parts, jointed to each other at the *steering-head*, the parts being called respectively the rear- and front-frames. One of these parts, that carrying the saddle, is usually much larger than the other (and is often called the frame, to the exclusion of the other part called *the fork*) ; the wheel—or wheels—mounted on the other (smaller) part of the frame is called the steering-wheel—or wheels. According to this definition, the driving-wheel of an 'Ordinary' is also the steering-wheel. In side-steering tricycles (see chap. xvi.) the frame is in three parts, and there are two steering-heads.

Cycles are *front-* or *rear-*steerers, according as the steering-wheel is mounted on the front- or rear-frame. All bicycles that have attained to any degree of public favour are front-steerers : The 'Ordinary,' the 'Kangaroo,' the 'Rover Safety,' the 'American Star,' and the 'Geared Ordinary.' A few successful tricycles have, however, been rear-steerers.

175. **Bicycle Steering.**—Let *a* (fig. 202) be the wheel fixed to the rear-frame, *b* the steering-wheel, and *d* the intersection of the steering-axis with the ground ; this, in most cases, is at or near the point of contact of the wheel with the ground, though in the 'Rover Safety,' with straight front forks, it occurs some little distance in *front*. Let the plan of the axes of the wheels *a* and *b* be produced to meet at *o*, then if the wheels roll, without slipping sideways, on the ground, the bicycle must move in a circle having

o as its centre. The steering-wheel, b, will describe an arc of larger radius than that described by the wheel a ; consequently if in making a sharp turn to avoid an obstacle the front wheel clears, the rear wheel will also clear.   In a rear-steering bicycle, on

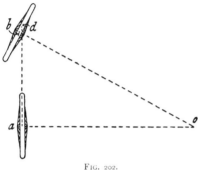

the other hand, it may happen that the rear wheel may foul an object which has been cleared by the front wheel.

The actual sequence of operations in steering a bicycle is not commonly understood. If a beginner turn the steering-wheel to one side before his body and the bicycle have attained the

FIG. 202.

necessary inclination, the balance will be lost. On the other hand, the beginner is often told to lean sideways in the direction he wants to steer. This operation cannot, however, be directly performed ; since, if he lean his body to the right, the bicycle will lean to the left, and the sideway motion of the mass-centre cannot be controlled in this way. It has been shown (sec. 162) that the path described by a bicycle, even when being ridden as straight as possible, is made up of a series of curves, the bicycle being inclined alternately to the right and to the left. If at the instant of resolving to steer suddenly to one side the bicyclist be inclined to that side, he simply delays turning the steering-wheel until his inclination has become comparatively large. The radius of curvature of the path corresponding to the large inclination being small, the steering-wheel can then be turned, and the bicycle will describe a curve of short radius. If, on the other hand, he be inclined to the opposite side, the steering-wheel is at first turned in the direction opposite to that in which he wishes to steer, so as to bring the bicycle vertical, and then change its inclination ; the further sequence of operations is the same as in the former case. Thus, to avoid an object it is often necessary to steer for a small fraction of a second towards it, then steer away from it ; this is probably the most difficult operation the beginner has to master. In steering, the rider's body should remain quite rigid in relation to the frame of the bicycle.

176. **Steering of Tricycles.**—The arrangement of the steering gear of a tricycle should be such that in rounding a corner the axes of the three wheels all intersect at the same point. In the 'Humber,' the 'Cripper,' and any tricycle with a pair of wheels mounted on one axle this condition is satisfied.

Let O be the intersection of the axes, a, b, c, of the three wheels. The tricycle as a whole rotating round O as a centre, the

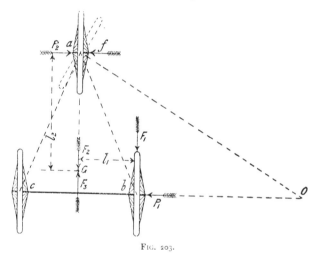

FIG. 203.

linear speed of the rim of wheel c will be greater than that of wheel b nearer the centre of rotation. If b and c are not driving-wheels, and are mounted independently on the axle, they will run automatically at the proper speeds. If b and c are driving-wheels, as in the 'Humber,' 'Cripper,' and 'Invincible' tricycles, some provision must be made to allow the wheel on the outside of the curve to travel faster than the inner. This is described in sections 188, 189.

177. **Weight on Steering-wheel.**—We have already seen that a considerable portion of the total weight of the machine must be

placed on the driving-wheel, so as to prevent skidding under the action of the driving effort. A certain amount of weight must also rest on the steering-wheel in order that it may perform its functions properly.

If the machine be moving at a high speed in a curve of short radius, the motion of the frame and rider can be expressed either as one of rotation about the point O, or as a translation equal to that of the mass-centre of the machine and rider, combined with a rotation about a vertical axis through the mass-centre G. If the rider should want to change from a straight to a curved course, the linear motion of the machine remains the same, but a rotation about an axis through the mass-centre must be impressed on it. To produce this a couple must act on the machine. The external forces, P_1 and P_2, constituting this couple can evidently only act at the points of contact of the wheel and the ground, and, presuming that the rolling friction may be neglected, can only be at right angles to the direction of rolling. The magnitudes of the forces P_1 and P_2 depend on the speed at which the cycle is running, and also on the general distribution of weight of the machine and rider—in mathematical language, on the moment of inertia of the system. The weight, w, on the steering-wheel must be equal to, or greater than, $\dfrac{P_1}{\mu}$, μ being the coefficient of friction. The moment of inertia, about its mass-centre, of a system consisting of a machine and two riders is very much greater than twice that of a system consisting of a machine and one rider; consequently the pressure required on the steering-wheels of tandems is much greater than twice that required on the steering-wheel of a single machine.

A simple analogy may help towards a better understanding of this. Suppose two persons of equal weight be seated at opposite ends of a see-saw, and that the up-and-down motion is imparted by a person standing on the ground, and applying force at one end of the see-saw. If now only one person be left on the see-saw, and he be placed at the middle exactly over the support, the person standing on the ground will have to supply a much smaller force than in the former case to produce swings of equal speed and amplitude. The swinging up and down of the see-saw corresponds to the change of steering of the cycle from left to right, the forces applied by the person standing on the ground to the forces, P_1 and P_2, of reaction of the ground on the wheels. The single person on the middle of the see-saw corresponds to a single rider on a cycle, the two persons at the ends to the riders on a tandem.

Sensitiveness of Steering.—We have continually spoken of the point of contact of a wheel with the ground, thereby meaning the geometrical point of contact of a circle of diameter equal to that of the wheel. The actual contact of a wheel with the ground takes place over a considerable surface, the lower portion of the tyre getting flattened out as shown, somewhat exaggerated in figure 204. The total pressure of the wheel on the ground is distributed over this area of contact. Considering tyres of the same thickness, it is evident that a wheel of large diameter will have the length of its surface of contact in the direction of the plane of the wheel greater than that of a wheel of smaller diameter.

FIG. 204.

Consider now the resistance to turning such a wheel, pivot-like, on the ground, as must be done in steering. Let A be the area of the surface of contact, and suppose the pressure of intensity, p, distributed uniformly over it, as will be very approximately the case with pneumatic tyres ; then

$$p = \frac{W}{A}.$$

Consider a small portion of the area of width, t, included between two concentric circular arcs of mean radius, r. Let a be the area of this piece, the total pressure on this will be $p\,a$, and the frictional resistance to spinning motion of this portion of the tyre on the ground will be $\mu\,p\,a$. The moment of this force about the geometrical centre, O, is

$$\mu\,p\,a\,r \quad . \quad . \quad . \quad . \quad . \quad . \quad . \quad (1)$$

and the total moment of resistance of the wheel to spinning on the ground is the sum of all such elements. If we consider the surface of contact to be a narrow rectangle, whose width is very small in comparison with its length, l, the *average* value of r in (1) will be $\dfrac{l}{4}$, and the total moment of resistance to spinning will be

$$\frac{\mu\,W\,l}{4} \quad . \quad . \quad . \quad . \quad . \quad . \quad . \quad . \quad (2)$$

Thus a greater pull will be required at the handle-bar to steer a large wheel than a small one ; in other words, a small steering-wheel is more sensitive than a large one. The assumption made above, that the width of the surface of contact is very small compared with its length, is not even approximately true for pneumatic tyres. The moment of resistance in this case will, however, increase with l, and, therefore, the conclusion as to the relative sensitiveness of small and large wheels holds.

The above expression gives the moment of resistance to turning the steering-wheel on the ground when the bicycle is at rest. This moment is quite considerable, and is much greater than the actual moment required to steer when the bicycle is in motion, as can be easily verified by experiment. The explanation of this phenomenon is practically of the same nature as the explanation, given in section 170, of the small force necessary to overcome friction in one direction, provided motion in a direction at right angles exists. In the present case the wheel is rotating about a horizontal axis during its forward motion ; the steering is effected by giving it a motion about a vertical axis. On account of the motion about a horizontal axis already existing, a comparatively small moment is sufficient to overcome the frictional resistance to motion about a vertical axis.

178. **Motion of Cycle Wheel.**—It is a popular notion that the motion of a vehicle wheel is one of pure rolling on the ground, but a little consideration will show that this is not always the case. So long as a tricycle moves in a straight line, the wheels merely roll on the ground, the instantaneous axis of rotation being a line through the point of contact of the wheel and ground, parallel to the axis. When the vehicle is moving in a curve, in addition to this rotation about a horizontal axis, the wheel possesses a motion round a vertical axis, and some parts of the tyre in contact with the ground slide over the ground, as described in section 177. The

instantaneous axis of rotation is now a line inclined to the ground.

Suppose that the plane of the wheel can be inclined to the vertical when the cycle is moving in a curve, as in the case of a bicycle or steering-wheel of a 'Cripper' tricycle. Let the axis of the wheel be produced to cut the ground at V, then if the cycle be at the instant turning about the point V as centre, the motion of the wheel on the ground will be one of pure rolling, no sliding being experienced by any point of the tyre in contact with the ground. The part of the wheel in contact with the ground may be considered part of a right circular cone, having its vertex at V. Such a cone would roll without slipping on a plane surface, the vertex, V, of the cone remaining always in the same position.

The intersection of the axis of the wheel with the ground is determined by the inclination of the wheel to the vertical. This inclination depends on the radius of the curve in which the bicycle is moving, and also its speed. For a curve of a given radius there is, therefore, one particular speed at which V will coincide with O, the centre of turning of the bicycle. At this speed there will be no spinning of the tyre on the ground, while at greater or less speeds spinning occurs to a greater or less degree.

179. Steering Without Hands.—In a front-driving bicycle, the saddle and crank-axle being carried by the rear- and front-frames respectively, there is theoretically no difficulty in steering without using the handle-bar. If it be desired to turn towards the right, a horizontal thrust at the left pedal as it passes its top position, or a pull at the right pedal as it passes its lowest position,

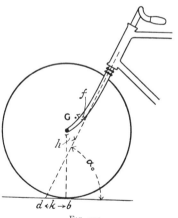

FIG. 205.

will effect the desired motion.

In a rear-driving bicycle, the saddle and crank-axle being

carried by the rear-frame, there is no direct connection between the rider and the steering-wheel axle except by the handle-bar.

Let a_0 be the angle the steering-axis makes with the horizontal when the bicycle is vertical (fig. 205) ; h the distance of the wheel centre from the steering-axis : k, the distance between b, the point of contact of the wheel with the ground, and d the point of intersection of the steering-axis with the ground, when the bicycle is vertical and the steering-wheel in its middle position ; f the

FIG. 206.

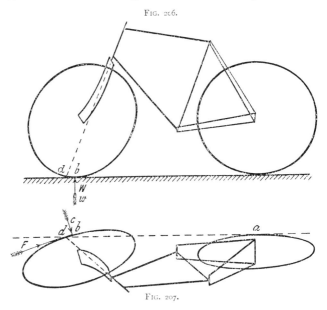

FIG. 207.

distance of the mass-centre of the steering-wheel and front-frame (including handle-bar, &c.) from the steering-axis ; θ the inclination of the middle plane of the rear-frame to the vertical ; ϕ the angle the handle-bar is moved from its middle position, *i.e.* the angle between the middle planes of the front and rear wheels ; and a the angle the steering-axis makes with the horizontal, corresponding to the values of θ and ϕ. Figs. 206 and 207 are elevation and plan of a bicycle heeling over. The forces acting on the front wheel and frame which may tend to turn it about the steering-axis are—the reaction of the ground, and the weight, w, of the front wheel and frame. The reactions at the ball-head intersect the steering-axis, and therefore cause no tendency to turn. The reaction of the ground can be resolved into three components—W, acting vertically upwards ; F, the resistance in the direction of motion of the wheel ; and C, the centripetal force at right angles to F. The line of action of F passes very near the steering-axis for all values of θ and ϕ, and since F is itself small in comparison with W and C, its moment may be neglected. Figs. 208 and 209 are elevation and plan enlarged from figs. 206 and 207, showing the relation of W to the steering-axis. $b\ d_1$ is the plan and $b_1{}^1\ d_1{}^1$ the elevation of the shortest line between W and the steering-axis. W can be resolved into a force, S, parallel to the steering-axis, and a force, T, at right angles to the plane containing S and the steering-axis. If $b_2{}^1\ b_1{}^1$ represent W to scale, $q^1\ b_1{}^1$ and $p^1\ b_1{}^1$ are the elevations of the forces T and S, while $Q\ b_1{}^1$ and $b_2{}^1\ Q$ show to scale the true magnitudes of T and S respectively ; *i.e.* $b_1{}^1\ b_2{}^1$ q^1 is the elevation of the force-triangle, and $b_1{}^1\ b_2{}^1\ Q$ is its true shape. Also it may be noticed that the line $b\ d_1$ in plan measures

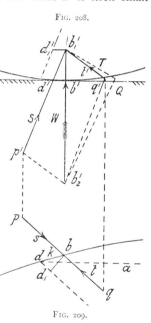

FIG. 208.

FIG. 209.

the true length of the perpendicular between W and the steering-axis ; and W tends to turn the steering-wheel still further, its moment about the steering-axis being $\overline{Q\ b_1{}^1} \times \overline{b\ d_1}$. The centripetal force C tends to turn the steering-wheel back into its middle position. The effect of the weight w in tending to turn the steering-wheel can be shown in exactly the same way as that of the vertical reaction W. The tendency is in general to increase the deviation of the steering-wheel, but when a straight fork is used the tendency is to reduce it, on account of the mass-centre of the handles being behind the steering-axis.

We shall now determine the analytical expressions for the moments of W, C, and w, assuming that the angles θ and ϕ are small, and that, therefore, we may use the approximations

$$sin\ \theta = \theta = tan\ \theta$$
$$sin\ \phi = \phi = tan\ \phi.$$

We have seen above that the moment of W is

$$\overline{Q\ b_1}^1 \times \overline{b\ d_1}.$$

Now $\overline{Q\ b_1}^1 = W\ cos\ a,$

also $sin\ a = sin\ a_0\ cos\ \theta.$

Therefore $\overline{Q\ b_1}^1 = W \sqrt{1 - sin^2\ a\ cos^2\ \theta}$

$$= W\ cos\ a_0 \text{ approximately.}$$

Now $\overline{b\ d_1} = \overline{b\ d}\ sin\ b\ d\ d_1$. The angle $b\ d\ d_1$ is made up of the two angles $a\ d\ b$ and $a\ d\ d_1$. The former is zero if ϕ is zero, and the latter is zero if θ is zero. For small values of θ and ϕ, the angle $a\ d\ b = \phi\ sin\ a_0$, and $a\ d\ d_1 = \theta\ tan\ a_0$.

Therefore $\overline{b\ d_1} = \overline{b\ d}\ sin\ (\theta\ tan\ a_0 + \phi\ sin\ a_0).$

Therefore, if we assume that $\overline{b\ d}$ remains constant, we have $\overline{b\ d_1} = k\ (\theta\ tan\ a_0 + \phi\ sin\ a_0)$ approximately, and moment of W is

$$W k\ sin\ a_0\ (\theta + \phi\ cos\ a_0) \quad . \quad . \quad . \quad . \quad . \quad (3)$$

The moment of C for small values of θ and ϕ will be approximately $C \times b\ d \times sin\ a_0$.

Now, if the angle ϕ remains constant

$$C = \frac{W v^2}{g\ R}, \quad R = \frac{l}{sin\ a\ d\ b} = \frac{l}{\phi\ sin\ a_0} \text{ approximately,}$$

v being the speed of the bicycle, R the radius of the circle described by the front wheel, and l the length of the wheel-base. Therefore the moment of C is

$$\frac{W v^2\ k\ \phi\ sin^2\ a_0}{g\ l} \cdot \quad . \quad . \quad . \quad . \quad . \quad (4)$$

The moment of w can be found as follows : Resolving w into two components parallel to and at right angles to the steering-axis, the latter is $w \cos a$. Figure 210 shows side and end elevations of the steering-axis and mass-centre, G. The perpendicular distance $B_1\, B_2$ between w and the steering-axis for a small value of θ is

$$\overline{G\,B_2} \times \theta = \frac{f\,\theta}{\cos a_0},$$

while for a small value of ϕ it is $f\,\phi$. Therefore moment of w is

$$w \cos a_0 \left(\frac{f\,\theta}{\cos a_0} + f\,\phi \right)$$

$$= w\,f\,(\theta + \phi \cos a_0) \quad . \quad \cdot \ \cdot \ \cdot \ \cdot \ \cdot \quad (5)$$

Hence, finally adding (3), (4), and (5), the moment tending to turn the steering-wheel still further from its middle position is

$$W\,k \sin a_0\,(\vartheta + \phi \cos a_0) - \frac{W\,k\,\phi\,v^2 \sin^2 a_0}{g\,l} + w\,f\,(\theta + \phi \cos a_0)$$

$$= (W\,k \sin a_0 + w\,f)\,(\vartheta - \phi \cos a_0) - \frac{W\,k\,\phi \sin^2 a_0}{g\,l}\,v^2 \quad . \quad (6)$$

To maintain equilibrium the expression (6) should have the value zero, to steer further to one side or other it should have a small positive value, and to steer straighter a small negative value.

For given values of v and ϕ there remains an element θ, the inclination of the rear-frame, at the command of the rider ; but even with a skilled rider the above moment varies probably so quickly that he could not adjust the inclination θ quickly enough to preserve equilibrium.

In the above expressions we have

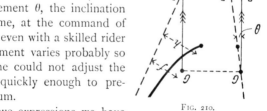

FIG. 210.

taken no account of the gyroscopic action of the wheel, though probably this is the most important factor in the problem Taking account of the gyroscopic action, the above moment about the steering-axis would produce a motion of precession about an axis at right angles to those of the ball-head and steering-wheel ; while to turn the steering-wheel about the steering-axis, a couple, with its axis at right angles to the steering-axis, would be required. This is produced by the side pressures on the steering tube; so that in steering without hands, if the rider wishes to turn to the right, he merely leans over slightly to the right, and the steering-wheel receives the required motion, provided the value of the expression (6) is small.

Example.—With the same data as in section 168, to turn the steering-wheel at the speed indicated, a couple of 2·42 foot-lbs. is required, *i.e.* if the ball-head be 8 inches long, side pressures of 3·63 lbs. would suffice to turn the front wheel at the speed indicated. To turn the steering-wheel more quickly, a greater side pressure must be exerted on the steering-head.

From section 168 the gyroscopic couple required is proportional to the square of the speed, and approximately proportional to the weight and to the diameter of the front wheel ; therefore, steering without hands should be easier the higher the speed, the larger the steering-wheel, and the heavier the rim of the steering-wheel. This agrees with the fact that a fair speed is necessary to perform the feat, that the feat is easier with pneumatic than with solid tyres, the former with rim being heavier than the latter ; it also accounts for the *easy* steering with large front wheels, and for the fact that the 'Bantam' is more difficult to steer without hands than the 'Ordinary.'

It may be noticed that if this explanation be correct, it should be possible to ride without hands a bicycle in which the steering-axis cuts the ground at the point of contact of the front wheel. M. Bourlet, who discusses the subject at considerable length, says this is impossible ; he also says that the mass-centre of the front wheel and frame must lie in front of the steering-axis ; but this would mean that a bicycle with straight forks could not be ridden without hands ; whereas some of the earliest 'Safety' bicycles, made with straight forks, were easily ridden without hands.

180. **Tendency of an Obstacle on the Road to Cause Swerving.**—If a bicycle run over a stone, the force exerted by the stone on the steering-wheel acts in a direction intersecting the steering-axis, and has thus no tendency to cause the steering-wheel to turn in either direction. In the same way, the steering-wheel of a

'Cripper' or 'Invincible' tricycle in running over a stone experiences no tendency to turn, and therefore no resistance need be applied by the rider at the handle-bar. The line of action of the force exerted on the machine cuts a vertical line through the mass-centre ; the force therefore only tends to reduce the speed of the machine, but not to deviate it from its path. If the obstacle meet one of the side wheels of a tricycle, the force exerted by the stone and the force of inertia of the rider form a couple tending to turn the machine and rider as a whole about their common mass-centre. In some tricycles the force exerted by the stone tends also to change the position of the steering gear, and so cause sudden swerving. A few of the chief types of tricycles are discussed in detail, with reference to these points, in the following sections.

181. **Cripper Tricycle** — Let one of the driving-wheels meet with an obstacle. Introducing at G, the mass-centre, two opposite forces, F_2 and F_3, each equal to F_1, no change is made in the static condition of the system. The force, F_1 (fig. 203), exerted by the stone on the machine is equivalent to an equal force, F_2, acting at the mass-centre of the machine and rider, and retarding the motion, and a couple formed by the forces F_1 and F_3 tending to turn the machine about its mass-centre, G. This turning is prevented by the side friction of the wheels on the ground. To actually turn about G, the driving-wheels must roll a little and the front steering-wheel slip sideways.

Let f be the resistance to slipping sideways of the front wheel, l_1 and l_2 the lengths of the perpendiculars from G on the lines of action of the forces F_1 and f, w the load on the steering-wheel, and μ the coefficient of friction between the steering-wheel and the ground. Then $f l_2$ must be equal to or greater than $F_1 l_1$. Also $f = \mu w$, therefore $\mu w l_2 \geqq F l$, or

$$w \geqq \frac{F l_1}{\mu l_2} \quad . \quad . \quad . \quad . \quad . \quad . \quad (7)$$

If, in the 'Cripper' tricycle, the steering-axis produced passes exactly through the point of contact of the steering-wheel with the ground (fig. 211), the reaction from the ground on the steering-wheel has no tendency to cause it to turn ; no resistance is necessary at the handle-bar when one of the driving-wheels strikes an obstacle. If, as in all modern tricycles, the steering-axis produced passes in front of the point of contact of the steering-wheel with the ground (fig. 212), the force, f, will tend to turn the steering-wheel sideways, and must be resisted by a force, F_4, at the handle-bar, such that $F_4 l_4 = f l_3$, l_3 being the length of the perpendicular

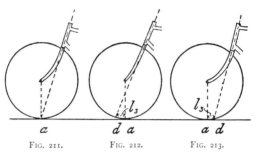

FIG. 211.　　　　FIG. 212.　　　　FIG. 213.

from the point of contact with the ground to the steering-axis, and l_4 the half-length of the handle-bar.

In a tricycle with a straight fork, the distance l_3, and therefore also the necessary force F_4, at the handle-bar to prevent swerving, is greater than with a curved fork (fig. 212).

182. **Royal Crescent Tricycle.**—In the 'Royal Crescent' tricycle (fig. 151), made by Messrs. Rudge & Co., the steering-axis intersected the ground at a point d (fig. 213), some distance behind the point of contact of the wheel. The force, f, would therefore tend to turn the steering-wheel about the steering-axis, in the opposite direction to that in the 'Cripper.' The distance, l_3, being much greater than in the 'Cripper,' the force, F_4, necessary at the handle-bar to prevent swerving was also greater. A spring control was used for the steering, so that a considerable force was necessary to move the steering-wheel from its middle position.

183. **Humber Tricycle.**—In a 'Humber' tricycle, an obstacle in front of one of the driving-wheels tends to turn the driving-axle round the steering-axis, a (fig. 214). This must be resisted by a force, F_4, applied by the rider at the handle-bar given by the equation $F_1 l_1 = F_4 l_4$, or the obstacle will change the direction of motion suddenly and a spill may occur. If the rider supply the

necessary force, F_4, the conditions as to the machine as a whole turning about the mass-centre G, and as to the weight necessary on the steering-wheel to prevent this turn-ing, are the same as discussed in section 181.

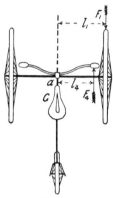

It will be seen from the above that the arrangement of the steering in the 'Humber' tricycle is less satisfactory than in some of the other types.

Any cycle in which there are a pair of independent wheels mounted on a com-mon axle, pivoted to the frame at its middle point, will be subject to the same defect of steering. Examples are afforded in figures 154, 155, and 182.

184. **Olympia Tricycle and Rudge Quadricycle.** — The wheel plan of an

FIG. 214.

'Olympia' tricycle is shown at figure 215. A single rear driving-wheel is used ; the two front wheels are side-steerers. In some of the earlier patterns of this tricycle made by Marriott &

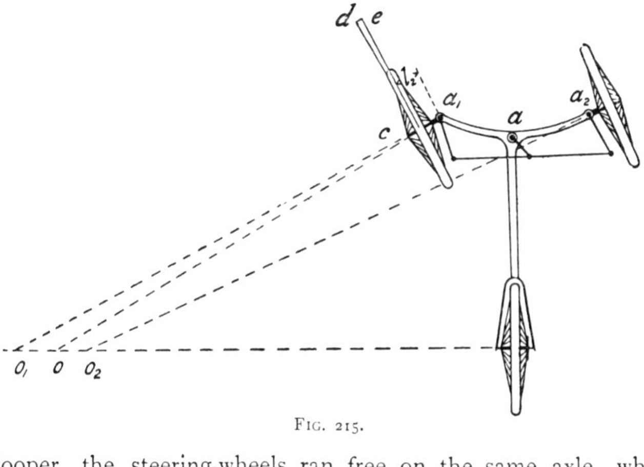

FIG. 215.

Cooper, the steering-wheels ran free on the same axle, which was pivoted at a to the rear-frame of the machine ; the action in steering was therefore the same as in the 'Humber' tricycle. In the modern patterns of the 'Olympia' tricycle the steering is effected by providing the steering-wheel spindles with separate steering-heads at a_1 and a_2. Short bell-cranks are formed on the spindles, and the ends of these cranks are connected by links to the end of a crank at the bottom of the steering-post a. The distance, l_2, between the steering-axis and the point of contact of the steering-wheel with the ground being much less than in the 'Humber' tricycle, the influence of an obstacle in causing swerving is correspondingly less, though in this respect the 'Olympia' is inferior to the 'Cripper.' The arrangement of this gear should be such that the axes of the steering-wheels in any position intersect at a point, O, situated somewhere on the axis of the driving-wheel. This cannot possibly be effected by any arrangement of linkwork, but the approximation to exactness may be practically all that can be desired for road riding. The gear should be arranged so that the bell-crank of the outer steering-wheel swings through a less angle from its middle position than that of the inner wheel.

If the axes of the wheels a_1 and a_2 intersect the axis of the driving-wheel at O_1 and O_2 (fig. 215), the machine as a whole may be supposed to turn about a point, O, somewhere between O_1 and O_2. Let c be the point of contact of wheel a_1 with the ground when the tricycle is moving round centre O, and let the linear velocity of a point on the frame vertically above c be represented by $c\,d$, drawn perpendicular to $O\,c$. From c draw $c\,e$ perpendicular, and from d draw $d\,e$ parallel, to the axis $O_1\,c$; these two lines intersecting at e, the actual velocity $c\,d$ is compounded of a velocity of rolling $c\,e$ of the wheel on the ground, and a velocity of side-slip, $e\,d$. The existence of this side-slip in running round curves necessitates careful arrangement of the steering mechanism, so that the centres O_1 and O_2 may never be widely separate. This side-slip must also add appreciably to the effort required to propel the 'Olympia' tricycle in a curved path, such as a racing track; and for such a purpose might possibly appreciably handicap it as compared with a 'Cripper.'

The steering gear of the 'Rudge' quadricycle is the same as that of the 'Olympia' tricycle.

185. **Rudge Coventry Rotary.**—In the 'Rudge Coventry Rotary' two-track tricycle, with single driving-wheel and two

steering-wheels (fig. 216), the reaction from the ground in driving being at F, there was continually a couple, $F l_1$, in action tending to turn the machine, and which was resisted by the reactions, f_1 and f_2, of the ground on the sides of the two side wheels. For equilibrium,

$$F l_1 = f l_2.$$

The steering-wheels were pivoted about axes passing through their points of contact with the ground and connected by short levers, connecting-rods, and a toothed-rack, to a toothed-wheel controlled by the rider. The arrangement, in this case, should again be such that in any position of the steering-gear the three axes intersect at a point O ; the machine would then turn about O as a centre.

If either of the steering-wheels pass over an obstacle, it is evident that

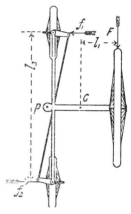

Fig. 216.

since the direction of the force acting on the wheel intersects the steering-axis there will be no tendency to turn the wheel, and therefore no resistance need be offered at the handle by the rider. The tendency of an obstacle to turn the machine as a whole about the mass-centre, G, is discussed in exactly the same way as for the 'Cripper' tricycle.

186. **Otto Dicycle.**—In the 'Otto' dicycle, the steering was effected by connecting each of the driving-wheels, by means of a smooth pulley and steel band, to the crank-axle. To run round a corner, the tension on one of the bands was reduced by the motion of the steering-handle, the band slipped on its pulley, and the other wheel being driven at a faster rate, the machine described the curve required. In a newer pattern with central gear (fig. 172) the motion was transmitted by a chain from the crank-axle to the common axle of the two wheels. The wheel-axle was divided into two portions, a differential gear being used, as explained in section 189. In steering, one of the driving-wheels was partially braked by a leather-lined metal strap, thereby making it more difficult to run than the other wheel ; one wheel was

thus driven faster than the other, and the machine described a curve.

If an obstacle met one of the wheels, its tendency was to retard the machine and to make it turn about its mass-centre. In performing this motion of rotation, neither of the wheels slipped sideways, and therefore no resistance was offered to the swerving ; consequently some other provision had to be made to prevent this motion. This was accomplished by locking the gear when running straight, so that the two driving-wheels were, for the time being, rigidly fixed to the axle, and ran at the same speed. If the horizontal force, F, actually caused the machine to swerve, one or other of the wheels actually slid on the ground. The frictional resistance to this sliding was $\mu \frac{W}{2}$, W being the weight of the machine and rider. If F was less than this, and the mechanism acted properly, the machine moved straight ahead over the obstacle.

187. **Single and Double-driving Tricycles.**—A tricycle, in which only one of the three wheels is driven, is said to be *single-driving*. The ' Rudge ' two-track and the ' Olympia ' are familiar examples. In single-driving tricycles the two idle wheels are supported independently, so that the three wheels have perfect freedom to rotate at different speeds.

If the two driving-wheels of a double-driving tricycle are (as is almost invariably the case) of the same diameter, while driving in a straight line they rotate at the same speed. They could, therefore, be rigidly fixed on the same axle, if only required to run straight ; but in running round a curve the outer wheel must rotate faster than the inner, unless one or other of the wheels skid, as well as roll, on the ground. Some arrangement of mechanism must be used to render possible the driving of the two wheels at different speeds.

188. **Clutch Gear for Tricycle Axles.**—Besides the ' Otto ' double-driving gear above described, two others, the clutch gear and the differential (or balance) gear, have been used to a considerable extent, though at present the differential gear is the only one used. In the ' Cheylesmore ' clutch gear (fig. 217), made by the Coventry Machinists Co., Limited, a sprocket wheel, w, in the

form of a shallow box, was mounted loosely near each end of the pedal crank-axle, and was connected by a chain to the corresponding driving-wheel. A cam, c, was fixed near each end of the crank-axle, and between the cam and the inner surface of the wheel, w, four balls, b, were placed ; the four spaces between the cam and the rim of the toothed-wheel being narrower at one end, and wider at the other, than the ball. In driving the axle in the direction of the arrow, the balls, b, were jammed between the wheel and the cam, the wheel consequently turned with the axle. If the axle were turned in the opposite direction, or if the wheel tended to move faster than the axle in the direction of the arrow,

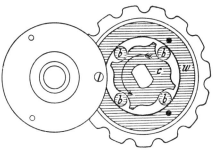

FIG. 217.

the balls, b, were liberated, and the cog-wheel revolved quite independently of the axle. While moving in a straight line both driving-wheels were driven ; but when running in a curve the inner wheel was driven by the clutch, while the outer wheel running faster than the inner overran the axle and liberated the balls, the outer wheel being thus left quite free to revolve at the required speed.

189. **Differential Gear for Tricycle Axle.**—Let two co-axial shafts, m and n (fig. 218), be geared to a shaft, k, the axis of which intersects that of the shafts, m and n, at right angles. The gearing may consist of three bevel wheels, a, b, and c, fixed respectively to shafts, m, k, and n. The three shafts are carried by bearings, m_1, k_1, and n_1 respectively. Let the shaft, k, be rotated in its bearings, it will communicate equal but opposite rotations to the shafts m and n. If ω_1 be the angular speed of the shaft m, that of n will be $-\omega_1$, and the relative angular speed of the shafts m and n will be $2 \omega_1$.

Now, let the shaft, k, carrying with it its bearings, k_1, be rotated about the axis, $m\,n$, with an angular speed, ω ; the teeth of the wheel, b, engaging with those of a and c, will cause the shafts, m and n, to rotate with the same speed, ω, about their common axis ; the shaft, k, being at rest relative to its bearings, k_1. If driving-wheels be mounted at the ends of the shafts, m and n, they will both be driven with the same angular speed ω about the axis $m\,n$.

Let now the shaft, k, be rotated in its bearings, giving a rotation ω_1 to the shaft m, and a rotation $-\omega_1$ to the shaft n, while k and its bearings are being simultaneously rotated about the axis $m\,n$

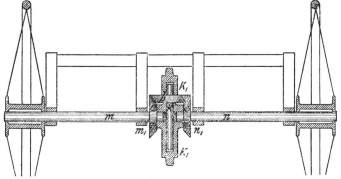

FIG. 218.

with the angular speed, ω. The resultant speed of the shaft m will be $(\omega+\omega_1)$, that of the shaft n will be $(\omega-\omega_1)$. Thus, finally, the average angular speed of the shafts m and n is the same as that of the bearings, k_1, while the difference of their angular speeds is quite independent of the angular speed of k_1. In Starley's *differential* tricycle gear, or *balance* gear, a chain-wheel is formed on the same piece of metal as the bearings, k_1, and is driven by a chain from the crank-axle. The driving effort of the rider is thus transmitted to the driving-wheels at the end of the shafts m and n. The shafts have still perfect freedom to rotate relatively to each other, and thus if in steering one wheel tends to go faster or slower than the other, there is nothing in the mechanism to prevent it.

In figure 218, the bevel-wheels, a and c, in gear with the wheel b are shown of equal size. In Starley's gear (fig. 219) a second wheel near the other end of the spindle, k, gears with those on the ends of the two half axles, so that the driving effort is transmitted at two points to each of these wheels. This forms, perhaps, the neatest possible gear, but a great variety could be made if necessary. Such a differential gear consists essentially of the chain-wheel, k_1, carrying a shaft, k, which gears in any manner with the shafts m and n. The particular form of gearing is optional; provided that it allows m and n to rotate relatively to each other. Thus in Singer's double-driving gear, the wheel, b, was a spur pinion, with

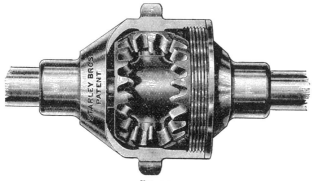

FIG. 219.

its axis parallel to $m\ n$, and engaging with a spur-wheel and an annular-wheel fixed respectively to the shafts, m and n. This gear had the slight disadvantage that equal efforts could not be communicated to the driving-wheels, that connected to the annular-wheel of the gear doing most of the work.

The balance gear being only used *differentially* for steering, the relative motion of the bevel-wheels, a, b, c (fig. 218), is very slow, and there is not the same absolute necessity for excessive accuracy as in toothed-wheel driving gear.

Example.—A tricycle with 28-in. driving-wheels, tracks 32 in. apart, being driven in a circle of 100 feet radius at a speed of 20 miles an hour, required the speed of the balance-gear.

While the centre of the machine moves in a circle 1200 inches radius, the inner and outer wheels move in circles $(1200 - 16)$ and $(1200 + 16)$ inches radii respectively. The circumferences of these circles are respectively $2\pi \times 1200$, $2\pi \times 1184$, and $2\pi \times 1216$ inches. While the centre of the machine moves over $2\pi \times 1200$ inches, the outer wheel moves over $2\pi \times 32$ inches more than the inner. The relative linear speed is therefore $\dfrac{2\pi \times 32}{2\pi \times 1200} \times 20$

$$= \cdot 5333 \text{ miles per hour}$$

$$= \frac{\cdot 5333 \times 5280 \times 12}{60} = 563\cdot 2 \text{ inches per minute.}$$

The circumference of a 28-in. wheel is $87\cdot 96$ in. The number of revolutions made by the outer part of the axle in excess of those made by the inner is therefore

$$\frac{563\cdot 2}{87\cdot 96} = 6\cdot 40 \text{ per minute.}$$

The number of revolutions of the axle divisions relative to the balance box, k, is therefore $3\cdot 20$ per minute.

CHAPTER XIX

MOTION OVER UNEVEN SURFACES

190. **Motion over a Stone.**—If a cycle be moving along a perfectly smooth, flat road, neglecting the slight horizontal sideway motion due to steering, the motion of every part of the frame of the machine is in a straight line. Suppose a bicycle to move over a stone which is so narrow that its top may be considered a point. The motion being in the direction of the arrow, the path of the centre of the driving-wheel will be a straight line $O A$ (fig. 220) parallel to the ground until the tyre

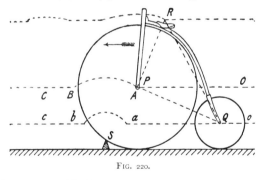

FIG. 220.

comes in contact with the obstacle at S, when the further motion of the wheel centre will be in a circular arc, $A B$, having S as centre. The further path of the wheel centre is the straight line, $B C$, parallel to the ground. The path of the centre of the rear wheel is of the same nature : a straight line, $o a$, until the tyre meets the obstacle S, the circular arc, $a b$, with S as centre, and then the straight line $b c$.

The motion of any point rigidly connected to the frame of

the bicycle can now be easily found. Let P and Q be the centres of the front and rear wheels respectively, and let it be required to find the form of the path of the point R lying on the saddle and rigidly connected to P and Q. Having drawn on the paper the paths of P and Q (fig. 220), take a small piece of tracing paper, and on it trace the triangle $P\,Q\,R$. Move this

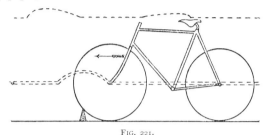

FIG. 221.

sheet of tracing paper over the drawing paper so that the points P and Q lie respectively on the curves $O\,A\,B\,C$ and $o\,a\,b\,c$. In this position prick through the point R, and a point on its path will be obtained. By repeating this process a number of points on the required path can be obtained sufficiently close together to draw a curve through them. Figures 220, 221, and 222

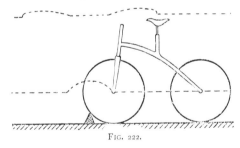

FIG. 222.

respectively show the curves described by a point a short distance above the saddle of an 'Ordinary,' of a 'Rear-driving Safety' with wheels 28 in. and 30 in. diameter, and of a 'Bantam' with both wheels 24 in. diameter, the point being midway between the wheel centres. A number of such curves are given and exhaustively discussed in R. P. Scott's 'Cycling Art, Energy, and Locomotion,' though it should be noticed that the curved portions of the saddle paths, due to the front and rear wheels passing over the obstruction, are shown placed in wrong positions.

191. **Influence of Size of Wheel.**—In figure 220 it will be noticed that the total heights of the curved portions of the paths of the wheel centres above the straight portions are the same, whatever be the diameter of the wheel; but the greater the diameter of the rolling wheel, the greater is the horizontal distance moved over by the wheel centre in passing over the stone. Thus with a large wheel the stone is mounted and passed over more gradually, and therefore with less shock, than with a small wheel. Therefore, other things being the same, large wheels are better than small for riding over loose stones lying on a good flat road.

192. **Influence of Saddle Position.**—The motion of the saddle may be conveniently resolved into vertical and horizontal components. In riding along a level road the vertical motion is zero and the horizontal motion uniform. When the front wheel meets an obstacle the motion of the frame may be expressed as a motion of translation equal to that of the rear wheel centre, Q, together with a motion of rotation of the frame about Q as centre. Let ω be the angular speed of this rotation at any instant. The linear motions of P and R relative to Q will be

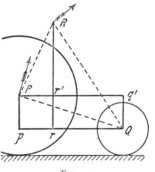

FIG. 223.

in directions at right angles to QP and QR respectively, and their speeds will be $\omega \times QP$ and $\omega \times QR$ respectively; the lines QP and QR (fig. 223) may therefore represent the magnitudes of the velocities, the directions being at right angles to these lines. Through Q draw a horizontal line, and to it draw perpendiculars Pp and Rr. Then Qp and Qr will represent the vertical components of the motions of P and Q respectively, Pp and Rr the horizontal components.

In the same way, if the front wheel be moving along the level, and the back wheel be passing over an obstacle, by drawing perpendiculars $R r^1$ and $Q q^1$ to a horizontal line through P, it can be shown that $P q^1$ and $P r^1$ represent the vertical components of the motions of Q and R respectively relative to P, $Q q^1$ and $R r^1$ the horizontal components.

Therefore, in a bicycle with equal wheels, the vertical 'jolting' communicated to the saddle by one of the wheels passing over an obstacle is proportional to the horizontal distance of the saddle from the centre of the other wheel, the horizontal 'pitching' to the vertical distance from the centre of the other wheel. With wheels of different sizes the average angular speeds ω are inversely proportional to the chords $A B$ and $a b$ (fig. 220); this ratio must be compounded with that mentioned above.

If the saddle of a tricycle be vertically over the centre of the wheel-base triangle, its vertical motion will be one-third that of one of the wheels passing over a stone. In the 'Rudge' quadricycle the vertical motion would be one-fourth, with similar conditions as to position of saddle.

From the above discussion it is readily seen that the most comfortable position for the saddle, as regards riding over rough roads, is midway between the wheel centres, the vertical motion of the saddle being then half that of a wheel going over a stone. In a tandem, with one seat outside the wheel centres, the vertical jolting of this seat is greater than that of the nearer wheel. Again, as regards horizontal pitching, the high bicycle compares unfavourably with the low; the rider on the top seat of the 'Eiffel' bicycle would have to hold on hard to avoid being pitched clean out of his seat while riding fast over a rough road. A long wheel-base is a decided advantage as regards horizontal pitching in riding over stones. The angular speed ω of the frame in mounting over a stone is, other conditions remaining the same, inversely proportional to the length of the wheel-base. Therefore, the pitching is also inversely proportional to the length of the wheel-base.

A curious point may be noticed in the case of the 'Ordinary.' From the saddle path shown (fig. 220) it will be seen that when the rear wheel, after surmounting the obstacle, is descending again to the level, the saddle actually moves backwards. This can only happen at slow speeds; at higher speeds the rear wheel actually leaves the stone before touching the ground, and the backward kink in the saddle path may be eliminated.

193. **Motion over Uneven Road.**--If the surface of the road be undulating, but free from loose stones, the paths of the wheel centres, P and Q, will be curves parallel to that of the road surface, and the path of any point rigidly fixed to the frame can be found by the same method. In a very bad case, the undulations being very close together (fig. 224), it may happen that the radius of curvature of one of the holes is less than the radius of a large bicycle wheel.

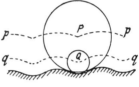

FIG. 224.

In this case the path, $p\,p$, of the large wheel will have abrupt angles, while that of the smaller wheel, $q\,q$, may be continuous, the large wheel being actually worse than the small one.

194. **Loss of Energy.**—If the motion of a wheel over an obstacle took place very slowly, there would theoretically be no loss of energy in passing over it, since the work done in raising

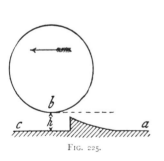

FIG. 225.

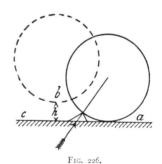

FIG. 226.

the weight would be restored as the weight descended; but at appreciable speeds the loss of energy by impact and shock may be considerable. Let a wheel moving in the direction of the arrow (fig. 225) pass over an obstacle of such a form that the wheel rises without sudden jerk or shock to a height h, the speed being so great that at its highest point the wheel is clear both of the obstacle and the ground. If W be the weight (including that of the wheel) resting on the axle, the energy lost will be Wh, since the kinetic energy in position b is this amount less than that in position a. The energy due to the fall from b to c is wasted in shock, there being no means of obtaining a forward effort from the work done during the descent.

If the wheel strike the obstacle suddenly (fig. 226) and then rises to the height h, clear of the ground and obstacle, the energy lost may be greater than Wh, the amount depending on the nature of the surface of the wheel tyre and the obstacle struck.

If the horizontal speed of the wheel be such that it does not leave contact with the obstacle in passing over it, the nature of the losses of energy can be shown as follows :

The centre of the wheel at the instant of coming into contact with the stone, S (fig. 227), is moving with velocity v in a hori-

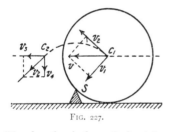

FIG. 227.

zontal direction. This can be resolved into a velocity v_1 in the direction $c_1 S$, joining the wheel centre to the stone, and a velocity v_2 at right angles to this direction. The velocity, v_1, is the velocity of impact of the wheel on the stone S, and the energy due to this velocity may be entirely lost.

If e be the index of elasticity, the velocity of rebound is $e v_1$, and with suitable elastic tyres the energy due to this velocity may be saved. The loss of energy due to the impact on the stone will be at least (sec. 69)

$$\left(1 - e^2\right) \frac{m v_1^2}{2 g} \quad . \quad . \quad . \quad . \quad . \quad . \quad . \quad . \quad (1)$$

and may be as great as

$$\frac{m v_1^2}{2 g} \quad . \quad . \quad . \quad . \quad . \quad . \quad . \quad . \quad (2)$$

where m is the weight of the portion of the machine rigidly connected to the wheel tyre.

The motion of the wheel continuing, the wheel centre mounts over the stone, describing a circle, $c_1 c_2$, with centre, S, and the tyre will again touch the ground on a point in front of the stone. If the speed of the machine be uniform, the velocity of the wheel centre as the wheel again just touches the ground may be equal in magnitude to v_2, the same as immediately after impact on the stone. This velocity, v_2 (fig. 227), can be resolved into horizontal and vertical components, v_3 and v_4. v_4 is the velocity of impact on the ground, and the energy due to it is either partially or entirely lost, and the final velocity of the wheel centre is v_3.

The assumption made above, that the speeds of the wheel centre, C, when in positions c_1 and c_2 are equal, is equivalent to assuming that the reactions of the stone on the wheel in any position before passing the vertical line through the stone is exactly equal to the reaction when at an equal distance past the stone ; or, briefly, the reactions as the wheel rolls on and off the stone are equal. With a hard unyielding tyre this is not even approximately true, except at very low speeds, consequently the positive forward effort exerted on the wheel as it rolls off the stone is less than the backward effort exerted as it rolls on, and the speed is seriously diminished. With a tyre that can adapt itself *instantaneously* to the inequalities of the road, the reactions during rolling on and off a stone are equal, and there is no loss of energy. The pneumatic tyre is the closest approximation to such an ideal tyre, while rubber is much better than iron.

If the road surface be undulating, the undulations being so long that the path of the wheel centre is a curve with no sudden discontinuities, there may be no loss of energy due to the undulations. If the undulations, however, be so short, and the speed of the machine so great, that the wheel after ascending an undulation actually leaves contact with the ground, there will be a loss of energy due to the impact on reaching the ground.

CHAPTER XX

195. **Expenditure of Power.**—The energy a cyclist generates while riding along a level road is expended in overcoming the various resistances to motion. These may be classed as follows : (1) Friction of bearings and gearing of the machine. (2) Rolling resistance of the wheels on the ground. (3) Resistance due to loss of energy by vibration. (4) Resistance of the air. The power expended in overcoming these resistances is the power actually communicated to the machine, and may be called the *brake* power of the rider. The power actually generated in the living heat-motor (the rider's body) may be called the *indicated* power ; the difference between the *indicated* and the brake powers will be the power spent in overcoming the frictional resistance of the motor—*i.e.* the friction of the rider's joints, muscles, and ligaments. At very high pedal speeds the brake power is small compared with the indicated ; in fact, by supporting the bicycle conveniently, taking off the chain, and pedalling as fast as he can, a rider may possibly develop more indicated power than when racing on a track, though the brake power is practically zero. The gearing of the bicycle, therefore, must not be made too low, or the greater part of the rider's energy will be spent in heating himself. The estimation of the work so wasted lies in the domain of the physiologist rather than in that of the engineer ; we proceed, therefore, to the consideration of the brake power and its expenditure.

196. **Resistance of Mechanism.**—The frictional resistance of the bearings is very small compared with the other resistances to be overcome ; the resistance due to friction of the bearings of a bicycle moving on a smooth track is practically the same at all

speeds. Professor Rankin estimated this at $\frac{1}{1000}$ part of the weight of the rider, but exact experiments are wanting.

The frictional resistance of the chain possibly varies with the pull on it, and as, other things being equal, the pull of the chain increases with the speed, the resistance will also vary with the speed. However, in comparison with the resistance due to rolling and with the air resistance, that of the chain is small, and may be included in the internal resistance of the machine, which we may say is approximately constant at all speeds.

197. **Rolling Resistance.**—The resistance to rolling is, according to the experiments of Morin, composed of two terms, one constant, the other proportional to the speed. With a pneumatic tyre on a smooth road the second term is negligible in comparison with the first, according to M. Bourlet. The rolling resistance is inversely proportional to the diameter of the wheel.

In 'Traité des Bicycles et Bicyclettes,' C. Bourlet says that the rolling resistance with pneumatic tyres is small, independent of the speed, and on a dry road it varies from

$$\cdot 005 \ W \text{ to } \cdot 01 \ W \quad . \quad . \quad . \quad . \quad . \quad . \quad (1)$$

while on a racing track the probable value for the resistance is $\cdot 004 \ W$, W being the total weight of machine and rider.

The resistance of a solid rubber tyre varies with the speed, and may possibly be expressible by a formula of the form

$$R = A + B v, \quad . \quad . \quad . \quad . \quad . \quad (2)$$

A and B being constants.

The power P required to overcome the rolling resistance $\cdot 005 \ W$ at the speed v is

$$P = \cdot 005 \ W v \text{ units} \quad . \quad . \quad . \quad . \quad (3)$$

If W be expressed in lbs. and v in miles per hour,

$$P = \cdot 44 \ W v \text{ foot-lbs. per min.} \quad . \quad . \quad (4)$$

198. **Loss of Energy by Vibration.**—One of the great advantages of a pneumatic tyre is that little or no vibration is communicated to the machine and rider. On a smooth road or track with pneumatic tyres the loss due to vibration is probably negligible ; but on a rough road it may be very large, and is

possibly proportional to the speed. With solid tyres, a considerable amount of energy is lost in vibration. Bourlet's experiments on the road show that the work wasted in vibration is about one-sixth of the total.

The use of a pneumatic tyre enables the tremulous vibration to be almost eliminated, no vibration being communicated to any part of the machine. For riding over very rough roads the introduction of springs into the wheel or frame may still further diminish vibration. The anti-vibrators should be placed so that they protect as great a portion of the machine from vibration as possible. In this respect a spring wheel should be better than a spring frame, and a spring frame, in turn, better than a spring saddle. The machine, as a whole, should be made sufficiently strong and rigid that none of its parts yield under the stresses to which they are subjected. Of course, when a spring yields and again extends, a certain amount of energy is lost ; it thus becomes a question as to when springs are advantageous or otherwise. Probably the rougher the road, the more can springs be used with advantage in the wheels, frame, and saddle ; whereas, on a smooth racing track, their continual motion would simply provide means of wasting a rider's energy.

199. **Resistance of the Air.**—M. Bourlet discusses the air resistance of a rider and machine, and concludes that it may be represented by a formula

$$R = k \, S \, v^2 \quad . \quad . \quad . \quad . \quad . \quad . \quad . \quad (5)$$

R being the air resistance, S the area of the surface exposed, v the speed, and k a constant. If the resistance be measured in kilogrammes, the area in square metres, and the speed in metres per second, $k = \cdot 06$. The area of surface exposed will depend on the size of the rider and his attitude on the bicycle. A mean value for S is $\cdot 5$ square metre ; then

$$R = \cdot 03 \, v^2 \quad . \quad . \quad . \quad . \quad . \quad . \quad . \quad (6)$$

If the resistance be measured in lbs., and the speed V in miles per hour,

$$R = \cdot 013 \, V^2 \quad . \quad . \quad . \quad . \quad . \quad . \quad (7)$$

The power required to overcome this resistance is

$$1 \cdot 144 \ V^3 \text{ foot-lbs. per minute} \quad . \quad . \quad . \quad . \quad . \quad (8)$$

Table X. gives the air resistances and the corresponding powers at different speeds calculated from these formula.

TABLE X.—AIR RESISTANCE TO 'SAFETY' BICYCLE AND RIDER.

Speed	Resistance	Power	Speed	Resistance	Power
Miles per hour	lbs.	Foot-lbs. per min.	Miles per hour	lbs.	Foot-lbs. per min.
5	·32	143	18	4·21	6,672
6	·47	247	19	4·69	7,846
7	·64	392	20	5·20	9,152
8	·83	586	21	5·73	10,600
9	1·05	834	22	6·29	12,180
10	1·30	1,144	23	6·88	13,920
11	1·57	1,522	24	7·49	15,820
12	1·87	1,977	25	8·12	17,870
13	2·20	2,513	26	8·79	20,100
14	2·55	3,139	27	9·48	22,520
15	2·92	3,861	28	10·19	25,110
16	3·33	4,685	29	10·93	27,900
17	3·76	5,620	30	11·70	30,890

If the wind be blowing exactly with or against the cyclist, his speed *relative to the air* must be used in the above formula. Thus, if the wind be blowing at the rate of 10 miles per hour, and the rider be moving at the rate of 20 miles per hour, while going *against* the wind, the air resistance is that due to a speed of 30 miles per hour, while going with the wind there is still a resistance due to a speed of 20 — 10 = 10 miles per hour.

If v be the speed of the cyclist, V that of the wind, while riding against the wind the relative speed is $(v + V)$. If the cyclist rides at a high speed, a very slight breeze against him may increase the air resistance considerably. Whilst riding with the wind the relative speed is $(v - V)$. In this case, if the speed of the wind be greater than that of the cyclist, there will be no resistance, but, on the contrary, assistance will be afforded by the wind. If the speed of the wind be less than that of the cyclist, there will be air resistance due to the speed $(v - V)$.

The power required to overcome air resistance in driving at v miles per hour against a wind blowing V miles an hour is

$$P = 1 \cdot 144 \, v \, (v + V)^2 \text{ foot-lbs. per minute} \quad . \quad . \quad (9)$$

that required in going with the wind,

$$P = 1 \cdot 144 \, v \, (v - V)^2 \text{ foot-lbs. per minute} \quad . \quad . \quad (10)$$

This equation gives also the power expended in overcoming air resistance by a rider behind pace-makers ; the principal beneficial effect of pace-makers being to create a current of wind of speed V assisting the rider.

With a side wind blowing, the air resistance is greater than that due to the relative speed. In moving through still air, or against a head wind, the cyclist drags with him a certain quantity of air. A side wind has the effect of changing very rapidly the actual particles dragged by the cyclist, so that in a given period of time the mass of air which has to be impressed with the rider's speed is greater than with a head wind of the same speed. Hence an increased resistance is experienced by the rider.

A consideration of the figures in Table X. will show that bicycle record-breaking depends more on pace-making arrangements than on any other single factor. For example, to ride unpaced at twenty-seven miles an hour requires the expenditure of more than two-thirds of a horse-power to overcome only the air resistance. Though an average speed of $27\frac{1}{2}$ miles per hour was kept up by Mr. R. Palmer and by Mr. F. D. Frost in the Bath Road Club 100-miles race, 1896, it is most improbable that they worked at anything like this rate during the whole period, the difference being due to the decrease in the air resistance caused by the pace-makers in front.

200. **Total Resistance.**—Summing up, the total resistance of the bicycle can be expressed by the formula

$$R = A + B v + C v^2 \quad . \quad . \quad . \quad . \quad . \quad (11)$$

and the power required to drive it by

$$P = A v + B v^2 + C v^3 . \quad . \quad . \quad . \quad . \quad (12)$$

A, B, and C being co-efficients depending on the nature of the mechanism and the condition of the road, but which are constant for the same machine on the same road at different speeds.

Figure 228 shows graphically the variation of the power required to propel a cycle as the speed increases. The speeds are set off as abscissæ. For any speed, $O\,S$, the power required to overcome the frictional resistance of the mechanism is set off as an ordinate $S\,M$; the power required to overcome rolling resistance is $M\,T$ (W being taken at 180 lbs.); the power required

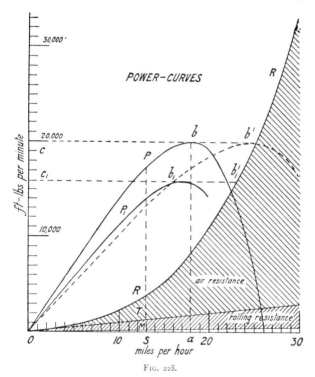

FIG. 228.

to overcome air resistance is $T\,R$; and the total power required is the ordinate $S\,R$. The curve M can be lowered by improvements in the mechanism, the curve T by improvements in the tyres and track-surface, and the curve R by improvements in pace-making.

Experiments on the total resistance of a cycle can be carried

out in two ways. Firstly, by towing the machine and rider along
a level road by means of another machine, the pull on the tow-
line being read off from a spring-balance. Secondly, by letting
the machine and rider run down a hill, the gradient of which is
known, until a uniform speed is attained; the ratio of the
resistance at the speed attained to the total weight of machine
and rider is the sine of the angle of inclination of the road. The
second method is not convenient for a series of experiments at
different speeds, since a number of hills of different gradients are
required; but since no extra assistance is required, a rider may
use it when unable to use the first method.

Table XI., taken from 'Engineering,' January 10, 1896, giving
results of experiments by Mr. H. M. Ravenshaw, serves to show
the variation of the resistance according to the state of the road.

TABLE XI.—RESISTANCE OF CYCLES ON COMMON ROADS.

Machine	Road	Total weight. Lbs.	Pounds per ton	Miles per hour
Tandem Tricycles, Pneumatic Tyres	Flint . . .	120	37	4
	,, . . .	290	31	4
	,, . . .	290	31	10·4
	,, . . .	290	31	7
	,, . . .	440	35	4
	,, . . .	440	35	8·3
	Asphalte pavement	290	31	4
	,, ,,	440	30	4
	,, ,,	440	30	6
	Heavy mud . .	290	73	4
	Wet mud . .	290	65	12
Tandem Bicycles, Pneumatic Tyres	Flint . . .	200	33	5
	,, . . .	370	30	5
	Heavy mud . .	200	95	5
	,, ,, .	370	78	5
	Flag pavement .	200	33	5
Single Tricycles, Solid Tyres . .	Flint . . .	220	60	4
	,, . . .	220	60	8
	Flag pavement .	220	60	5
	Heavy mud . .	200	146	4

CHAPTER XXI

201. **A Machine** is a collection of bodies designed to transmit and modify motion and force. The moving parts of a machine are so connected, that a change in the position of one piece involves, in general, a certain definite change in the position of the others. A bicycle or tricycle is a machine in which work done by the rider's muscles is utilised in changing the position of the machine and rider. Coming to narrower limits, we may say a cycle is a machine by which the oscillatory movement of the rider's legs is converted into motion of rotation of a wheel or wheels rolling along the ground, on which is mounted a frame carrying the rider. Still more narrowly, we may consider a cycle as a mechanism for converting the motion of the pedals, which may be either oscillatory or circular, into motion of rotation of the driving wheel.

202. **Higher and Lower Pairs.**—Each part of a machine must be in contact with at least one other part ; two parts of a mechanism in contact and which may have relative motion forming a *pair*. If the two parts have contact over a surface, as is necessary when heavy pressures are transmitted, the pair is said to be *lower*. From this definition there can only be three kinds of lower pairs—turning pairs, sliding pairs, and screw pairs ; as in a shaft and its journal, a cylinder and piston, a bolt and its nut, respectively. If the elements of a pair do not have contact over a surface, or if one of the elements is not rigid, the pair is said to be *higher*, the relative motion of the pair being, as a rule, much more complex than that of lower pairs. A pair of toothed-wheels in contact, a flexible band and drum, a ball and its bearing-case, are examples of higher pairs.

Link or Connector.—Two elements of consecutive pairs may be connected together by a *link*. An assemblage of pairs connected by links constitute a *kinematic chain*, or a *mechanism*, or a *gear*. The simplest kinematic chain contains four pairs connected by four links ; it is therefore called a four-link mechanism. If one link be fixed, a motion given to a second link will produce a determinable motion of the two remaining links. Three pairs united by three links constitute a rigid triangle, while a five-link chain requires further constraint for movement of a definite character to be produced. The four-link kinematic chain is the basis of probably 99 per cent. of all linkwork mechanisms.

203. **Classification of Gearing**.—Professor Rankine defines an elementary combination in mechanism as a pair of primary moving pieces so connected that one transmits motion to the other ; that whose motion is the cause is called the driver, the other the follower. The connection between the driver and follower may be :

(1) By rolling contact of their surfaces, as in toothless wheels.

(2) By sliding contact of their surfaces, as in toothed-wheels and cams, &c.

(3) By flexible bands, such as belts, cords, and gearing chains.

(4) By linkwork, such as connecting-rods, &c.

(5) By reduplication of cords, as in the case of ropes and pulleys.

(6) By an intervening fluid.

The driving gear of cycles has been made from classes (2), (3), and (4), each of which will form the subject of a separate chapter. An example of (1) is found in the 'Rotherham' cyclometer, the wheel of which is driven by rolling contact from the tyre of the front wheel. The pump of a pneumatic tyre is an example of (6). We cannot recollect an example in cycle construction corresponding to (5), though it would be easy to design one to work in connection with a pedal clutch gear, such as the 'Merlin.'

204. **Efficiency of a Machine**.—If the pairs of a mechanism could perform their relative motion without friction, the work done by the prime mover at the driving end of the machine would be transmitted intact to the driven end ; in other words, the work got out of the machine would be equal to that put into

it. But however skilfully the parts be designed to reduce friction to the lowest possible amount, there is always *some* frictional resistance which consumes energy, so that the work got out of the machine is less than that put into it, by the amount of work spent in overcoming the frictional resistance of the pairs.

The ratio of the work transmitted by the machine to that supplied to it is called the *efficiency of the machine.* The efficiency of a machine will be higher according as the number of its pairs is small ; an increase in the number of pairs increases the opportunities for work to be wasted away. Thus, in general, the simpler the mechanism used, the better will be the results obtained.

It seems perhaps unnecessary to say that no advantage can be derived from mere complexity of mechanism, but the number of driving gears for cycles that are being patented shows either that the perpetual motion inventor has plenty of vitality, or that the technical common sense of a large number of cycle purchasers is not of a very high standard.

205. **Power.**—We have already seen that the work done by an agent is the product of the applied force, into the distance through which the point of application of the force is moved in the direction of the applied force. The *power* of an agent is equal to the *rate* of doing work—that is, power may be defined as the *work done per unit of time.* If E be the work done in t seconds, and P the power of the agent, then

$$P = \frac{E}{t}.$$

But E is equal to Fs, where F is the force acting and s the distance moved ; therefore

$$P = \frac{Fs}{t}.$$

But $\frac{s}{t}$ is equal to the speed ; therefore

$$P = Fv \quad . \quad . \quad . \quad . \quad . \quad . \quad . \quad (1)$$

That is, the power of the agent is equal to the product of the acting force and the speed of its point of application. The same principle is expressed in the maxim, 'What is gained in power is lost in speed'; the word 'power' in this maxim having the meaning we have associated with 'force' throughout this book.

In a frictionless machine the power is transmitted without loss. The above equation shows that any given horse-power may be transmitted by any force F, however small, provided the speed v can be made sufficiently great. On the other hand, if the speed of transmission be very small, a very large force, F, may correspond to a very small transmission of power. An example of the former case occurs in transmitting power to great distances by means of wire rope. Here the speed of the rope is made as large as it is found practicable to run the pulleys, so that a rope of comparative small diameter may transmit a considerable amount of power. An example of the latter case occurs in a hydraulic forging press, where the pressure exerted on the ram is, in many cases, 10,000 tons; but the speed of the ram being small—only a few inches per minute—the horse-power required to work such a press may be comparatively small.

These principles are of direct application to the gearing of cycles.

Example I.— Suppose two rear-driving bicycles each to have 28-inch driving-wheels geared to 56 inches; let the bicycles be equal in every respect, except that in one the numbers of teeth in the wheels on the crank-axle and hub are 16 and 8 respectively, while in the other the numbers are 18 and 9 respectively. When going along the same gradient at the same speed, the speeds of the chain relative to the machine are in the ratio of 8 to 9; consequently, the pulls on the chain will be in the ratio 9 to 8, that on the chain of the bicycle having the smaller wheels being the greater.

Example II.— Let two bicycles be the same in every respect, except that in one the cranks are 6 inches long, in the other 7 inches. When running along the same road at the same speed, the work done in overcoming the resistance will be the same in the two cases, and, therefore, the work done by the pressure of the feet on the pedals is the same in both cases. But the pedals' speeds are in the ratio of 6 to 7, therefore the average pressures to be applied to the pedals are in the ratio 7 to 6, the shorter crank requiring the greater pressure.

Example III.—Suppose two Safety bicycles to be equal in every respect, except that one is geared to 56 inches, the other to 63 inches. With equal riders, running along the same road at the same speed, the work done in both cases will be equal. But the distances moved over during one revolution of the crank are in the ratio of 56 to 63, that is, 8 to 9. The numbers of revolutions required to move over a given distance will therefore be in the ratio of the reciprocals of the distance—that is, 9 to 8. Consequently, the average pressures to be applied to the pedals in the two cases will be in the ratio of 8 to 9, the bicycle with the low gear requiring the smaller pressure on the pedals

The whole question of gear for a bicycle thus resolves itself into a question of what will suit best the convenience of the rider. Assuming that the maximum power of two riders is exactly the same, one may be able to develop his maximum power by a comparatively light pressure on the pedals and a high speed of revolution of the cranks, the other may develop his maximum power with a heavier pressure and a smaller speed of revolution of the crank-axle. The former would therefore do his best work on a lower geared machine than the latter. The question of length of crank depends also on the same general principles, different riders being able to develop their maximum powers on different lengths of crank.

The maximum power a rider can develop by pedalling a crank-axle is probably at low speeds proportional to the speed of driving ; at higher speeds the power does not increase so rapidly as the speed, and soon reaches an absolute maximum ; at still higher speeds the rapidity of pedalling is too great, and the power actually communicated to the crank-axle rapidly falls to zero. These variations of the power with the speed are graphically represented by the curves P and P_1 (fig. 228), P_1 being for longer sustained effort than P ; a certain speed of the crank-axle corresponding to a definite speed of the cycle on the path, so long as the gearing remains unaltered. The height of the ordinates will depend on the duration of the ride, and the maximum power $a\,b$ for an effort of short duration may be developed at a less axle speed than the maximum $a_1 b_1$ for a longer effort. By increasing the amount of gearing-up, the abscissæ of the curve would be all proportionately increased, while the ordinates remain as before. The best gearing-up possible for the rider will be such that the power curve of the machine intersects the rider's power curve at the highest point of the latter. From b, the highest point of the rider's power curve with a certain gearing-up, draw $b b^1$ to intersect at b^1 the power curve R of the machine, then the rider will develop the greatest speed $c b^1$ on the machine if the gearing-up be increased in the ratio of $c b$ to $c b^1$. If, as seems to the author most probable, the ratio $\dfrac{c b^1}{c b}$ for the shorter effort is greater than the ratio $\dfrac{c_1 b^1_1}{c_1 b_1}$ for the longer effort, the gearing-up should be greater for the former than for the latter. That is, to attain in all races his highest possible speed, the shorter the distance the higher should be the gear used by the rider.

Very little is known as to the maximum power that can be developed by a cyclist, no accurate experiments, to the author's knowledge, having been made. Rankine gives 4,350 foot-lbs. per minute as the average power of a man working eight hours raising his own weight up a staircase or ladder, and 17,200 foot-lbs. per minute in turning a winch for two minutes. Possibly racing cyclists of the front rank develop for short periods two-thirds of a horse-power—*i.e.* 22,000 foot-lbs. per minute. If this estimate and that of the air resistance (sec. 199) be correct, from figure 228 it is evident that a speed of 28 miles per hour could not be attained on a single bicycle, in still air, without pace-makers, even though the mechanism and the tyres were theoretically perfect. It should be noted that the conventional horse power, 33,000 foot-lbs. per minute, introduced by Watt, and employed by engineers as the unit of power, is considerably in excess of the average power of a draught horse.

206. **Variable-speed Gear.**—The maximum power of any rider is exerted at a particular speed of pedal and with a particular length of crank. The best results on all kinds and conditions of roads would probably be attained if the pedal could always be kept moving at this particular speed whatever the resistance ; the

gearing would then have to vary the distance travelled over per stroke of pedal, until equilibrium between the effort and resistance was established. An ideal variable gear would be one which could be altered continuously and automatically, so that when going uphill a low gear was in operation, and when going down hill a high gear. A number of two-speed gears have been used with success, and are described in chapter xxvii., but no continuously varying gear has been used for a cycle driving gear, though such a combination is well known in other branches of applied mechanics.

207. **Perpetual Motion.**—Many inventors and schemers do not appreciate the importance of the principle of 'what is gained in force, or effort, is lost in speed.' Since for a given power the effort or force can be increased indefinitely by suitable gearing, and likewise the speed, they appear to reason that by a suitably devised mechanism it may be possible to increase both together, and thus get more power from the machine than is put into it. A crank of variable length, the leverage being greater on the down than on the up-stroke, is a favourite device. The Simpson lever-chain is another device having the same object in view. The angular speeds of the crank-axle and back hub are inversely proportional to their numbers of teeth ; with an ordinary chain the distances of the lines of action from the centres are directly proportional to these numbers. By driving the back hub chain-wheel from pins on the chain links at a greater distance from the wheel centre, it was claimed that an increased leverage was obtained, and that the lever-chain was therefore greatly superior to the ordinary. It is possible, by using an algebraic fallacy which may easily escape the notice of anyone not sufficiently skilled in mathematics, to prove that $2 \times 2 = 5$; but though the human understanding may be deceived by the mechanical and algebraic paradoxes, in neither case are the laws of Nature altered or suspended. When once the doctrine of the 'conservation of energy' is thoroughly appreciated, plausible mechanical devices for *creating* energy will receive no more attention than they deserve.

208. **Downward Pressure.**—In all pedomotive cycles the *general* direction of the pressure exerted by the rider on the pedals

is vertically downwards. If P be the average vertical pressure and d the vertical distance between the highest and lowest points of the pedal's path, the work done by the rider per stroke of pedal is Pd. This is quite independent of the form of the pedal path.

209. **Cranks and Levers.**—If the pedals are fixed to the ends of cranks revolving uniformly, the vertical component of the pedal's motion will be a simple harmonic motion, and, neglecting ankle action, the motion of the rider's knee will be approximately simple harmonic motion along a circular arc.

When the crank is vertical, its direction coincides with that of the vertical pressure, and consequently no pressure, however great, will tend to drive the crank in either direction. The crank is then said to be on a 'dead-centre.' In steam-engines, and mechanisms in which the crank is employed to convert oscillating into circular motion, a fly-wheel is used to carry the crank over the dead-centre. In cycles, when speed has been got up, the whole mass of the machine and rider tends to continue the motion, and thus acts as a fly-wheel carrying the crank over the dead-centre, so that in riding at moderate or high speeds the existence of the dead-centre is hardly suspected. In riding at a very slow speed, however, the existence of the dead-centre is more manifest. If two cranks are placed at right angles to each other on the same shaft, while one is on the dead-centre the other is in the best position for exerting the downward effort, and there is no tendency of the shaft to stop.

In the above discussion we have assumed that the connecting-rod which drives the crank can only transmit a simple thrust or pull ; if, in addition to this, the connecting-rod can transmit a transverse effort there may be no dead-centre. In turning the handle of a winch by hand, the arm acts as a connecting-rod which can transmit, thrust, pull, and transverse effort, so that no dead-centre exists. In Fleming & Ferguson's marine-engine two cylinders are connected by piston-rods and intermediate links to two corners of a triangular connecting-rod, the third corner of which is at the crank ; with this arrangement there is no dead-centre, the single crank and triangular connecting-rod being in this respect equivalent to two cranks at right angles.

The existence of the dead-centre is supposed by some to be

a disadvantage inherent to the crank, but the efficiency of the mechanism is not in any way directly affected by it.

210. **Variable Leverage Cranks.**—One favourite notion of those inventors who have no clear and exact ideas of mechanical principles, is to have a crank of variable length arranged so that the leverage may be great during the down-stroke of the pedal and small during the up-stroke ; their idea evidently being to obtain all the mechanical advantages of a long crank, and yet only make the foot travel through a distance corresponding to a short crank. We have shown above that, presuming the pressure is vertical, the work done per stroke of pedal depends only on the pressure applied, and the vertical distance between the highest and lowest points of the pedal path ; the distance of the pedal from the centre of the crank-spindle having no *direct* influence whatever. The pedal path in most of the variable crank gears that have appeared from time to time is simply an epicycloidal curve which does not differ very much in shape from a circle, but which is placed nearer the front of the machine than an equal circle concentric with the crank-axle. Thus, the gear only accomplishes in a clumsy manner what could be done by a simple crank, having its axle placed a little further forward than that of the variable crank.

Let O (fig. 229) be the centre of a variable crank, and $c\,d$ the pedal path during the up-stroke. Let the length of the crank become greater, the path of the pedal during this extension being $d\,a$, and let the arc $a\,b$ be the pedal path during the down-stroke. The crank will then shorten, $b\,c$ being the pedal path. If the pressure be vertically downward, work will be done only while the pedal moves from a to b, and the angle of driving will be the small angle $a\,o\,b$. Thus while with a variable crank a greater turning effort may be exerted than with a fixed crank, the arc of action is correspondingly less.

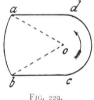

FIG. 229.

211. **Speed of Knee-joint during Pedalling.**—Regarding that part of the leg between the knee and the foot as a connecting-rod, that between the knee and the hip-joint as a lever vibrating about a fixed centre, the speed of the knee corresponding to a uniform speed of the pedal can easily be determined by the method of

section 33. Figure 23 is a polar curve showing the varying speed of the knee for different positions of the crank. From this curve it will be seen that on the down-stroke the maximum speed is attained when the crank is nearly horizontal, but on the up-stroke the maximum speed is not attained till the crank is nearly 45° above the horizontal. The speed then rapidly diminishes, and is nearly zero when the crank is vertical. The shorter the crank, in comparison with the rider's leg, the more closely does the motion of the knee approximate to simple harmonic motion ; with simple harmonic motion the polar curve is two circles.

In any gear in which a crank connected to the driving-wheel is used, the speed of the knee-joint will vary approximately as above described—*i.e.* it will gradually come to rest as it approaches its highest and lowest positions, then gradually increase in speed until a maximum is attained.

212. **Pedal-clutch Mechanism.**—Instead of cranks, clutch gears have been used for the driving mechanism. In these a cylindrical drum is placed at each side of the axle and runs freely on it. A long strap, with one end firmly fixed to the drum, is coiled once or twice round it, the other end is fastened to the pedal lever. When the pedal is depressed, the drum is automatically clutched rigidly to the shaft ; when the pressure is removed from the pedal, the pedal lever is raised by a spring and the drum released from the axle. One of the most successful clutch gears was that used on the 'Merlin' bicycles (fig. 176) and tricycles made by the Brixton Cycle Company.

The general advantage which a clutch gear was supposed to have as compared with a crank was that any length of stroke could be taken from a pat of an inch up to the full throw of the gear. However, even supposing that the clutches which lock the drums to the axle and the springs which lift the pedal levers are perfect in action, the gear has the serious defect that the down-stroke of the pedal begins quite suddenly and is performed at a constant speed ; thus the legs must have a considerable speed imparted suddenly to them. At moderate and high speeds this is a decided disadvantage as against the gradual motion required for the crank-geared cycle. There is the further serious practical disadvantage that no clutch that has been hitherto designed is perfectly instantaneous in its action of engaging and disengaging. When a clutch is used for continual driving, as in the clutch driving gears of some of the early tricycles, and where no great importance need be attached to the delay of a second or two in the action of the clutch gear, the case is quite different. Mr. Scott, in 'Cycling Art, Energy, and Locomotion,' has put the comparison between the crank gear and clutch gear for pedals in a nutshell thus : " In the crank-clutch cycle, as in other uses, the immediate solid grip is a matter of very little concern ; if a half turn of the parts takes place before clutching, it does very little harm, since it is so small a fraction of the entire number of revolutions to be made before the grip is released. But if a grip is to be taken at every down-stroke of the foot, as in a lever-clutch cycle, the least slip or lost motion is fatal."

These two objections are so weighty, that in spite of the immense advantage of providing a simple variable gear, pedal-clutch gears have never been much used.

213. **Diagrams of Crank Effort.**—Though the pressure on the pedal may be constant during the down-stroke, the effort tending to turn the crank will vary with the varying crank position. The actual pressure on the pedal may be resolved into two components, parallel and at right angles to the crank ; the former, the radial component, merely causes pressure on the bearing, and, since no motion takes place in its direction, no work is done by it ; the latter, the tangential component, constitutes the active effort tending to turn the crank. If $O C$ (fig. 230) be the crank in any position, and P the total pressure on the pedal, the radial and tangential components, R and T, are equal to the

FIG. 230.

projections of P respectively parallel to, and at right angles to the crank $O C$. If the tangential component T be set off along the corresponding crank direction, a *polar* curve of crank effort will be obtained.

If the pressure, P, be constant during the down-stroke, and be directed vertically downwards, the polar curve of crank effort will be a circle. Let p be the effort exerted by the rider at any instant at his knee-joint in the direction of the motion of the latter, let t be the corresponding tangential effort on the pedal, let s be a very small space moved through by the pedal, and s^1 the corresponding space moved through by the knee-joint. Then the work done at the knee-joint is $p\,s^1$, the corresponding work done at the pedal $t\,s$; these two must be equal, presuming there is no appreciable loss in the transmission. Therefore

$$t = \frac{s^1}{s}\,p \qquad . \qquad . \quad . \quad . \quad . \quad . \quad .(2)$$

But $\dfrac{s^1}{s}$ is the ratio of the speeds of the knee-joint and pedal respectively, and is represented by the intercept Df (fig. 21). If, therefore, the effort at the knee-joint be constant during the down-stroke of the pedal, figure 23 is the curve of crank effort as well as the speed curve of the knee.

If, starting from any position, the distance moved through by the pedal relative to the machine be set off along a horizontal

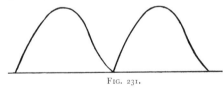

FIG. 231.

line, and the corresponding tangential effort on the crank be erected as an ordinate, a *rectangular* curve of crank effort will be obtained.

Corresponding to the circle as the polar curve of crank effort, the rectangular curve will be a *curve of sines.* Figure 231 shows the rectangular curve corresponding to the down-stroke polar curve in figure 23.

The area included between the base line and the rectangular curve of crank effort represents the amount of work done. The mean height of the rectangular curve therefore represents the mean tangential effort to be applied at the end of the crank in order to overcome the resistance of the cycle.

214. **Actual Pressure on Pedals.**—The actual pressure on the pedal during the motion of the cycle is not even approximately constant. Mr. R. P. Scott investigated the actual pressure on the pedal by means of an instrument which he calls the 'Cyclograph,' the description of which we take from 'Cycling

Art, Energy, and Locomotion.' "A frame, *A A* (fig. 232), is provided with means to attach it to the pedal of any machine. A table, *B*, supported by springs, *E E*, has a vertical movement through the frame *A A*, and car-

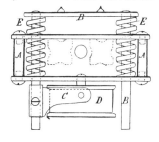

ries a marker, *C*. The frame carries a drum, *D*, containing within mechanism which causes it to revolve regularly upon its axis. The cylindrical surface of this drum *D* is wrapped with a slip of registering paper removable at will. When we wish to take the total foot pressure, the cyclograph is placed upon the pedal and the foot upon

FIG. 232.

the table. The drum having been wound and supplied with the registering slip, and the marker *C* with a pencil bearing against the slip, we are ready to throw the trigger and start the drum, by means of a string attached to the trigger, which is held by the rider so that he can start the apparatus at just such time as he desires a record of the pressure."

Figure 233 shows a cyclograph from a 52-inch 'Ordinary' on a race track, speed 18 miles per hour ; figure 234 that from the same

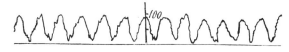

FIG. 233.

machine ascending a gradient 1 in 10, speed 4 miles an hour ; and figure 235 is from the same machine back-pedalling down a

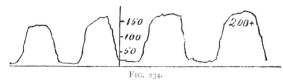

FIG. 234.

gradient 1 in 12. Figure 236 is from a rear-driver geared to 54 inches up a gradient 1 in 20 at a speed of 9 miles an hour ; and figure 237 is from the same machine going up a gradient of 1 in 7 at a speed of 10 miles per hour. The figures on the diagrams

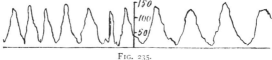

FIG. 235.

are lbs. pressure on the pedal. These curves and many others are discussed in the work above referred to.

These curves give no notion as to the varying *tangential* effort

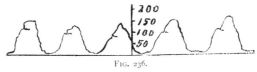

FIG. 236.

on the crank, which is, of course, of more importance than the total pressure. Mallard & Bardon's dynamometric pedal, referred

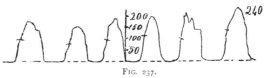

FIG. 237.

to by C. Bourlet, is an instrument in which the tangential component of the pedal pressure is measured and recorded.

215. **Pedalling.**—A vertical push during the down-stroke of the pedal is the most intense effort that the cyclist can communicate, and unfortunately it is the only one that many cyclists are capable of exerting. From Scott's cyclograph diagrams it will be seen that in only one case is the pedal pressure zero during the up-stroke. The first improvement, therefore, that should be made in pedalling is to lift the foot during the up-stroke, though not actually allowing it to get out of contact with the pedal. Toe-clips will be of advantage in acquiring this.

Next, just before the crank reaches its upper dead-centre a horizontal push should be exerted on the pedal, and before it reaches the lower dead-centre the pedal should be clawed backwards. These motions, if performed satisfactorily, will considerably extend the arc of driving.

Ankle Action.—To perform these motions satisfactorily the ankle must be bent inwards when the pedal is near the top, and fully extended when near the bottom. Figures 238, 239, and 240, from a booklet describing the 'Sunbeam' cycles issued by Mr. John Marston, show the positions of the ankle when the crank is at the top, the middle of the down-stroke, and the bottom respectively. The method of acquiring a good ankle action is well described in the 'Sunbeam' booklet and in Macredy's 'The Art and Pastime of Cycling.' Besides increasing

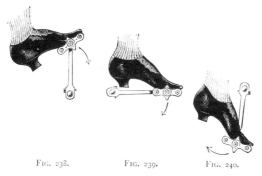

Fig. 238.   Fig. 239.   Fig. 240.

the arc of driving, ankle action has the further advantage of diminishing the extent of the motion of the leg. With a good ankle action the speed curves shown in figures 23, 501, and 511 may be considerably modified ; in fact, the addition of a fifth link (between the foot and ankle) to the kinematic chain in figure 22 makes the motion of the leg indeterminate.

If the shoe of the rider be fastened to the pedal an upward pull may be exerted, and the action of pedalling becomes more like that of turning a crank by hand, the arc of action being extended to the complete revolution. With *pulling pedals* more work is thrown on the flexor muscles of the legs, to the corresponding relief of the extensors.

216. **Manumotive Cycles.**—A few cycles, principally tricycles, have been designed to be driven by the action of the hand and arms.

Singers' 'Velociman' has been for a number of years the best example of this type of machine. Figure 241 shows an up-to-date example. The effort is applied by the hands to two long levers, which, by sliding joints in place of connecting-rods, drive cranks at opposite ends of an axle ; this axle is connected by chain

FIG. 241.

gearing to the balance gear on the driving-axle. The steering is done by the back pressing against a cushion supported at the end of a long steering bar.

217. Auxiliary Hand-power Mechanisms.—A number of cycles have been made from time to time with gearing operated by hand, having the intention of supplementing the effort communicated by the pedals. The idea of the inventors is that the greater the number of muscles concerned in the propulsion, the greater will be the speed, or a given speed will be obtained with less fatigue ; but though this may be true for extraordinary efforts of short duration, it is probably quite erroneous for long-continued efforts. Whatever set of muscles be employed to do work, a man has only one heart and one pair of lungs to perform the functions required of them. It is a matter of everyday experience that the cyclist can tax his heart and lungs to their utmost, using only pedals and cranks ; so that, unless inventors can provide a method of stimulating these organs to do more than they are at present capable of, it seems worse than useless to complicate the machine with auxiliary hand-power mechanism. Regarded as a motor, the human body may be compared to a number of engines deriving steam from one boiler, supplied with feed-water by one feed-pump. If one engine is capable of using all the steam generated in the boiler, no additional, but rather less, useful work will be obtained by setting additional engines running. It is a fact well known to engineers that a steam-engine works most economically when running under its heaviest load. One engine, therefore, will utilise the steam generated in the boiler more efficiently than several. The lungs may be compared to the furnace of the boiler, the blood to the feed-water, the heart to the feed-pump which circulates the feed-water, the muscles of the legs to an engine capable of utilising all the energy supplied by the combustion of the fuel in the furnace, the arms to a small engine. If the analogy can be pushed so far, less work will be got from the body by using both legs and arms simultaneously than by using the legs only ; and this quite independently of the frictional resistance of the additional mechanism.

The ' Road-sculler ' and ' Oarsman ' tricycles were designed so that the rider might exercise the muscles of his legs, back, chest, and arms, as in rowing. The speed attained was less than in the crank-driven tricycle, the mechanism being more complex and therefore less efficient, while from the foregoing discussion it seems probable that the rider, though using more muscles, actually developed less indicated power.

PART III
DETAILS

PART III

DETAILS

CHAPTER XXII

THE FRAME (DESCRIPTIVE)

218. **Frames in General.**—The frame of a bicycle forms practically a beam which carries a load—the weight of the rider—and is supported at two points, the wheel centres. In order to allow of steering, this beam is divided into two parts connected by a hinge joint—the steering-head. The two parts are sometimes referred to as the 'front-frame' and the 'rear-frame'; the front-frame of a 'Safety' including the front fork, head-tube, and the handle-bar. The rear-frame has assumed many forms, which will be discussed in some detail. In all bicycles that have attained to any degree of success the rear-frame has been the larger of the two; hence sometimes when 'the frame' is mentioned without any further qualification, the rear-frame is meant. It is usually evident from the context whether 'the frame' means the rear-frame or the complete frame.

219. **Frames of Front-drivers.**—The 'Ordinary' has the simplest, structurally, of all cycle frames, consisting of a single tube, called the backbone, forked at its lower extremity for the reception of the hind wheel, and hinged to the top of the fork carrying the front wheel. The frame of the 'Geared Ordinary' is the same as that of the 'Ordinary,' the distance between the seat and the top of the driving-wheel being too small to admit of bracing the structure. With the further reduction of the size of the driving-wheel, and the greater distance obtained between the saddle and top of the driving-wheel, it becomes possible to use a braced frame. Figure 242

shows a front-driving frame made by the Abingdon Works Company (Limited). Here the weight of the rider is taken up

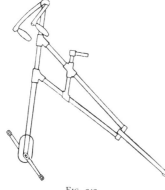

by the two straight tubes, each of which will be subject to bending-moment due to half the total weight.

Figure 132 shows one form of frame used by the Crypto Works Company (Limited), in their 'Bantam.' The bracing in this is more apparent than real, since the weight of the rider is transferred to the middle of a straight tube of very little less length than the total distance

FIG. 242.

between the wheel centres. This tube must, therefore, be made strong enough to resist the bending-moment.

FIG. 243.

Figure 243 shows the frame of the 'Bantamette,' made by the same company, and which can be ridden by a lady with skirts.

Here, of course, the backbone is subjected to bending-stresses, and a very strong tube must be used for it. Figure 291 shows a properly braced front-driving frame designed by the author, which is practically equivalent to a triangular truss. The short tube joining the steering-head to the seat-lug is made stout enough to resist the bending due to the saddle-pin attachment, while the seat-struts are subjected only to compression, and the lower stays to tension.

220. **Frames of Rear-drivers.**—The rear-driving chain-driven ' Safety ' introduced in 1885 is kinematically the same as the popular machine of the present day. The greatest difference between them lies in the design of the frame. So many designs of frame have been used that we can only notice a few general types here.

The original ' Humber ' frame (fig. 128) has a general resemblance to the present-day diamond-frame, though from a structural point of view, the want of a tube joining the saddle-pillar to the crank-axle makes it greatly different as regards strength.

Figure 244 shows the ' Pioneer ' dwarf Safety, made by H. J. Pausey, 1885. This is of the cross-frame type, and consists practically of two members, one joining the driving-wheel spindle to the steering-head, the other running from the saddle to the crank-axle. It will be noticed that the frame is not braced or stayed in any manner, so that the whole weight of the rider is transferred to the back-

FIG. 244.

bone. When driving, the pull of the chain tends to bring the crank-axle and driving-wheel centres nearer together, and there being no direct struts to resist this action, the frame is structurally weak. In this respect it is much worse than the ' Humber ' frame (fig. 128).

Figure 126 shows the ' Rover ' Safety made by Messrs. Starley & Sutton, 1885. The frame is of the open diamond type, the front fork is vertical, and the steering is not direct, but the handle-bar is mounted on a secondary spindle connected by short links to the front fork.

Figure 245 shows a Safety made by the Birmingham Small Arms and Metal Co., Limited. The principal difference between

FIG. 245.

this frame and that of figure 244 consists in the substitution of indirect for direct steering.

Figure 127 shows the 'Rover' Safety, made by Messrs. Starley

FIG. 246.

& Sutton in 1886. The frame is of the open diamond type, with curved tubes, and direct steering is used. The approximation to the present type of frame is closer than in any of the previous examples.

Figure 246 shows the 'Swift' Safety, made by the Coventry

FIG. 247.

Machinists Co., 1887. This frame is of the open diamond type ; the top and bottom tubes from the steering head are curved.

The first improvement on the elementary cross-frame (fig. 244) was to insert struts, or a lower fork, between the crank-bracket

FIG. 248.

and driving-wheel spindle (fig. 247), so that the pull of the chain could be properly resisted. Another improvement was to connect the steering-head and the top of the saddle-post by a light stay (fig. 248). In the 'Ivel' Safety of 1887 (fig. 249) a stay ran from the steering-head to the crank-bracket, but the chain-struts were omitted. The 'Humber' Safety of this period (fig. 250) had the crank-bracket stay and chain-struts. The 'Invincible' Safety, made by the Surrey Machinists Co. in 1888 (fig. 251), had, in

FIG. 249.

addition, a stay between the steering-head and top of saddle-post; while a later machine (fig. 252), by the same firm, had stay-rods from the driving-wheel spindle to the top of saddle-post. This

FIG. 250.

bicycle was made forkless, the wheel-spindles being supported only at one end; but in this respect the design is not to be recommended.

The frame of the 'Sparkbrook' Safety, 1887 (fig. 253), may be noticed. It is a kind of compromise between the cross-frame and the open diamond ; the crank-bracket and driving-wheel spindle

FIG. 251.

are directly connected, but the crank-axle is connected to a point about the middle of the upper tube of the frame. The bending-

FIG. 252.

moment, which attains nearly its maximum value at this point, is resisted by this single tube, which consequently must be rather heavy.

The frame of the 'Quadrant' bicycle (fig. 254) differs essentially from either the diamond- or cross-frame. In this bicycle the main frame is continued forward on each side of the steering-

FIG. 253.

wheel ; the spindle of the steering-wheel is not held in a fork, but its ends are mounted on cases which roll on curved guides or 'quadrants.' From each case a light coupling-rod gives connec-

FIG. 254.

tion to a double lever at the bottom of the steering-pillar. The frame in front of this steering-pillar consists practically of two tubes with no bracing, while the bracing of the rear portion is very imperfect. This arrangement for controlling the motion of the steering-wheel is the same as used in the 'Quadrant' tricycle.

The frame of the 'Rover' Safety of 1888 (fig. 255) shows a

FIG. 255.

great advance on any of the earlier frames above described. It may be described as a combination of the cross- and diamond-

FIG. 256.

frames. The main tube from the steering-head is joined on about the middle of the down-tube from the saddle to the crank-bracket, which thus may be considered to be supported at its ends and loaded in the middle, and must therefore be fairly heavy to resist

the bending-moment on it. Another weak point in the design is
the making of the top tube curved instead of straight.

The 'Referee' frame (fig. 256) was one of the earliest with

Fig. 257.

practically perfect bracing. The crank-bracket being kept as near
as possible to the rim of the driving-wheel, the diamond was
stiffened by a curved down-tube. A short vertical saddle-tube was

Fig. 258.

continued above the top tube, thus allowing the saddle and pin to
be turned forward or backward—a good point which has been
abandoned in later frames. Ball-socket steering was used.

Figure 257 shows the Safety made by Singer & Co., 1888, the frame of which differs very little essentially from that of figure 255.

FIG. 259.

Figure 258 shows the 'Singer' Safety of 1889, the frame of which differs considerably from all types hitherto described. The

FIG. 260.

remarks applied to the design of the frame in figure 255 may also be applied to this frame.

The 'Ormonde' (fig. 259) and the 'Mohawk' (fig. 260) frames may be noticed, the latter having the down-tube from saddle to crank-bracket in duplicate.

Figure 130 shows the 'Humber' Safety of 1889. This frame gives the first close approximation to the present almost universally used 'Humber' frame.

In 1890 the 'Humber' Safety, with extended wheel-base, was introduced. In this machine the distance between the crank-axle and the driving-wheel was increased, thereby increasing the distance between the points of contact of the two wheels with the ground. With this increased distance it was possible to join the seat-lug to the crank-bracket by a straight down-tube, thereby giving the well-known 'Humber' frame (fig. 131). The stem of the saddle-pin goes inside this tube, and a neater appearance is obtained thereby. This frame is not a perfectly braced structure,

FIG. 261

the introduction of a tube to form one of the diagonals of the diamond being necessary to convert it into a perfectly framed structure. This has been done in the 'Girder' Safety frame (fig. 296). With a well designed 'Humber' frame, however, the possible bending-moment on the tubes, due to the omission of the diagonal, is so small that it is practically not worth while to introduce the extra tube.

Quite recently a 'pyramid'-frame (fig. 261) has been introduced in America. It remains to be seen whether the excessive rake of the steering head, necessary with this design, will allow of the easy steering we are accustomed to with the diamond-frame.

Bamboo Frames.—From the discussion of the stresses on the frame (chap. xxiii.) it will be seen that when the frame is properly braced, and its members so arranged that the stresses on them are along their axes, the maximum tensile or compressive stress on the material is small. If a steel tube were made as light as possible, with merely sufficient sectional area to resist these principal stresses, it would be so thin that it would be unable to resist rough handling, and would speedily become indented locally. A lighter material with greater thickness, though of less strength, would resist these local forces better. The bamboo frame (fig. 262) is an effort in this direction, the

FIG. 262.

bamboo tubes being stronger locally than steel tubes of equal weight and external diameter.

Aluminium Frames.—From the extreme lightness of aluminium compared with iron or steel, many attempts have been made to employ it in cycle construction. The tenacity of the pure metal is, however, very low, and its ductility still lower, compared with steel ; while no alloy containing a large percentage of aluminium, and therefore very light, has been found to combine the strength and ductility necessary for it to compete favourably with steel. Of course, for parts which are not subjected to severe stresses it may probably be used with advantage.

221. **Frames of Ladies' Safeties.**—The design of the frame of a Ladies' Safety is more difficult than the design of the frame for a man's Safety. In the early Ladies' Safeties the frame was usually of the single tube type, and may be represented by the ' Rover' Ladies' Safety (fig. 263). The single tube from the crank-bracket to the steering-head is subjected to the entire bending-moment, and must therefore be of fairly large section. If the lady rider wears skirts, the top tube, as used in a man's bicycle, must be omitted ; and if a second bracing tube be introduced, it must be very low down. Figure 264 shows one of the usual

forms of Ladies' Safety, a tube being taken from the top of the steering-head to a point on the down-tube a few inches above the crank-bracket. By this arrangement, of course, the down-

FIG. 263.

tube is subjected to a bending stress, while the frame, as a whole, is weakest in the neighbourhood of the crank-axle. Since the bending-moment on the frame diminishes from the crank-axle

FIG. 264.

towards the front wheel centre, it is better to have the two tubes from the steering-head diverging (fig. 265) instead of converging as they approach the crank-axle ; the depth of the frame would then vary proportionally to the depth of the bending-moment diagram, and the bending stresses on the members of the frame

Fig. 265.

would be least. Such an attempt at bracing the frame of a Ladies' Safety, as is illustrated in figure 266, is useless, since at the point P the depth of the frame is zero, and the only improvement is that the bending at the point P is resisted by two tubes instead of one.

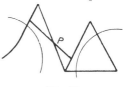

Fig. 266.

222. **Tandem Frames.** — A great variety of frames are in use at present, the processes of natural selection not having gone on for such a long time as is the case with frames for single machines. A frame (fig. 267), resembling that of the ordinary diamond-frame, with the addition of a central parallelogram, has been used. It will be noticed at once that the middle portion is not arranged to the best advantage for resisting shearing-force,

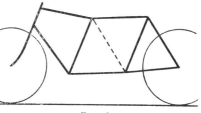

Fig. 267.

so that as regards strength, the middle portion of the frame is simply equivalent to two tubes lying side by side and subjected to bending.

Figure 268 shows a tandem frame, by the New Howe Machine Company, in which three tubes resist the bending on any vertical section ; and figure 269 shows a frame, by the Coventry Machin-

FIG. 268.

ists' Company, with the front seat arranged for a lady. Both these frames should be stronger, weight for weight, than that in figure 267, but they are not perfectly braced structures, and the bending-moment on the tubes will be considerable.

FIG. 269.

The addition of a diagonal to the central parallelogram, indicated by the dotted line (fig. 267), converts the frame into a braced structure, and the strength is proportionately increased.

The front quadrilateral of the frame (fig. 267) requires a diagonal to make the frame a perfectly braced structure, and, though riding along a level road, it is possible, by properly disposing the top and bottom tubes, to insure that there shall be no bending on them, it would seem advisable to provide against contingencies by inserting this diagonal in tandem frames. Such is done in the ' Thompson & James's' frame (fig. 140).

Figure 270 shows a tandem frame, made by Messrs. J. H. Brooks & Co., intended to take a lady on either the front or back seat. On the side of the machine on which the chain is placed, instead of a single fork-tube two tubes are used, one above and one below the chain, and both lying in the plane of the chain.

FIG. 270.

Thus the lower part of the frame constitutes a beam to resist the bending-moment, and the upper portions are used merely to support the saddles.

Figure 271 shows a tandem frame also intended to take a lady on either the front or back seat, designed by the author. The frame is dropped below the axle—the lower part is, in fact, a braced structure of exactly the same nature as that in figure 267. The crank-bracket is held by a pair of levers, the lower ends of which are hinged on the pin at the lower point of junction of the frame tubes. The upper ends can be clamped in position on the tubes which form the chain-struts. The driving-wheel spindle is thus permanently fastened to the frame, and therefore remains always in track. A single screw is used to adjust the crank-bracket, on releasing the top clamping screws of the supporting levers. Although this is a reversion to the hanging crank-bracket, it may

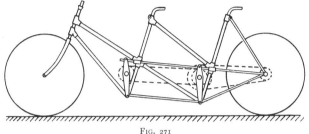

FIG. 271

be pointed out that it is connected rigidly to the frame at four points, and may therefore be depended upon not to work loose.

223. **Tricycle Frames.**—In the early tricycles Y-shaped

FIG. 272.

frames for front-driving rear-steerers and loop-frames for front-steerers were usually employed, while in side-drivers, such as the Coventry Rotary, the frame was T-shaped, the top of the T being in a longitudinal direction. The frame of the 'Cripper' tricycle (fig. 150) was also T-shaped, the top of the T forming a bridge supporting the axle, and the vertical branch of the ⊤ running forward from the middle of the axle to the steering-head and supporting the crank-axle and seat. These frames were almost entirely unbraced, and their strength depended only on the diameter and thickness of the tubes.

The diamond-frame for tricycles, on the same general lines as the diamond-frame used in bicycles, marks a great improvement

FIG. 273.

in this respect, figure 272 illustrating the frame of the 'Ivel' tricycle. Figure 152 illustrates a tricycle with diamond-frame made by the Premier Cycle Company (Limited). It will be noticed that the frame is the same as that for a bicycle, with the addition of a bridge and four brackets supporting the axle. The next improvement, as regards the proper bracing of the frame, is the spreading of the seat-struts, so that they run towards the ends of the bridge, the bending stresses on the axle-bridge being slightly reduced by this arrangement. Figure 273 shows a tricycle with this arrangement, by Messrs. Singer & Co., but with the front part dropped, so that it may be ridden by a lady.

In nearly all modern tricycles the driving-axle has been supported by four bearings, two near the chain-wheel, so that the pull of the chain can be resisted as directly as possible, and two at the outer ends, as close to the driving-wheels as possible, each bearing being held in a bracket from the bridge. The whole arrangement of driving-axle, bridge, and brackets looks rather complex, while the chain-struts are subjected to the same severe bending stresses as those of a bicycle (sec. 238). A great improvement is Starley's combined bridge and axle, the bridge being a tube concentric with, and outside, the axle. Figure 274 is

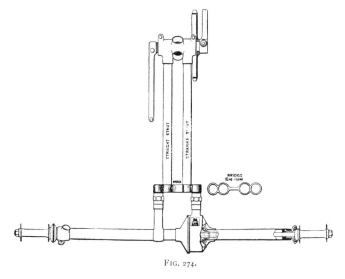

FIG. 274.

a plan showing the arrangement of the combined bridge and axle, crank-bracket and chain-struts, as made by the Abingdon Company, the lug for the seat-strut being shown at the left-hand side of the figure. The driving cog-wheel on the axle is inclosed in an enlarged portion of the outer tube, in which two spaces are made to allow the chain to pass out and in. The chain adjustment is got by lengthening or shortening the chain-struts by means of a right- and left-handed screw, the hexagonal tubular nuts being clearly shown in the figure, an arrangement patented by the author in 1889. Messrs. Starley Brothers have still further improved the tricycle frame by making the chain gear exactly central, so that the design of the frame is simplified by using only one tube as a chain-strut, while the bending stresses caused by the pull of the chain are eliminated. The crank-bracket (fig. 153) is enlarged at the middle to form a box encircling a chain-wheel, two openings being provided for the chain to pass in and out, as in the axle-box, while three lugs are made on the outside of the box to take the three frame tubes. The narrowest possible tread is thus obtained. This, in the author's opinion, marks the highest level attained in the design of frame for a double-rear-driving tricycle.

224. **Spring-frames.**—In the days of solid tyres many attempts were made to support the frame of a bicycle or tricycle on springs,

FIG. 275.

so that joltings due to the inequalities of the road might not be transmitted to the frame. The universal adoption of pneumatic tyres has led to the almost total abandonment of spring-frames. The springs should be so disposed that the distances between saddle, handle-bar, and crank-axle remain unaltered. In Harrington's vibration check, which was typical of a number of appliances that could be fitted to the non-driving wheel of a bicycle, the wheel spindle was not fixed direct to the fork ends, but to a pair of short arms fastened to the fork ends and controlled by springs. This allowed the front wheel to move over an obstruction without communicating all the vertical motion to the frame.

Figure 275 shows the ' British Star ' spring-frame Safety, made by Messrs. Guest & Barrow, the rear wheel being isolated by a powerful spring from the part of the frame carrying the saddle.

FIG. 276.

Figure 276 shows the 'Cremorne' spring-frame Safety, the springs being introduced near the spindle of the driving-wheel. In

FIG. 277.

both these spring-frames the lower fork is jointed to the frame at or near the crank-bracket. In the 'Elland' spring-frame, made by Cooper, Kitchen & Co., the spring was introduced just below the seat lug, and the lower fork was hinged to the crank-bracket.

Figure 277 shows the 'Whippet' spring-frame bicycle, the most popular of the type, in which the driving-wheel and steering-wheel forks are carried in a rigid frame. The portion of the frame carrying saddle, crank-axle, and handle-bar is suspended from the main frame by a powerful spiral spring and a system of jointed bars, the arrangements of which are shown clearly in the drawing.

Figure 278 shows a spring-frame bicycle now made by Messrs. Humber & Co. (Limited), the rear fork being jointed to the frame

FIG. 278.

at the crank-bracket, and the front wheel being suspended by a pair of anti-vibrators. The rear fork is subjected to a considerable bending moment, and must therefore be made heavy; in this respect the design is inferior to many of the earlier spring-frames.

225. **The Front-frame.**—The front-frames of bicycles and tricycles show great uniformity in general design, any differences between those of different makers being in the details. The front-frame consists of the fork sides, which are now usually tubes of oval section tapered towards the wheel centre; the fork crown; the steering-tube; and the handle-bar. The double-plate fork

crown (fig. 279) is now almost universally used. The fork sides are brazed to the crown-plates. In the best work it is usual to insert a liner at the foot of the steering-tube, shown projecting in figure 279, so as to strengthen the part. The fork tubes are again

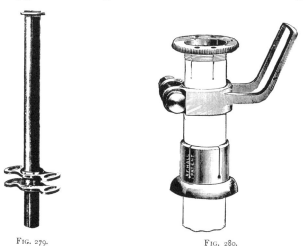

FIG. 279.

FIG. 280.

strengthened by a liner, the top of which also forms a convenient finish for the fork crown.

The top adjustment cone (fig. 280) of the ball-head is slipped on near the top of the steering-tube, the latter having been previously placed in position through the ball-head. The end of the tube is screwed, to provide the necessary adjustment of the cone. The end of the tube and the tubular portion of the adjustment cone are slit, and the handle-bar having been fixed in the necessary position, the three are clamped together by a split ring and tightening screw. The lamp-bracket is often made a projection from this tightening ring, as shown in figure 280. Figure 280 illustrates the ball-head made by the Cycle Components Manufacturing Company (Limited), and shows the adjustment cone, lamp-bracket, and the adjusting nut apart on the steering-tube, while figure 281 shows the ball-head complete, with the parts assembled in position.

The steering-head of the 'Falcon' bicycle, made by the Yost Manufacturing Company, Toledo, U.S.A., differs from that by the Cycle Components Company, in having the adjusting cone screwed on the steering-tube. The top bearing cup is butted

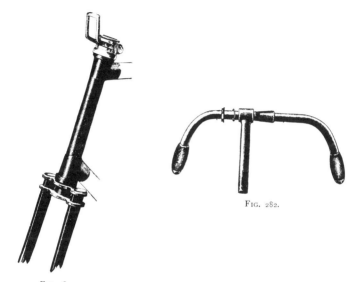

FIG. 282.

FIG. 281.

against the frame tube of the steering-head, the top lug embracing, and being brazed to, both.

It is becoming more usual not only to make the handle-pillar adjustable in the steering-tube, but also to make the handle-bar adjustable in the socket at the head of the handle-pillar. One of the best designs for accomplishing this (fig. 282) is that used in the 'Dayton' bicycles, made by the Davis Sewing Machine Company, Dayton, U.S.A A conical surface is formed on the handle-bar, and fits a corresponding surface on the socket at the top of the handle-pillar. A short portion of the handle-bar is screwed on the exterior ; the handle-bar is fixed in the required position by screwing up a thin nut, and thus wedging the two conical surfaces together.

The handle-bar is most severely stressed at its junction with the handle-pillar. A handle-bar liner (fig. 283), as made by the

FIG. 283.

Cycle Components Manufacturing Company, is used to strengthen it.

The front-frame of the usual type of the present day is essentially a beam subjected to bending, showing in this respect no improvement on that of the earliest tricycles. In tandems and

FIG. 284.

triplets many accidents have resulted from the collapse of the front frame; additional strength is therefore desirable for this, generally the weakest part of these machines. This can be attained by making the steering-tube and fork sides of sufficient section, and also by entirely new designs for the front-frame.

The '*Referee*' front-frame (fig. 284) is made by continuing the fork sides up through the crown to the top of, and outside, the steering-head. The maximum bending-moment is thus resisted by the two fork sides and the steering-tube, instead of only by the latter, as in the ordinary pattern. There should be no possibility, therefore, of the steering-tube giving way.

Duplex fork sides (fig. 285) continued to the top of the steering-head are a still further improvement in the same direction; the forward tube acting as a strut, the rear tube as a tie, though both are subjected, in some degree, to direct bending.

A braced front frame has been made in the 'Furore' tandem.

In a bicycle designed by the author in 1888, with the object of eliminating, as far as possible, all bending stresses on the frame tubes, the steering-head was behind the steering-wheel, and consequently the latter could be supported by a trussed frame. The complete frame (fig. 286) had a general resemblance to a

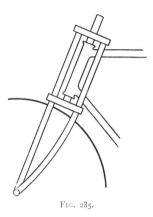

FIG. 285.

queen-post roof-truss. This design answered all requirements as regards lightness and strength ; but as an expert rider experienced

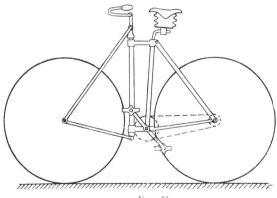

FIG. 286.

almost as much difficulty in learning to ride this machine as a novice in learning to ride one of the usual type, it was abandoned.

In tandems steered by the rear rider, the front-frame could be immensely strengthened by taking stay-tubes from the ends of

the front wheel spindle to a double-armed lever near the bottom
of the steering-pillar (fig. 287). These stay-tubes would have to
be bent, as shown in plan (fig. 288), to clear the steering-wheel

FIG. 287.

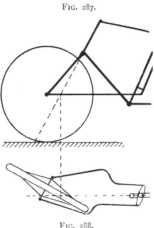

FIG. 288.

when turning a corner. The front fork would then be made
straight, as it would act as a strut, while the stress would be
almost entirely removed from the steering-tube.

CHAPTER XXIII

THE FRAME (STRESSES)

226. **Frames of Front-drivers**.—*a b c* (fig. 290) shows the bending-moment curve on the frame of an 'Ordinary' (fig. 289) due to the weight, *W*, of the rider. The weight of the rider does not come on the backbone at one point, but, by the arrange

ment of the saddle spring, at two points, p_1 and p_2. If perpendiculars be drawn from p_1 and p_2 to meet this curve at d_1 and d_2, d_1 *b* d_2 will be the bending-moment curve of the spring, and the remainder *a* d_1 d_2 *c* of the original bending-moment curve will give the bending-moment on the backbone and rear fork. The bending-moment on the backbone is greatest near the head, and diminishes towards the lower end. Accordingly, the backbones of 'Ordinaries' were invariably tapered.

In the 'Ordinary' the front fork was vertical, and consequently the bending-moment on the frame just

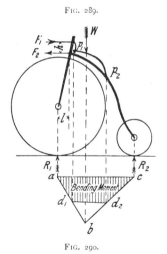

Fig. 289.

Fig. 290.

at the steering-head was zero. In the 'Rational,' however, the front fork was sloped, and a bending-moment, $R_1 l$, existed at the steering-head, *l* being the horizontal distance of the steering-head behind the front wheel centre. There would consequently be two equal forces, F_1 and F_2, acting at right angles to the head, at the top and bottom centres, such that

$$F_1 h = R_1 l \quad . \quad . \quad . \quad . \quad . \quad . \quad . \quad . \quad (1)$$

where h is the distance between the top and bottom centres. The greater the distance h, the smaller would be the force F_1, and thus a long head might be expected to work more smoothly and easily than a short one.

It is easily seen that the side pressure on the steering-head of a 'Safety' bicycle or 'Cripper' tricycle arises in exactly the same way. The arrangement of the frame of a 'Safety' is such as permits of a much longer steering-head than can be used in the 'Ordinary,' and as the pressure on the front wheel is much less than in the 'Ordinary,' the side pressure on the steering-head is also very much smaller.

Example I.— In a 'Geared Ordinary' the rake of the front fork is 4 inches, the distance between the top and bottom rows of balls in the head is 3 inches, the weight of the rider is 150 lbs., and the saddle is so placed that two-thirds of the weight rest on the front wheel ; find the side pressure on the ball-head. The reaction, R_1 (fig. 289), in this case is

$$\frac{2}{3} \times 150 = 100 \text{ lbs.,}$$

the bending-moment at the head is

$$100 \times 4 = 400 \text{ inch-lbs.,}$$

the force F is therefore

$$\frac{400}{3} = 133\cdot3 \text{ lbs.}$$

Example II.—In a 'Safety' bicycle the ball steering-head is $9\frac{1}{2}$ inches long, the horizontal distance of the middle of the head behind the front wheel centre is 9 inches, the rider weighs 150 lbs., and one-fourth of his weight rests on the front wheel ; find the side pressure on the steering-head. In this case, the reaction R_1 is

$$\frac{1}{4} \times 150 = 37\cdot5 \text{ lbs.,}$$

the bending-moment on the head is

$$37 \cdot 5 \times 9 = 337 \cdot 5 \text{ inch-lbs.,}$$

the side pressure on the steering-head is therefore

$$\frac{337 \cdot 5}{9\frac{1}{2}} = 35 \cdot 5 \text{ lbs.}$$

Example III.—In Example I., if the point p_1 (fig. 289) of maximum bending-moment on the backbone be 6 inches behind the front wheel centre, find the necessary section of the backbone. The bending-moment at p_1 will be

$$100 \times 6 = 600 \text{ inch-lbs.}$$

If the maximum tensile stress be taken 15,000 lbs. per sq. in., substituting in the formula $M = Zf$, we get

$$Z = \frac{600}{15,000} = \cdot 04 \text{ inch-units.}$$

From Table IV., p. 112, a tube $1\frac{1}{4}$ in. diameter, number 20 W.G. would be sufficient.

The section necessary at any other point of the backbone may be found in the same manner, but where the total weight of the part is small, it is usual to make the section at which the straining action is greatest sufficiently strong, and if the section be kept uniform throughout, all the other parts will have an excess of strength. In the backbone of an 'Ordinary,' the section should not diminish by the tapering so quickly as the bending-moment.

Example IV.—In a front-driver in which the load and the relative position of the wheel centres and seat are as shown in figure 291, the stresses can be easily calculated as follows :

Taking the moment about the centre of the rear wheel, we get $R_1 \times 36 = (120 \times 23) + (30 \times 36)$; therefore

$$R_1 = 106 \cdot 7 \text{ lbs., } R_2 = 43 \cdot 3 \text{ lbs.}$$

The maximum bending-moment (fig. 293) on the frame occurs on the section passing through the seat, and is—

$$M = 43 \cdot 3 \times 23 = 996 \text{ inch-lbs.}$$

If the frame simply consists of a backbone formed by a tube $1\frac{1}{4}$ in. diameter, 18 W.G., we find from Table IV., p. 112—

$$Z = \cdot 0525, \text{ and } f = \frac{M}{Z} = \frac{996}{\cdot 0525} = 19,000 \text{ lbs. per sq. in.}$$

Braced Frame for Front-driver.—A simple form of braced frame is shown diagrammatically in figure 291. The short tube

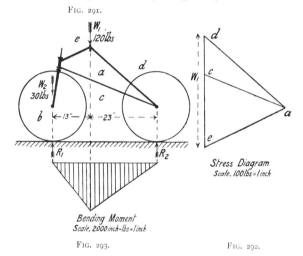

FIG. 291.

Stress Diagram
Scale, 100 lbs = 1 inch

Bending Moment
Scale, 2000 inch-lbs = 1 inch

FIG. 293.   FIG. 292.

from the steering-head to the seat-lug is made stout enough to resist the bending-moment due to the saddle adjustment, while the seat-struts are subjected to pure compression, and the lower stays to pure tension. Figure 292 shows the stress-diagram for this braced structure ; from which the thrust on the seat-struts, $a\,d$, is 116 lbs. If they are made of two tubes $\frac{5}{8}$ in. diameter, 28 W.G., from Table IV., p. 112, $A = 2 \times \cdot 0284 = \cdot 057$, and the compressive stress is

$$f = \frac{116}{\cdot 057} = 2,000 \text{ lbs. per sq. in.}$$

The pull on the lower stay, $a\,c$, is 95 lbs., and if the stays are made of tubes of the same diameter and thickness as the seat-struts, the tensile stress will be correspondingly low.

The greatest stress on the top-tube will be due to the saddle adjustment. With the horizontal branch of the ∟ pin 3 in. long, a total horizontal adjustment of 6 in. can be provided ; the maximum bending-moment on the tube will be

$$M = 150 \times 3 = 450 \text{ inch-lbs.}$$

If the tube be 1 in. diameter, 20 W.G., $Z = \cdot 0253$, and the maximum stress on the tube will be

$$f = \frac{M}{Z} = \frac{450}{\cdot 0253} = 17,800 \text{ lbs. per sq. in.}$$

227. **Rear-driving Safety Frame.**—The bending-moment curve for the frame (taken as a whole) of any bicycle is independent of the shape of the frame, and depends on the weight to be carried, and the position of the mass-centre relative to the centres of the wheels. The actual stresses on the individual members of the frame, however, depend on the shape of the frame. The frame of a rear-driving chain-driven Safety must provide supports for the wheel spindle at W, the crank-axle at C, the saddle at S, and the steering-head at H_1 and H_2 (fig. 296). Two principal types of frame are to be distinguished. In the *cross-frame* the point H_1 and H_2 were very close together, and the opposite corners of the quadrilateral $W C H S$ were united by tubes. In the *diamond-frame*, adjacent corners of the pentagon, $H_1 H_2 C W S$, are united by tubes. In both the diamond- and the cross-frames *additional* ties and struts are inserted, the object being to make the frame as rigid as possible, and, of course, to reduce its weight to the lowest possible consistent with strength The weight of a bar necessary to resist a given straining action depends on the magnitude of the straining action and its direction in relation to the bar. We have already seen that a force applied transversely to a bar and causing bending, to be effectually resisted, will require a bar of much greater sectional area than if the force be either direct compression or tension. It may thus be laid down as a guiding principle in designing cycle frames, that *the various members should be so disposed, that as far as possible they are all subjected to direct compression or tension, but not bending,* It follows from this that each member should be attached to

other members at only two points. A bar on which forces can only be applied at two points—its ends—cannot possibly be subjected to bending. If a third 'support' be added, the possibility and probability of subjecting the bar to bending arises. The early Safety frames and some Tandem frames of the present day show many examples of bad design, a long tube often being 'supported' at one or more intermediate points, the result being to throw a transverse strain on it, and therefore weaken, instead of strengthen the structure.

228. **The Ideal Braced Safety Frame.**—In a Safety rear-frame the external forces act on five points ; the weight of the

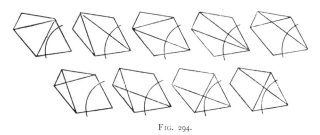

FIG. 294.

rider being applied partially at the saddle S, and at the crank-axle C, the reaction of the back wheel at W, and the pressure on the steering-head at H_1 and H_2 (fig. 296). If the five points

FIG. 295.

H_1, H_2, S, W, and C be joined by bars or tubes dividing the space into triangles, the frame will be perfectly braced, and there will be only direct tensile or compressive stresses on the bars. Figure 296 shows the arrangement used in the 'Girder' Safety frame, while figure 294 shows a number of possible arrangements of perfectly braced rear-frames. Figure 295 shows another perfectly braced rear-frame, in which the lower back fork between the crank-axle and driving wheel spindle is omitted. Comparing this with figure 320, it will be seen that a very narrow tread may be obtained with this frame, a saving of at least the diameter of one lower fork tube being effected.

Example.—The rider weighs 150 lbs., 30 lbs. of which is applied at the crank-axle, the remainder, 120 lbs., at the saddle S.

From the given dimensions of the machine (fig. 296), the reaction, R_1 and R_2, on the front and back wheels can be calculated.
Considering the complete frame, and with the dimensions
marked, taking moments about the centre of the front wheel, we
have

$$(30 \times 23) + (120 \times 33) = R_2 \times 42$$

from which

$$R_2 = 110 \cdot 7 \text{ lbs.}$$

and

$$R_1 = 150 - 110 \cdot 7 = 39 \cdot 3 \text{ lbs.}$$

Consider now the front-frame consisting of the fork and steering-
tube ; it is acted on by three forces, the reaction, R_1, upwards, and

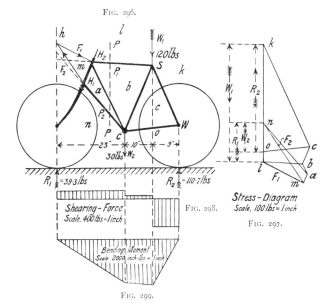

FIG. 296.

FIG. 298.

Stress - Diagram
Scale, 100 lbs = 1 inch

FIG. 297.

Shearing - Force
Scale, 400 lbs = 1 inch

Bending Moment
Scale 2000 inch-lbs = 1 inch

FIG. 299.

the reactions between it and the rear-frame, at H_1 and H_2.
These three forces must therefore (see sec. 45) pass through the
same point. With the ordinary arrangement of ball-head the
vertical pressure of the front portion of the frame acts on the rear

portion at H_1, and the resultant force at H_2 may be assumed at right-angles to the steering-head, H_1 H_2. Therefore, from H_2 draw H_2 h, intersecting the vertical through the front wheel centre at h ; join H_1 h, giving the direction of the force at H_1.

The stress-diagram (fig. 297) can now be constructed by the method of section 83. From this diagram it is easily seen that the lower back fork is in tension, and also the tube from the lower end of the steering-head ; the other members of the frame are in compression. By measurement from figure 297 the thrust k c along the seat-struts is 117 lbs. ; that along the down-tube, b c, about 19 lbs. ; along the top-tube, b l, about 41 lbs. ; along the steering-head, a m, 12 lbs. ; along the diagonal, a b, from the top of the steering-head to the crank-bracket, 7 lbs. ; and the tension on the lower back fork, c o, is 58 lbs. ; on the bottom-tube, a n, 64 lbs. These values will, of course, vary slightly according to the dimensions of the frame.

Taking a working stress of 10,000 lbs. per square inch, the sectional area of the two tubes constituting the seat-struts would only require to be $\frac{117}{10000} = $ ·0117 sq. in., provided the diameter was great enough to resist buckling. The section of the other tubes would be correspondingly small. We shall see later, however, that many of the frame tubes are subjected to bending, and that the maximum stresses due to such bending are much greater than those considered above.

229. **Humber Diamond-frame.**—The force on the tube between the spaces a and b (fig. 296) is very small, and by careful design may be made zero (see sec. 230). In the 'Humber' diamond-frame this tube is suppressed, and thus if the frame tubes were connected by pin-joints at C, S, H_1 and H_2, the frame would be no longer able to retain its form when subjected to the applied forces. That the frame actually retains its shape is due to the fact that the frame joints are rigid, and that the individual members are capable of resistance to bending. If all the frame joints are rigid, the stress in any member cannot be determined by statical methods, but the elasticity and deformation of the parts under stress must be considered. However, by making certain assumptions, results which may be approximately true can be obtained by statical methods.

Example I.—Suppose the tubes $C\ H_1$ and $H_1\ H_2$ to be fastened together at H_1, so as to form one rigid structure, which we may consider connected by pin-joints to the frame at C and H_2, the other joints of the frame being pin-joints. The distance of H_1 from the axis of the suppressed member, $H_2\ C$, is 6 in. The bending-moment at H_1 on the part $C\ H_1\ H_2$ is therefore 7 lbs. × 6 in. = 42 inch-lbs.; the bending-moment diminishes towards zero at C and H. If the tube $H_1\ C$ be 1 in. diameter, and a working stress of 10,000 lbs. per sq. in. be allowed, substituting in formula (3), section 101, we get

$$10{,}000 = \frac{64}{A} + \frac{4 \times 42}{A \times 1}$$

or,
$$A = {\cdot}0232 \text{ sq. in.}$$

Consulting Table IV., p. 112, we see that the thinnest there given, No. 32 W.G., has an excess of strength. If the tube $H_2\ C$ had been retained, the sectional area of the tube $H_1\ C$ need only have been

$$A = \frac{64}{10{,}000} = {\cdot}0064 \text{ sq. in.}$$

Example II.—Suppose the tubes $H_1\ H_2$ and $H_2\ S$ rigidly fastened at H_2, and connected at H_1 and S by a pin-joint to the rest of the frame. The part $H_1\ H_2\ S$ may then be considered as a beam carrying a load of 7 lbs. at H_2. The perpendicular distances of H_1 and S from the line of action of this force are 6 in. and 18 in. respectively. The bending-moment at H_2 is therefore (see sec. 87)

$$\frac{7 \times 6 \times 18}{24} = 31{\cdot}5 \text{ inch-lbs.}$$

The compressions along $H_1\ H_2$ and $H_2\ S$ will be increased by the components of the original force, 7 lbs., along the suppressed bar at $H_2\ C$. Similarly, the forces along $C\ H_1$ and $C\ S$ will be altered. The thickness of tube required can be worked out as in Example I. above.

230. Diamond-frame, with no Bending on the Frame Tubes.
—Consider the complete frame divided by a plane, $P\ P$ (fig. 296) immediately behind the steering-head, $H_1\ H_2$. If the frame

tubes $H_2 S$ and $H_1 C$ are not subjected to bending, the forces exerted by the front part of the frame on the rear part must be in the direction of the tubes. The forces acting at P_1 and P_2 on the front portion of the frame are equal in magnitude but reversed in direction. The only other external force acting on the front portion of the frame is the reaction, R_1, of the wheel on the spindle; these three forces are in equilibrium, and therefore must all pass through the same point. The condition then that the tubes in a diamond-frame should be subjected to no bending is that *the axes of the top- and bottom-tubes should, if produced, intersect at a point vertically over the front wheel centre.* This is very nearly the case in figure 296; if it was exactly, the force $b a$ along $H_2 C$ would be zero.

231. **Open Diamond-frame.**—The open diamond-frame (figs. 127, 246), though in external appearance very like the 'Humber' frame, is subjected to totally different straining actions. In the first place, if the joints at C, H_1, H_2, S and W be pin-joints, under the action of the forces the frame would at once collapse. Practically, the top-tube, $H_2 S$, and the seat-struts, $S W$, form one rigid beam, which must be strong enough to resist the bending-moment due to the load at S. Taking the same dimensions as in figure 296, the distances of H_2 and W from the line of action of the load at S are 21 in. and 9 in. respectively, and the weight of the rider 150 lbs., the bending-moment at S will be

$$\frac{150 \times 21 \times 9}{30} = 945 \text{ inch-lbs.}$$

Taking $f = 20,000$ lbs. per sq. in., and substituting in the formula $M = Z f$,

$$Z = \frac{945}{20,000} = \cdot 0472 \text{ in.}^3$$

From Table IV., p. 112, a tube 1 in. diam., 14 W.G., would be required; or a tube $1\frac{1}{8}$ diam., 17 W.G.

When the rider is going easily his whole weight rests on the saddle, and must be supported by the beam $H_2 S W$. On the other hand, when working hard, as in riding up a steep hill, his whole weight may be applied to the pedals, and, therefore, will come on the frame at C. The bottom-tube, $H_1 C$, and the lower back fork, $C\ W$, must be rigidly jointed together at C, and form a beam sufficiently strong to resist this bending-moment. Taking the same dimensions as in figure 296, the bending-moment at C is

$$\frac{150 \times 15 \times 19}{34} = 1{,}257 \text{ inch-lbs.}$$

$$Z = \frac{M}{f} = \frac{1{,}257}{20{,}000} = \cdot0628 \text{ in.}$$

and a 1 in. tube, 11 W.G., or a $1\frac{1}{8}$ in. tube, 14 W.G., would be required. A comparison of these results with those of sections 228-9 will reveal the weakness of the open diamond-frame.

232. **Cross-frame.**—In the cross-frame (fig. 300), the forked backbone a runs straight from the steering-head to the back

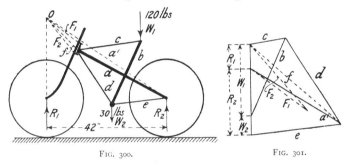

FIG. 300.   FIG. 301.

wheel spindle. The crank-bracket and seat-lug are connected by the down-tube b. The earlier cross-frames consisted only of these two members ; but in the later ones, bottom stays e, from the crank-bracket to the back wheel spindle, and stays c and d, from the steering-head to the crank-bracket and seat-lug respectively, were added. With this arrangement, the down-tube b is subjected to thrust, the stays c, d, and e to tension, and the backbone a to thrust, combined with bending, due to the forces acting on the steering-head.

The stress-diagram can be drawn as follows : The loads W_1 and W_2 at the seat-lug and crank-axle respectively being given, the reactions R_1 and R_2 at the wheel centres can be calculated, as in section 89. At the back wheel spindle three forces act. The magnitude and direction of R_2 are known, the pull of the lower fork e is in the direction of its axis, and is, therefore, known, but the thrust on the backbone a is not along its axis ; its direction is not known, and we cannot, therefore, begin the stress-diagram with the forces acting at this point. If the initial tension on the stays c and d be assumed such that there is no straining action at the junction of the down-tube b and the backbone a, the former will be subjected to a thrust along its axis, and, therefore, the directions of the three forces acting at the seat-lug are known. Draw W_1 (fig. 301), equal and parallel to the weight acting at the seat-lug, and complete the force-triangle $W_1 \, b \, c$, the sides being parallel to the correspondingly lettered members of the frame (fig. 300). Proceeding to the crank-bracket, the forces acting are the weight W_2, the thrust of the down-tube b, and the pulls of the stays d and e, the directions of which are known ; the force-polygon $b \, W_2 \, e \, d$ can therefore be drawn. Proceeding now to the back wheel spindle, the pull of the stay e and the upward reaction R_2 are known. Setting off R_2 (fig. 301) from the extremity of the side e, and joining the other extremities of e and R_2, the direction and magnitude a^1 of the thrust on the backbone are obtained. This thrust does not act along the axis of the backbone, which is, therefore, in addition to a thrust along its axis, subjected to a bending-moment varying from zero at the back wheel spindle to a maximum $P \, l$ at the steering centre, P being the thrust measured from figure 301, and l the distance of a perpendicular on the line of action of the thrust from the centre of the backbone at the steering-head.

The forces acting on the backbone a are : the pull of the lower stay e and the reaction R_2, having the resultant a^1 ; the pulls c and d, with resultant f acting, of course, through the point of intersection of c and d ; the pressures F_1 and F_2 on the steering centres. Since R_1, F_1 and F_2 reversed are the only forces acting on the front-frame, the resultant of F_1 and F_2 must be R_1. Thus the forces acting on the backbone a are equivalent to the three forces R_1, a^1 and f, which must, therefore, all pass through the point O. A check on the accuracy of the stress-diagram is thus obtained.

The point O being determined, by joining it to the top and bottom steering centres the directions of F_1 and F_2 are obtained, and their magnitudes by drawing the force-triangle $R_1 F_1 F_2$ (fig. 301).

233. **Frame of Ladies' Safety.**—Figure 302 is the frame-diagram of a Ladies' Safety. Having given the loads at the seat-lug and crank-axle, the reaction of the wheels can be calculated, as in section 89. The stresses on the seat-struts a and the back fork e may be found in exactly the same manner as for the diamond frame ; viz. by drawing the force-triangle $R_2 a e$ (fig. 303). The best arrangement of the two tubes c and d from the steering-head

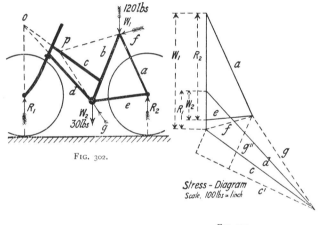

FIG. 302.

Stress - Diagram
Scale, 100 lbs = 1 inch

FIG. 303.

will be when their axes intersect at a point o vertically over the front wheel centre. Assuming that the forces on these two tubes are parallel to their axes, they are determined by drawing the force-triangle $R_1 d c$ for the three forces (fig. 303) acting on the front-frame of the machine. The down-tube b is acted on by three forces—(1) The thrust of the tube c ; (2) the resultant f of the thrust a and load W_1 acting at the seat-lug ; (3) the resultant g of the pulls d and e and the load W_2, acting at the crank-axle. These three forces form the force-triangle $c f g$ (fig. 303). A check on the accuracy of the work is obtained by the fact that the forces f and g (fig. 302) must intersect at a point p on the axis of the tube c.

From figure 303, the thrust of c on the down-tube b is 145 lbs., while its component c^1 at right angles to b is 142 lbs. The down-tube b is 22 in. long, divided by c into two segments, 7 and 15 in. The greatest bending-moment on it is therefore

$$\frac{142 \times 7 \times 15}{22} = 667 \text{ inch-lbs.}$$

The lower part of the down-tube is subjected to a thrust g'' (the component of the force g parallel to the down-tube) of 62 lbs.

234. **Curved Tubes.**—About the years 1890–2 a great number of Safety frames were made with the individual tubes curved in various ways. The curving of the tubes was made on æsthetic grounds, and possibly the tremendous increase in the maximum stress due to this curving was not appreciated. The maximum stress on a curved tube subjected to compression or tension at its ends is discussed in section 101.

Example.—Let a tube be bent so that its middle point is a distance equal to four diameters from the straight line joining its ends ; the maximum stress is, by (4) section 101,

$$\frac{P}{A} + \frac{4 \times P \times 4\,d}{A\,d} = \frac{17\,P}{A}$$

The tube would, therefore, have to be seventeen times the sectional area of a straight tube subjected to the same thrust.

235. **Influence of Saddle Adjustment.**—So far we have considered the mass-centre of the rider to be vertically over the point S (fig. 296) ; this is approximately the case when the saddle is fixed direct, as in some racing machines, to the top of the back fork without the intervention of an adjustable saddle-pin. But when an adjustable saddle-pin is used, the weight on the saddle acts at a distance l, usually from 3 to 6 inches behind the point S. The weight W acting at this distance is equivalent to an equal weight acting at S, together with a couple $W\,l$, producing a bending-moment $W_1\,l$ at S. From the manner in which the adjustable pillar is usually fixed at S, this bending-moment is generally transmitted to the down-tube $S\,C$, which must therefore be stout enough to resist it. Since, however, the

joint at S is rigid, a small part of this bending-moment may be transmitted to the tube $S H_2$.

Example.—Taking $l = 5$ in., and $W = 120$ lbs., as in figure 296, $M = 120 \times 5 = 600$ inch-lbs. Taking the direct thrust along $S C$ 19 lbs., as in section 228, and a working stress 20,000 lbs. per sq. in., the diameter of the tube 1 in., and substituting in (3) section 101,

$$20,000 = \frac{19}{A} + \frac{4 \times 600}{A},$$

we find $A = \cdot 1210$ sq. in. From Table IV., p. 113, the tube would require to be 18 W.G.

If the saddle were placed vertically over S, and no bending came on the tube $S C$, its sectional area would be $\dfrac{19}{20,000}$ $= \cdot 00095$ sq. in. : one-hundredth part of the section necessary with the saddle placed sideways from S.

This example is typical of the enormous additions which must be made to the weights of the tubes of a frame when the forces do not act exactly along the axes of the tubes.

By the use of the T-shaped seat-pillar (fig. 304) the range of horizontal adjustment can be increased without increasing unduly

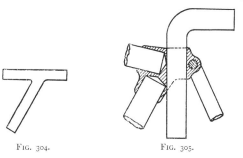

F<small>IG</small>. 304.   F<small>IG</small>. 305.

the stresses due to bending ; or, for a given range of horizontal adjustment, the bending stresses are lower with the T-shaped than with the L-shaped seat-pillar. The adjustment got by an L pin with horizontal and vertical limbs is much better (figs. 256, 260), since by turning the L pin round, the saddle may be adjusted either before or behind the seat-lug S. Thus, for a horizontal adjustment of 6 inches, the maximum eccentricity *l* need not be greater than 3 inches. By combining such an **L** pin with the 'Humber' frame it would be possible to further reduce the stresses on the frame. Figure 305 shows a seat-lug for this purpose, designed by the author.

For racing machines of the very lightest type possible the best result is obtained by fastening the saddle direct at *S*; this, of course, does not allow of any adjustment, and a machine that might suit one rider admirably might not be suitable for others.

236. **Influence of Chain Adjustment.** — In chain-driven Safeties it is found that chains stretch, no matter how carefully made,

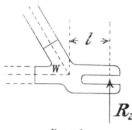

FIG. 306

after being some time in use, and therefore some provision must be made for taking up the slack. This is usually done by making the distance between the centres of the crank-axle and the driving-wheel adjustable. Figure 306 shows a common faulty design for the stamping at the driving-wheel spindle. The force R_2 (fig. 306) is equivalent to an equal force acting at *W* plus a bending-moment $R_2 l$, which is transmitted to the upper and lower forks.

Example. — If the distance *l* (fig. 306) be $\frac{1}{2}$ inch, and R_2 be, as in the example of section 228, 111 lbs., the bending-moment transmitted to the forks is 55·5 inch-lbs. The direct compression along the seat-struts *S W* is 41 lbs. (fig. 297), that along the lower fork *W C* is 58 lbs. Taking 10,000 lbs. per sq. in. as the working stress of the material, a section of ·0041 sq. in. for the top fork, and ·0058 sq. in. for the bottom fork would be sufficient, if they were not subjected to bending. Suppose the bending to be taken up entirely by the lower fork, made of two tubes $\frac{3}{4}$ in. diameter, and of total area *A* ; then, when subjected to bending as well as to direct compression or tension, the maximum stress to which they are subjected is given by the formula (3) of section 101. Substituting the above numerical values of *f*, *P*, *M*, and *d*, we have

$$10,000 = \frac{58}{A} + \frac{4 \times 55.5}{A \times .75},$$

or $A = \cdot 035$ sq. in. Thus the maximum stress, when the force R_2 is applied $\frac{1}{2}$ in. from the point of intersection of the forks, is nearly seven times as great as when it is applied in the best possible position.

Swinging Back Fork.—The centre of the driving-wheel may be always at the intersection of the top and bottom forks if the top fork be attached to the frame at S by a bolt—the bolt used for tightening the saddle-pin may serve for this purpose—and its lower ends be provided with eye-holes for the reception of the spindle of the driving-wheel. This arrangement, now almost universal, was first designed by the author in 1889. The lower fork may then be provided with a plain straight slot (fig. 307), along which the wheel spindle can be pulled by an adjusting screw. During a small adjustment of this nature the angle

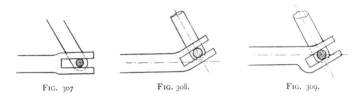

Fig. 307.   Fig. 308.   Fig. 309.

$S\,C\,W$ (fig. 296) will vary slightly, so that theoretically the lower forks should be attached by a pin-joint at C; but practically the elasticity of the tubes is sufficiently great to allow of the use of a rigid joint. In the form of this adjustment used by Messrs. Humber & Co. the slot is not in the direction of the axis of the lower fork, but curved (fig. 308) to a circular arc struck from S as centre. In this way there is no tendency to alter the angle $S\,C\,W$ (fig. 296); but the fact that the centre of the wheel spindle does not always lie on the axis of the lower fork $C\,W$ throws a combined tension and bending on it, the bending-moment being equal to $P\,l$, where P is the direct force on the lower fork parallel to its axis, and l is the distance of the centre of the wheel from the axis of the lower fork.

Example.—Let $l = \dfrac{d}{2}$, that is, the centre of the wheel is just on a line with the top of the tube of the lower fork.

Substituting in (3), section 101,

$$f = \frac{P}{A} + {}_4\frac{P\,d}{2\,A\,d} = {}_3\frac{P}{A}.$$

If the centre of the wheel lay on the axis of the tube the stress would be uniformly distributed and equal to $\frac{P}{A}$. Thus the stress on the lower fork is increased by the eccentricity of the force acting on it to three times its value with no eccentricity.

A better arrangement for the slot would be that shown in figure 309, where the spindle is adjusted equally above and below the centre line of the lower fork tubes.

237. **Influence of Pedal Pressure.**—In the foregoing discussion we have considered the forces to be applied in the middle plane of the bicycle frame ; but the rider applies pressure on the pedals at a considerable distance from the middle plane, and thus additional transverse straining actions are introduced. We now proceed to investigate the corresponding stresses.

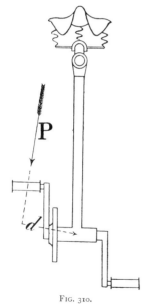

FIG. 310.

Figure 310 is a transverse sectional elevation, showing the pedals, cranks, crank-bracket, saddle, and down-tube, to the foot of which the crank-bracket is fixed. A force, P, applied to the pedal will cause a bending of the crank-bracket, which will be transmitted to the down-tube. From the arrangement of the lower fork in relation to the crank-bracket it is seen that practically none of this bending-moment can be transmitted to the lower fork. A small portion of the bending-moment may be transmitted to the bottom-tube $H_1\,C$ (fig. 296), but the greater part will be transmitted to the down-tube.

The magnitude of the bending-moment is $P\,d$, d being the length of the perpendicular from the centre of the crank-bracket on to the line of action of the force P. The narrower the tread the smaller will be d, and therefore the smaller the transverse stresses on the frame. Hence the importance of obtaining a narrow tread.

Example I.—Let P be 150 lbs., the tread, *i.e.* the distance from centre to centre of the pedals measured parallel to the crank-axle, 11 inches. The distance d may be taken equal to half the tread, *i.e.* $5\frac{1}{2}$ inches. The bending-moment on the foot of the down-tube will be $150 \times 5\frac{1}{2} = 825$ inch-lbs. Let the down-tube be $1\frac{1}{8}$ in. diameter, 20 W.G. From Table IV., p. 113, the Z for the section is ·0325 in.³; substituting these values in the formula $M = Zf$, we get

$$825 = ·0325 f,$$

i.e. $f = 25,400$ lbs. per sq. in.

Compared with the result on page 310, got by considering the forces applied in the middle plane of the frame, it is seen that on the down-tube the stress due to transverse bending is the most important.

In the double diamond-frame the single down-tube of figure 310 is replaced by the two tubes which support the crank-bracket near its ends (fig. 311). This gives a much better construction to resist the transverse stresses, but unfortunately it is not so neat in appearance as the single tube, and its use has been practically abandoned of recent years. The maximum stress produced in this case can be easily calculated and may be illustrated by an example.

Example II.—Let the tubes be $\frac{3}{4}$ in. diameter, 20 W.G., with their ends 3 in. apart. Under the action of the force P the nearer tube will

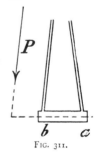

FIG. 311.

be subjected to tension, the further one to compression. Taking moments about a (fig. 311), the point of attachment of the further tube to the crank-bracket, we get

$$7\,P = 3\,F, \ i.e. \ F = \frac{7}{3}\,P = 350 \text{ lbs.}$$

The sectional area of the tube, from Table IV., p. 113, is ·0807 sq. in., therefore the stress on the tube is

$$f = \frac{350}{·0807} = 4,336 \text{ lbs. per sq. in.}$$

238. **Influence of Pull of Chain on Chain-struts.**—In riding easily along a level road, when very little effort is being exerted, the tension on the chain is small, and the stresses on the lower back fork, or chain-struts, will be as discussed in section 228. But when considerable effort is being exerted on the pedal, the tension on the chain is considerable, and since the chain does not lie in the middle plane of the frame, additional straining actions are introduced.

The tension F on the chain (fig. 312) can be easily found by considering the single rigid body formed by the pedal-pins, cranks,

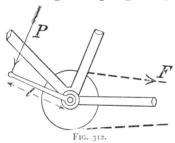

FIG. 312.

crank-axle, and chain-wheel. This rigid body is free to turn about the geometric axis of the crank-bracket, and it is acted on by three forces: P the pressure on the pedal-pin, the pull F of the chain, and the reaction of the balls on the crank-axle. Taking moments about the geometric-axis of the axle, that of the latter force vanishes, and we get $P l = F r$; l being the length of crank, and r the radius of the sprocket-wheel.

Example I.—Let $P = 150$ lbs., $l = 6\frac{1}{2}$ in., and let the chain-wheel have eighteen teeth to fit the 'Humber' pattern chain. From Table XV., p. 405, we get $r = 2·87$ in.; therefore

$$2·87 \, F = 6\frac{1}{2} P, \text{ and } F = \frac{6·5}{2·87} \times 150 = 340 \text{ lbs}$$

Figure 313 is a plan showing the crank-bracket and the lower back fork. Consider the horizontal components of the forces acting on the crank-bracket. If the pressure on the pedals be vertical there will be no horizontal component due to it, and we are left with the force F_1, the horizontal component of the pull on the chain. This is equilibrated by the horizontal components

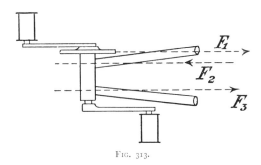

FIG. 313.

of the reactions at the bearings, therefore the crank-bracket i acted on by the forces at the bearings and the forces F_2 and F_3 exerted by the ends of the lower back fork.

Example II.—Let the chain-line be $2\frac{1}{8}$ in. (*i.e.* the distance from the centre of the chain to the centre of the fork is $2\frac{1}{8}$ in.), let the fork ends at the crank-bracket be 3 in. apart ; then the forces to be considered are shown in figure 314.

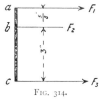

FIG. 314.

To find the pull F_3, take moments about b.

$$\tfrac{5}{8} F_1 = 3 F_3, \text{ therefore } F_3 = \frac{\cdot 625}{3} 340 = 70 \cdot 8 \text{ lbs.}$$

To find the compression F_2 on the near tube, take moments about c, and we get $3\frac{5}{8} F = 3 F_2$,

$$\therefore F_2 = \frac{3 \cdot 625}{3} 340 = 410 \text{ lbs.}$$

Comparing with the results of section 228, the compression on the near tube of the fork is much greater than the tension due to the weight of the rider applied centrally. The near tube, therefore, must be designed to resist compression.

Bending of Chain-struts.—The sides of the lower back

fork, the crank-bracket, and the back wheel spindle together form an open quadrilateral without bracing (fig. 315), $a\,b$ and $d\,c$ being the fork sides, $b\,c$ the crank-bracket, and $a\,d$ the

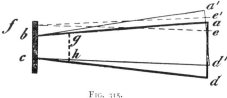

FIG. 315.

wheel spindle. If this structure be acted on by forces there will be in general a tendency to distortion. The tension of the chain, ef, is such that the points e and f on the spindle and crank-bracket respectively, in the plane of the chain, tend to approach each other, and the structure is distorted into the position $a^1\,b\,c\,d^1$. The action can be easily imagined by supposing the structure jointed at the corners a, b, c, and d. In the actual structure this distortion is only resisted by the stiffness of the joints, and the bending-moment can be investigated thus : Consider the equilibrium of the wheel spindle $a\,d$ (fig. 316). It is acted on by the pressure

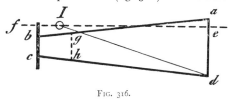

FIG. 316.

on the two bearings (the resultant of which is the pull of the chain ef), and the forces exerted by the fork sides at the points a and d respectively. The spindle is acted on by three forces, which, being in equilibrium, must all pass through the same point I, lying somewhere on the line $e\,f$ produced indefinitely in both directions. Thus, the force acting on the fork side $a\,b$ is in the direction $a\,I$. If F_2 be the magnitude of this force, and l_2 the perpendicular from b on $a\,I$, there will be a bending-moment $M_2 = F_2\,l_2$. With a similar notation for the fork side $c\,d$, there will be a bending-moment $M_3 = F_3\,l_3$ at the point c of the fork side. If I coincided with the point of intersection of $a\,b$ and $e\,f$, M_2 would be zero, but M_3 would be very great.

Example III.—We might assume such a position for I that M_2

and M_3 would be approximately equal. In this position, taking the data of the above examples, l_2 would be about $\frac{3}{8}$ in., and

$$M_2 = F_2\, l_2 = 410 \times \tfrac{3}{8} = 154 \text{ inch-lbs.}$$

If the lower fork be of round tube $\frac{3}{4}$ in. diameter, 20 W.G., we find, from Table IV., p 113, $Z = \cdot 0137$ in.³ Substituting in the formula $M = Zf$ we get

$$f = \frac{154}{\cdot 0137} = 11,200 \text{ lbs. per sq. in.}$$

The sectional area of the tube, from Table IV., p. 113, is $\cdot 0807$ sq. in. ; therefore the stress due to the compression of 410 lbs. is

$$f = \frac{410}{\cdot 0807} = 5,080 \text{ lbs. per sq. in.}$$

Thus, the maximum compressive stress on the fork at b is

$$f = 11,200 + \cdot 5,080 = 16,280 \text{ lbs. per sq. in.}$$

Section of Chain-struts.—The tubes from which the chain-struts are made are usually of round section. Occasionally tubes of oval section are used, the larger diameter of the tube being placed vertically. Since the plane of bending of the fork tubes is horizontal, if the fullest advantage be desired the oval tubes should be placed with the larger diameter horizontal. But the horizontal diameter is limited by the necessity of getting a narrow tread. For a given sectional area (or weight) of tube, and horizontal diameter, the bending resistance will be greater, the greater the vertical diameter and the less the thickness of the tube ; since a larger proportion of the material will be at the greatest distance from the neutral axis.

FIG. 317.

D tubes have also been used with the flat side vertical. The discussion in section 98 has shown a difference of about one per cent. in favour of the **D** tube consisting of a semicircle and its diameter. Square or rectangular tubes have not been used to any great extent for the chain-struts, but the discussion in section 99 shows that for equal sectional area and diameter they are much stronger than the round tube. If the horizontal diameter b be constant, and the vertical unlimited, a rectangular tube with great vertical diameter will be stronger, weight for weight, than a square tube ; Z approaching the value $\dfrac{A\,b}{2}$, corresponding to the whole sectional area being concentrated at the two sides parallel to the neutral axis, the other two sides being indefinitely thin.

A still more economical section for the lower fork tubes would be a hollow rectangle, the vertical sides being longer and thicker than the horizontal. This might be attained by drawing a thin rectangular tube of uniform thickness, and brazing two flat strips on its wider faces (fig. 317).

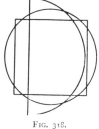

Figure 318 shows the sections of round, D, and square tubes of equal perimeter.

Lower Fork with Bridge Bracket.—If the cog-wheel on the crank-axle be placed between the two bearings, as in the ' Ormonde ' bicycle (fig. 259), the chain will run between the two lower fork sides (fig. 319), and there

FIG. 318.

will be no bending stresses on the fork tubes due to the pull of the chain. The objection to this arrangement is that the tread must

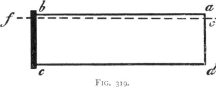

FIG. 319.

be increased considerably in order to have a bearing outside the cogwheel on the crank-axle.

' *Referee* ' *Lower Back Fork.*—In the ' Referee ' bicycle the bending on the fork

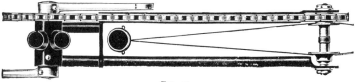

FIG. 320.

sides is eliminated by an ingenious arrangement shown in figure

320. The fork tubes are parallel to the plane of the chain, and instead of running forward to the crank-bracket, they end at an intermediate bridge piece connected to the crank-bracket by two parallel tubes lying closer together than the fork sides. If the end lugs to which the ends of the driving-wheel spindle are fastened were central with the tubes, the bending stresses might be entirely confined to the bridge piece.

239. **Tandem Bicycle Frames.**—The design of tandem frames is much more difficult than that of single bicycle frames, since

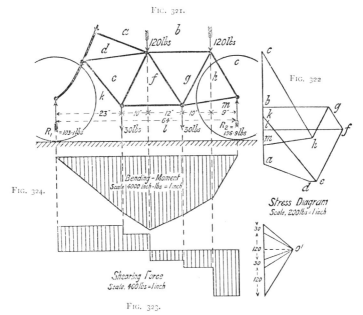

FIG. 321.

FIG. 322.

FIG. 324.

Bending - Moment
Scale 4000 inch-lbs = 1 inch

Stress Diagram
Scale, 200 lbs = 1 inch

Shearing Force
Scale, 400 lbs = 1 inch

FIG. 323.

the weight to be carried is double, and the span of the present popular type of tandem from centre to centre of wheels is also greater than that of the single machine. The maximum bending-moment on a tandem frame is therefore much greater than that on the frame of a single bicycle. If, however, one of the riders overhangs the wheel centre, the maximum bending-moment on the frame may actually be less than on that of the single machine.

In the 'Rucker' tandem bicycle (fig. 135) each rider was nearly vertically over the centre of his driving-wheel, and the maximum bending-moment on the backbone was not very great.

Example.—With 120 lbs. applied at the rear saddle, with an overhang of 10 in., the maximum bending-moment was

$$M = 120 \times 10 = 1,200 \text{ inch-lbs.}$$

If the maximum stress on the backbone had not to exceed 20,000 lbs.,

$$Z = \frac{1,200}{20,000} = \cdot 060 \text{ in.}^3$$

A tube $1\frac{1}{4}$ in. diameter, 17 W.G., would have been sufficient. It may be noticed that with one rider overhanging the wheel-base the bending-moment changes sign about the middle of the frame—*i.e.* if the backbone were originally straight, while carrying the riders the rear portion would be slightly bent with its centre of curvature downwards, the front portion with its centre of curvature upwards.

Figure 321 shows the frame of a rear-driving tandem Safety with both riders between the wheel centres, similar to that of figure 296. The top- and bottom-tubes of the forward portion of the frame should be arranged so that they intersect on the vertical through the front wheel centre, but in order to make the stress-diagram more general they are not so shown in figure 321. Figure 322 shows the stress-diagram, regarding the frame as a plane structure, while figures 323 and 324 are the shearing-force and bending-moment diagrams respectively.

The scale of the stress-diagram, 200 lbs. to an inch, has been chosen half that of the stress-diagram of the single machine (fig. 297), and a few comparisons may be made. The thrusts on the top-tubes, *a d* and *b g*, of the tandem are respectively about $2\frac{1}{2}$ and $3\frac{1}{2}$ times that on the top-tube of the single machine. The pull on the front bottom-tube, *e k*, of the tandem is about $2\frac{1}{2}$ times that for the single. The thrusts on the diagonal, *f g*, and the front down-tube, *e f*, are respectively $3\frac{1}{2}$ and $6\frac{1}{2}$ times that on the down-tube of the single machine ; while the *pull* on the rear down-tube, *g h*, of the tandem is about four times the thrust on the down-tube of the single machine. The pulls on the lower back fork, *m h*, and on the middle chain-struts, $f\,l$, are respectively about 2 and $3\frac{1}{2}$ times that on the lower back fork of the single machine.

In making the above comparisons it should be remembered that the single frame (fig. 296) is relatively higher than the tandem frame (fig. 321) illustrated. If the latter were higher, the stresses on its members would be less.

The scale of the bending-moment diagram (fig. 324) is 4,000 inch-lbs. to an inch, twice that for the single machine (fig. 299). The maximum bending-moment is more than three times that for the single machine.

A glance at the shearing-force diagrams (figs. 298 and 323) shows that on a vertical section passing through the rear down-tube of the tandem the shear is negative, while at the down-tube of the single machine the shear is positive. Hence the stress on the rear down-tube is tensile. This can also be shown by a glance at the force-polygon, $l\,m\,h\,g\,f$ (fig. 322), for the five forces acting at the rear crank-bracket (fig. 321); the force $h\,g$, being directed away from the bracket, indicates a pull on the down-tube.

The thrust on the tube $d\,e$ is small, and vanishes when the front top- and bottom-tubes intersect vertically above the front wheel centre. The thrust on the diagonal tube, $f\,g$, of the middle parallelogram is 60 lbs., smaller than the thrust or pull on any other member of the frame. This explains why the frame with open parallelogram (fig. 267) and those with no proper diagonal bracing are able to stand for any time under the loads to which they are subjected.

The maximum stresses on the members of the frame due to the vertical loads will be largely increased by the stresses due to the pull of the chain, the thrust of the pedals, and the seat adjustment, as already discussed. The magnitudes of these stresses will be proportionately greater in the tandem than in the single frame.

Tandem frames may be also subjected to considerable twisting strains. If the front and rear riders sit on opposite sides of the central plane of the machine, the middle part will be subjected to torsion. This torsion can be best resisted by one tube of large diameter ; *no arrangement of bracing in a plane can strengthen a tandem frame against twisting.*

240. **Stresses on Tricycle Frames.**—Nearly all the frames of early tricycles were unbraced, and their strength depended entirely on the thickness and diameter of the tubes used, one exception being that of the 'Coventry Rotary' (fig. 144), the side portion of which formed practically a triangular truss ; another, that of the 'Invincible,' a central portion of which was fairly well braced.

In the early 'Cripper' tricycles the frame was usually of T shape, and consisted of a *bridge* supporting the axle, and a *backbone* supporting the saddle and crank-axle. With the usual arrangements of wheels and saddle, about three-eighths of the weight of the rider rested on each driving-wheel. The strength of the bridge can easily be calculated thus :

Example I.—If the weight transferred to the middle of the bridge be 120 lbs., the track of the wheels be 30 inches apart, the middle of the bridge is subjected to a bending-moment

$$M = \frac{Wl}{4} = \frac{120 \times 30}{4} = 900 \text{ inch-lbs.}$$

If the maximum stress be 20,000 lbs. per sq. in.,

$$Z = \frac{M}{f} = \frac{900}{20,000} = \cdot 045 \text{ in.}^3$$

A tube $1\frac{1}{8}$ in. diameter, 17 W.G. (see Table IV.), will be sufficient.

In calculating the strength of the backbone the worst case will be when the total weight of the rider is applied at the crank-axle. Taking the relative distances as in the Safety bicycle (fig. 296),

$$M = \frac{150 \times 23 \times 19}{42} = 1,560 \text{ inch-lbs.}$$

$$Z = \frac{1,560}{20,000} = \cdot 078 \text{ in.}^3$$

A tube $1\frac{3}{8}$ in. diameter, 16 W.G., will be sufficient.

With frames made on the same general design as that of the Safety bicycle the stresses will be calculated as already discussed for the bicycle, the only important additional part being the bridge supporting the axle. Its strength may be calculated as in the above example. The stresses on the axle-bridge are diminished by taking the seat-struts to the outer end of the bridge, as in

'Starley's' frame (fig. 153), and in the 'Singer' frame (fig. 273). Figure 325 is plan and elevation of the rear portion of 'Starley's' frame. At the outer end of the bridge, which in this case is a tube concentric with the axle, there are three forces acting, which, however, do not all lie in the same plane. These are the re action of the wheel R, the thrust T along the seat-strut, and the pull A along the bridge. These forces in the plan are denoted by the corresponding small letters, and in the elevation

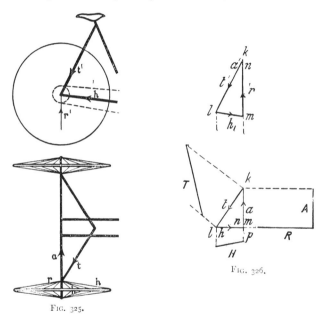

FIG. 326.

FIG. 325.

with the corresponding small letter with a dash (^{1}) attached. If a force, H, parallel to the chain-struts be applied at the end of the axle, the four forces H, A, R, and T will be in equilibrium, and may be represented by four successive edges of a tetrahedron respectively parallel to the direction of the forces. The plan and elevation of this tetrahedron, $k\,l\,m\,n$, is drawn in figure 326, the length of the side corresponding to the force R being drawn to any convenient scale. The magnitudes of the forces H, A, and

T can be measured off from the true lengths of the corresponding edges of the tetrahedron. These are shown in the plan.

Example II.—Suppose $R = 60$ lbs., and the direction of the tubes is such that $H = 30$ lbs., the resultant of the three forces R A, and T is equal and opposite to H; thus the bridge is subjected to a bending in the plane of the chain-strut. If the distance from the end to the centre of the bridge be 14 in.,

$$M = 30 \times 14 = 420 \text{ inch-lbs.}$$

If the bridge be 1 in. diameter and 20 W.G.,

$$Z = \cdot 0253 \text{ in.}^3, \text{ and } f = \frac{420}{\cdot 0253} = 16{,}600 \text{ lbs. per sq. in.}$$

The axle will also be subjected to a bending-moment in a vertical plane, due to the fact that the centre of the wheel is overhung some distance from the end of the bridge. If the overhang be 3 in., the bending-moment $= 60 \times 3 = 180$ inch-lbs., a smaller value than that found above.

241. The Front-frame.—The front-frame (fig. 327) is acted on by three forces—the reaction R_1 of the front wheel on its spindle,

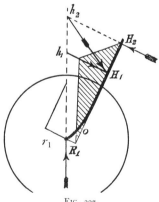

and the reactions H_1 and H_2 of the ball-head on the steering-tube. Since the front-frame is in equilibrium under the action of these forces, they must all pass through a point h, situated somewhere on the vertical line passing through the wheel centre. If we assume that the direction of the force H_2 at the upper bearing of the ball-head is at right angles to the head, the point h will be determined; the magnitudes of H_1 and H_2 can then easily be determined by an application of the triangle of forces.

FIG. 327.

Let r_1, h_1, h_2 be the components of the forces R_1, H_1, H_2 at right angles to the ball-head, then the front-frame is subjected to a bending-moment due to these three forces, and the bending-moment diagram may be represented by the shaded triangle (fig. 327).

Example I.—If $R_1 = 40$ lbs., the slope of the ball head be such that $r = 20$ lbs., and the distance between the lines of action of r_1 and h_1 be 17 in., the greatest bending-moment will be

$$M = 20 \times 17 = 340 \text{ inch-lbs.}$$

If the steering-tube be 1 in. diameter, 20 W.G., we get from Table IV., p. 113, $Z = \cdot0253$, and the maximum stress on the tube will be

$$f = \frac{M}{Z} = \frac{340}{\cdot0253} = 13,440 \text{ lbs. per sq. in.}$$

It is now becoming usual to strengthen the steering-tube by a liner at its lower end. For the nearest approximation to uniform strength throughout its length it is evident, from the shape of the bending-moment diagram, that its section should vary uniformly from top to bottom. If the liner extend half the length of the ball-head the tube will be of equal strength at the bottom and the middle, and will have an excess of strength at other points.

In a tandem bicycle the nature of the forces on the front-frame are exactly the same as above discussed, but are greater in magnitude. If in a tandem $R_1 = 100$ lbs., with the same dimensions as given above, M will be 850 inch-lbs.

Example II.—If the steering-tube be 1 in. diameter, 18 W.G., and be reinforced by a liner, 18 W.G., the combined thickness of tube and liner is ·096 inches, a little greater than that of a tube 13 W.G. The Z of a 1-in. tube 13 W.G. is ·055, therefore the maximum stress on the tube is

$$f = \frac{850}{\cdot055} = 15,460 \text{ lbs. per sq. in.}$$

The Fork Sides, at their junction to the crown, have to resist nearly the maximum bending-moment (fig. 327). They are usually made of tubes of oval section, tapering towards the wheel centre. The discussion of tubes of oval and rectangular sections (secs. 97 and 99) has shown the latter form to be the superior ; and, as there is no limitation of space to be considered in designing the front fork, the sides may with advantage be made of rectangular tube. If the rectangular tube be of uniform thickness, it has been stated (sec. 99) that for the greatest strength its depth should be three times its width. A still greater economy can be got by thickening the sides of the tube parallel to the neutral axis, either by brazing strips to a tube of uniform thickness (fig. 328), or during the process of drawing.

Pressure on Crown-plates.—The forces acting on the fork (fig. 329) are R_1, F_1, and F_2, the reactions of the crown-plates.

Example III.—Let the crown-plates be $\frac{3}{4}$ in. apart. Taking the components of these forces at right angles to the steering-head, and taking moments about the centre of the upper plate, we get

FIG. 328.

$$f_1 = \frac{16\cdot5}{\cdot75} \times 20 = 440 \text{ lbs.}, \text{ i.e. } 220 \text{ lbs. on each side.}$$

In the same way we get

$$f_2 = 420 \text{ lbs.}, \text{ i.e. } 210 \text{ lbs. on each side.}$$

The great advantage of the plate crown over the old solid crown is that the forces f_1 and f_2 are made to act as far apart as possible with a given depth of crown, whereas with the older solid crown the pressure was distributed over quite an appreciable distance, so that the distance between the resultant pressures f_1 and f_2 was small; the forces f_1 and f_2 were therefore correspondingly larger, since the moment to be resisted was the same.

In some recent designs of crowns the two plates are united by short tubes *outside* of the fork sides. As regards the attachment of the fork sides, this arrangement is therefore practically equivalent to the old solid crown. If any strengthening is desired, it should be done by an inside liner. Triple crown-plates have been used for tandems; but, as far as we can see, the middle plate contributes nothing to the strength of the joint, and may with advantage be omitted

FIG. 329.

Handle bar.—The handle-bar, when pulled upwards by the

rider with a force P at each handle, is subjected to a bending-moment Pl, l being the distance from the handle to the centre of the handle-pillar.

Example IV.—If $l = 12$ in., then $M = 12\,P$ inch-lbs. Let the handle-bar be $\frac{7}{8}$ in. diameter, 18 W.G., $Z = \cdot 0244$ in.³, and let the maximum stress on the handle-bar, f, be 20,000 lbs. per sq. in.; substituting in the formula $M = Z f$, we get

$$12\,P = \cdot 0244 \times 20,000.$$
$$\therefore \quad P = 41 \text{ lbs.}$$

That is, a total upward pull of 82 lbs. will produce a maximum stress of 20,000 lbs. per sq. in.

If the handles be bent backwards, the handle-bar is also subjected to a twisting-moment, which, however, usually produces smaller stresses than the bending-moment. For example, if the handle be bent 4 in. backwards, the twisting-moment $T = 4\,P$. The modulus of resistance to torsion of a $\frac{7}{8}$ in. tube, 18 W.G., is, from Table IV., p. 113, $\cdot 0488$ in.³ ; and therefore, with the same value for P as in the above example, we get $4 \times 41 = \cdot 0488\,f$,

or $\qquad\qquad f = 3,360$ lbs. per sq. in.

242. **General Considerations Relating to Design of Frame.**—The importance of having the forces acting on a tie or strut exactly central cannot be over-estimated ; the few examples already given above show how the maximum stress is enormously increased by a very slight deviation of the applied force from the axis. In iron bridge or roof building, this point is thoroughly appreciated by engineers ; but in bicycle building the forces acting on each tube of a frame are, as a rule, so small that tubes of the smallest section theoretically possible cannot be conveniently made. The tubes on the market are so much greater in sectional area than those of minimum theoretical section that they are strong enough to resist the increased stresses due to eccentricity of application of the forces ; and thus little or no attention has been paid to this important point of design.

The consideration of the shearing-force and bending-moment diagrams simultaneously with the outline of the frame is instructive, and reveals at a glance some weak points in various types of frames. The vertical section at any point of a properly braced frame will cut three members ; the moment of the horizontal components of the forces acting on these members will be equal to the bending-moment at the section, while the sum of the vertical components will be equal to the shearing-force. Therefore, in general, any part of a frame in which the vertical depth is small will be a place of weakness. The Ladies' Safety frames (figs. 264 and 265) have already been discussed. That shown in figure 266 is weakest at the point of crossing of the two tubes to the steering-head, the depth of the frame being zero at this point, so that only the *bending* resistance of these tubes can be relied on. The cross-frame (fig. 249) is very weak in the backbone, just behind the point where the down-tube crosses it. The Sparkbrook frame (fig. 253) is weakest at a point on the top-tube, just in front of the point of attachment of the tube from the crank-bracket. The frames (figs. 244 and 245) are practically equivalent to a single tube unbraced. The frames shown in figures 247–251 are weakest just behind the steering-head.

The consideration of the shearing-force curve shows the necessity for the provision of the diagonal of the central parallelogram in a tandem frame. The top- and bottom-tubes are nearly horizontal, so that if they were acted on by forces parallel to their axes they could not resist the shearing-force. The shearing-force must therefore be resisted by an inclined member of the frame, or, failing this, the forces on the top- and bottom-tubes cannot be parallel to their axes, and they must be subjected to bending. The same remarks apply to a frame formed by the duplication of either the top- or bottom-tubes without the provision of a diagonal, as in figure 268.

243. **Introductory.**—Wheels may be divided into two classes—rolling wheels and non-rolling wheels. In rolling wheels the instantaneous axis of rotation is at the circumference; examples are, bicycle wheels, vehicle wheels, railway carriage wheels, &c. Such rolling wheels have, in general, a *fixed* axis of rotation relative to the frame, which has a motion of translation when the wheel rolls. Non-rolling wheels are those not included in the above class; they may be mounted on fixed axes, their circumferences being free, or in contact with other wheels. Such are fly-wheels, gear-wheels, rope- or belt-pulleys, &c.

Wheels may again be subdivided, from a structural point of view, into solid wheels, wheels with arms, nave, and rim, cast or stamped in one piece, and built-up wheels. In a solid rolling wheel, the load applied at the centre of the wheel is transmitted by compression of the material of the wheel to its point of contact with the ground.

244. **Compression-spoke Wheels.**—A built-up wheel usually consists of three portions—the *hub* (nave, or boss), at the centre of the wheel; the *rim* or *periphery* of the wheel; and the *spokes* or *arms*, connecting the rim to the hub. Built-up wheels may be divided again into two classes, according to the method of action of the spokes. A wheel may be conceived to be made without a rim, consisting only of nave and spokes (fig. 330). In this case the load applied at the centre of the wheel is evidently transmitted by compression of the spoke, which is at the instant in contact with the ground. If the spokes are numerous, the rolling motion over a hard surface may be made fairly regular. In the ordinary

wooden cart or carriage wheel (fig. 331), the ends of the spokes are connected by wooden felloes, *f*, the felloes being mortised to receive the spoke ends, and an iron tyre, *t*, encircles the whole. This iron tyre is usually shrunk on when hot, and in cooling it compresses the felloes and spokes. This construction is very simple, since only one piece—the iron tyre—is required

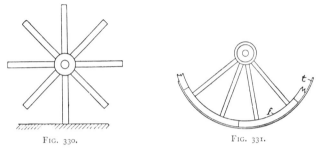

FIG. 330.   FIG. 331.

to bind the whole structure together. The compression wheel compares favourably in this respect with the tension wheel. On the other hand, the sectional area of the spokes must be great, in order to resist buckling under the compression ; very light wheels cannot, therefore, be made with compression spokes. The method of transmitting the load from the centre of the wheel to the ground is practically the same as in figure 330.

245. **Tension-spoke Wheels.**—The initial stresses in a bicycle wheel of the usual construction are exactly the reverse of those

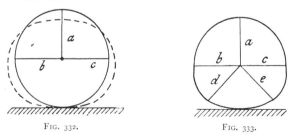

FIG. 332.   FIG. 333.

on the compression-spoke wheel (fig. 331). The method of action of the tension-spoke wheel may be shown as follows. Suppose the hub connected by a single wire, *a*, to a point on the top of the rim, a load applied at the centre of the wheel would be transferred to the top of the rim and would tend to flatten it, the sides would tend to bulge outwards, and the rim to assume the shape shown by the dotted lines (fig. 332). This horizontal bulging might be prevented by connecting the hub to the rim by two additional spokes, b and c. If, now, a load were applied at the centre of the wheel, the three spokes, a, b, and c, would be subjected to tension, and if the rim were not very stiff it would tend to flatten at its lower part, as indicated in figure 333. Additional spokes, d and e, would restrain this bulging. Thus, by using a sufficient number of spokes capable of resisting tension, the load applied at the centre of the wheel can be transmitted to the ground without appreciable distortion of the rim.

246. **Initial Compression in Rim.**—In building a bicycle-wheel the spokes are always screwed up until they are fairly tight.

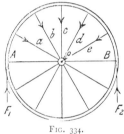

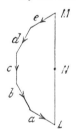

FIG. 334.   FIG. 335

The tension on all the spokes should be, of course, the same. This tightening up of the spokes will throw an initial compression on the rim, which may be determined as follows. Suppose the rim cut by a plane, $A\,O\,B$, passing through the centre of the wheel (fig. 334). Consider the equilibrium of the upper portion of the rim of the wheel : it is acted on by the pulls of the spokes a, b, c, d . . . and the reactions F_1 and F_2 of the lower part of the rim at A and B. If the tension t be the same in all the spokes, the force-polygon a, b, c, d . . . (fig. 335) will be half of a regular polygon. The sum of the forces F at A and B will be equal to the closing side, $L\,M$, of the force-polygon.

If the number of spokes in the wheel be great, the force-polygon (fig. 335) may be considered a circle. Then, if n be the number of spokes in the wheel, the circumference $L M$ (fig. 335) is equal to $\dfrac{n\,t}{2}$, the diameter $L\,M$ to $\dfrac{n\,t}{\pi}$.

But
$$2\,F = L\,M = \frac{n\,t}{\pi} \; ;$$

therefore
$$F = \frac{n\,t}{2\,\pi} \quad . \quad . \quad . \quad . \quad . \quad . \quad . \quad . \quad (1)$$

Example.—The driving-wheel of a Safety has 40 spokes, No. 14 W.G., which are screwed up to a tension of 10,000 lbs. per sq. in. Find the compression on the rim.

From Table XII., page 346, the sectional area of each spoke is ·00503 sq. in. ; the pull t is therefore ·00503 × 10,000 = 50·3 lbs. Substituting in (1),

$$F = \frac{40 \times 50\cdot3}{2 \times 3\cdot1416} = 320 \text{ lbs.}$$

247. Direct-spoke Driving-wheel.—The mode of transmission of the load from the centre of a bicycle wheel to the ground having been explained, it remains to show how the driving effort

FIG. 336.

is transmitted from the hub to the rim. In a large gear-wheel the arms are rigidly fixed to the nave, and while a driving effort is being exerted, the arms press on the rim of the wheel in a tangential direction. Thus each arm may be considered as a beam rigidly fixed to the nave and loaded by a force at its end near the rim. The spokes of a bicycle wheel are not stiff enough to transmit in this manner forces transverse to their axis, being to all intents and purposes perfectly flexible. When a driving force is exerted the hub turns through a small angle without moving the rim, so that the spokes whose axes initially all passed through the centre of the wheel now touch a circle, s (fig. 336). Let r be the radius of this circle, and P the pull of the driving chain which is exerted at a radius R. Considering the equilibrium of the hub, the moment of the force P about the centre is $P R$;

the moment of the forces due to the pull of the spokes on the hub is

$$n\,t\,r.$$

Thus, $PR = n\,t\,r,$

and

$$r = \frac{PR}{n\,t} \quad . \quad . \quad . \quad . \quad . \quad (2)$$

Example.—Let the driving-wheel have 40 spokes, each with an initial tension of 50 lbs. ; let the pull of the chain be 300 lbs., and be exerted at a radius of $1\frac{1}{2}$ in. Find the size of the circle s, and the angle of displacement of the hub.

Substituting in (2)

$$r = \frac{300 \times 1\frac{1}{2}}{40 \times 50} = \cdot225''.$$

Figure 337 is a drawing showing the displacement of the hub. Let $c\,d$ be the radius of the circle touched by the spokes, $b\,a$ the initial position of a spoke, $b^1\,a^1$ the displaced position, and let the distance of the point of attachment of the spokes from the centre of the hub be $\frac{7}{8}$ in. ; the angle of displacement of the hub, $a\,c\,a^1$ will be approximately

$$\frac{\cdot225}{\cdot875} = \cdot257 \text{ radians,}$$

or,

$$\frac{\cdot257 \times 180}{\pi} = 14\cdot7 \text{ deg.}$$

FIG. 337.

If the driving effort be reversed, as in back-pedalling, the hub will first return to its original position relative to the rim, and then be displaced in the opposite direction before the reversed driving effort can be transmitted.

Thus, a direct-spoke bicycle wheel is not a rigid structure, but has quite a perceptible amount of tangential flexibility between the hub and the rim.

Lever Tension Driving-wheels.—In the early days of the 'Ordinary,' wheels were often made with a pair of long levers projecting from the hub, from the ends of which wires went off to the rim. These tangential wires were adjustable, and by tightening them the rim was moved round relative to the hub, and thus the tension on the spokes could be adjusted. The tangential driving effort was also supposed to be transferred from the hub to the rim by the lever and tangent wires, while the

FIG. 338.

radial spokes only transmitted the weight from the hub to the rim. Figure 338 shows the 'Ariel' bicycle with a pair of lever tension wheels.

248. Tangent-spoke Wheels.—In a tangent wheel the spokes are not arranged radially, but touch a circle concentric with the

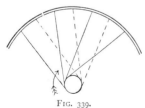

FIG. 339.

hub (fig. 339). The pull on the tangent-spokes indicated by the full lines would tend to make the hub turn in the direction of the arrow. Another set of spokes, represented by the dotted lines, must be laid inclined in the opposite direction, so that the hub may be in equilibrium. The initial tension should be the same on all the spokes.

Let a driving effort in the direction of the arrow be applied at the hub. This will have the effect of increasing the tension on one half of the spokes and diminishing the tension on the other half. If r be the radius of the circle to which the spokes are tangential, t_1 and t_2 the tensions on the tight and slack spokes respectively, the total tangential pull of the spokes at the hub is

$$\frac{n}{2}(t_1 - t_2).$$

Therefore

$$P\,R = \frac{n\,r}{2}(t_1 - t_2),$$

from which

$$t_1 - t_2 = \frac{2\,P\,R}{n\,r}\ .\ \ .\ \ .\ \ .\ \ .\ \ .\ \ (3)$$

Example.—Let r be $\frac{7}{8}$ in., the spokes 15 W.G., the modulus of elasticity of the spokes 10,000 tons per sq. in.; then, taking the rest of the data as in section 247, find the angle of displacement of the hub relative to the rim under the driving effort.

Substituting in (3),

$$t_1 - t_2 = \frac{2 \times 300 \times 1\frac{1}{2}}{40 \times \frac{7}{8}} = 25\cdot 7\ \text{lbs.}$$

The sectional area of each spoke (Table XII.) is ·00407 sq. in. ; the increase or diminution of the tension due to the pull of the chain is therefore

$$\frac{25\cdot 7}{2 \times \cdot 00407} = 3,156\ \text{lbs. per sq. in.} = 1\cdot 41\ \text{tons per sq. in.}$$

The extension of one set of spokes and the contraction of the other set will thus be $\frac{1\cdot 41}{10,000}$ th part of their original length, which length in a 28-in. driving-wheel is about 12 in. The displacement of a point on the circle of radius $\frac{7}{8}$ in. is thus

$$\frac{1\cdot 41 \times 12}{10,000} = \cdot 00169\ \text{in.}$$

The angle the hub is displaced relative to the rim will be

$$\frac{\cdot 00169 \times 180}{\frac{7}{8} \times \pi} = \cdot 11\ \text{deg.}$$

Comparing this example with that of section 247, the superiority of the tangent wheel in tangential stiffness is apparent. In this example it should be noted that the initial pull on the spokes does not enter into the calculation. Consequently, the initial pull on tangent-spokes may with advantage be less than that on direct-spokes.

249. **Direct-spokes.**—The spokes of a direct-spoke wheel are usually of the form shown in figure 340, the conical head at the end engaging in the rim, and the

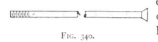

FIG. 340.

other end being screwed into the hub. For the sake of preserving the spoke of equal strength throughout, its end is often *butted* before being screwed (fig. 341), the section at the bottom of the thread in this case being

FIG. 341.

at least as great as at the middle of the spoke.

In figure 339 the spoke is shown making an acute angle with the hub. As a matter of fact, under the action of a driving effort the spokes near the hub will be bent, as shown exaggerated in

FIG. 342.

figure 342. The continual flexure under the driving effort weakens and ultimately causes breakage of direct spokes, unless made of greater sectional area than would be necessary if they could be connected to the hub by some form of pin-joint. The conical head lies loosely in the rim, and being quite free to adjust itself to any alteration of direction, the spoke near the rim is not subjected to such severe straining actions as at the hub.

250. **Tangent-spokes.** — Tangent-spokes cannot be conveniently screwed into the hub, but are threaded through holes in a flange of the hub, the end of the

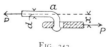

FIG. 343.

spoke being made as indicated in figure 343. This sharp bend of the spoke seriously affects its strength. Let P be the pull on the spoke, and d its diameter. On the section of the spoke at a there will be a bending-moment $P x$, x being the distance between the middle of the section a and the hub flange; this distance may be taken approximately equal to d. The bending-moment is then $P\,d, Z = \dfrac{\pi\,d^{\,3}}{3^2}$, and the maximum stress, f, due to bending will be found by substitution in the formula $M = Z\,f$. Therefore

$$P\,d = \frac{\pi\,d^{\,3}}{3^2}f$$

and

$$f = \frac{3^2\,P\,d}{\pi\,d^{\,3}} = \frac{3^2\,P}{\pi\,d^{\,2}}.$$

The tensile stress on the middle of the spoke is

$$\frac{4\,P}{\pi\,d^{\,2}}$$

Thus the stress due to bending on the section at the corner is eight times that on the body of the spoke due to a straight pull.

Figure 344 shows a tangent-spoke strengthened at the end by butting.

The ends of tangent-spokes must be fastened to the rim by means of nuts or nipples. The nipple has its inner surface screwed to fit the screw on the end of the spoke, has a conical head which lies in a corresponding counter-sunk hole in the rim, and a square or hexagonal body threaded through

F$_{IG}$. 344.

the hole in the rim for screwing up by means of a small spanner.

A piece of wire threaded through the hub flange (fig. 345), and its ends fastened to the rim by nipples in the usual way, is often used to form a pair of tangent spokes. The objection to the spoke shown in figure 343 still holds with regard to this form; but the fact that no head has to be formed at the hub probably makes it slightly stronger than a single spoke of the same diameter headed at the end.

F$_{IG}$. 345.

Figure 346 shows the form of tangent-spoke used by the St. George's Engineering Co. in the 'Rapid' cycle wheels. The

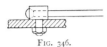

spoke is quite straight from end to end, and is fastened to the rim in the usual way by a nipple. It is fastened to the hub by means of

FIG. 346.

a short stud projecting from the hub flange, a small hole being drilled in the projecting head of the stud, and the spoke threaded through it. The headed end of the spoke is pulled up against the stud. Spokes of this form are not subjected to bending, and are therefore much stronger than tangent-spokes of the usual form of the same gauge.

TABLE XII.—SECTIONAL AREAS AND WEIGHTS PER 100 FT. LENGTH OF STEEL SPOKES.

Imperial standard wire gauge	Diameter	Sectional area	Weight of 100 ft.
	In.	Sq. in.	Lbs.
6	·192	·02895	10·005
7	·176	·02433	8·409
8	·160	·02011	6·950
9	·144	·01629	5·629
10	·128	·01287	4·447
11	·116	·01057	3·652
12	·104	·00849	2·936
13	·092	·00665	2·298
14	·080	·00503	1·738
15	·072	·00407	1·407
16	·064	·00322	1·112
17	·056	·0024€	·850
18	·048	·00181	·625
19	·040	·00126	·434
20	·036	·00102	·352

251. **Sharp's Tangent Wheel.**—The distinctive features of this wheel, invented by the author, are illustrated in figure 347. The hub is suspended from the rim by a series of wire loops, one loop forming a pair of spokes. In figure 347, for the sake of clearness of illustration, one loop or pair of spokes is shown thickened. The ends are fastened to the rim by nuts or nipples in the usual way. There is no fastening of the spokes to the hub,

beyond that due to friction. Figure 348 represents the appearance of the spokes in contact with the hub. The arc of contact
of the spoke and hub is a spiral, so that all the ends of the
spokes on one side of the middle plane of the wheel begin contact
with the hub at the same distance from the middle, the other ends
all leaving the hub nearer the middle plane. A wheel could be
made with loops of wire having circular contact with the hub, but
it would not be symmetrical, and the spokes would not all be of

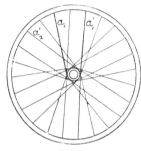

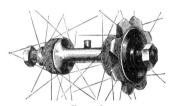

Fig. 347.   Fig. 348.

the same length. By making the spokes have a spiral arc of contact with the hub, the positions of all the spokes relative to the
hub are exactly similar, the wheel is symmetrical, and the spokes
are all of the same length. It will be noticed that there are no
sudden bends in the spokes, so that they are much stronger than
in the ordinary tangent wheel, no additional bending stresses being
introduced. For non-driving cycle wheels there can be no question as to the sufficiency of the hub fastening, but it may at
first sight seem startling that the mere friction of the spokes
on the hub should be sufficient to transmit the driving effort to
the rim, though it is well known that by coiling a rope round a
smooth drum almost any amount of friction can be obtained.
This system of construction is applicable to all types of built-up
metal wheels, and has been applied with success to fly-wheels and
belt-pulleys, and to the 'Biggest Wheel on Earth'—the gigantic
pleasure-wheel at Earl's Court.

Let t be the initial tension on the spokes; then, while the
driving effort is being exerted, the tension on one half of each loop

rises to t_1, and on the other half falls to t_2. If t_1 be very much greater than t_2 there will not be sufficient friction between the hub

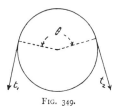

FIG. 349.

and the wire, and slipping will occur. Let θ be the angle of contact (fig. 349) and μ the coefficient of friction between the spoke and hub. Then, when slipping takes place,

$$\frac{t_1}{t_2} = \varepsilon^{\mu\,\theta} \quad . \quad . \quad . \quad . \quad . \quad . \quad (4)$$

If $\frac{t_1}{t_2}$ is less than determined by (4), slipping will not occur.

Equation (4) may be written in the form,

$$log_\epsilon \frac{t_1}{t_2} = \mu\,\theta,$$

the symbol $log_\epsilon \frac{t_1}{t_2}$ denoting the logarithm, to the 'Naperian' or natural base, of the number $\frac{t_1}{t_2}$. Using a table of 'common' logarithms, a more convenient form is —

$$log \frac{t_1}{t_2} = ·4343\,\mu\,\theta \quad . \quad . \quad . \quad . \quad (5)$$

Example I.—A driving-wheel 28 in. diameter, on this system, has 40 spokes wrapped round a cylindrical portion of the hub $1\frac{1}{2}$ in. diameter, the initial tension on each spoke is 60 lbs., the pull on the chain is 300 lbs., and is exerted at a radius of $1\frac{1}{2}$ in. Find whether slipping will take place or not.

Let the arc of contact be half a turn, as shown approximately in figure 347, then $\theta = \pi$, the coefficient of friction μ for metal on metal dry surface will be about from ·2 to ·35, but assuming that oil from the bearing may get between the surfaces, we may take a low value, say o·15 ; substituting in (5)

$$log \frac{t_1}{t_2} = ·4343 \times ·15 \times 3·1416 = ·2046,$$

from which, consulting a table of logarithms,

$$\frac{t_1}{t_2} = 1·602$$

when slipping takes place. But from (3)

$$t_1 - t_2 = \frac{2\,P\,R}{n\,r} = \frac{2 \times 300 \times 1\frac{1}{2}}{40 \times \frac{3}{4}} = 30 \text{ lbs.}$$

Therefore
$$t_1 = 60 + 15 = 75$$
$$t_2 = 60 - 15 = 45$$
and
$$\frac{t_1}{t_2} = 1\cdot5.$$

Thus, with the above conditions, slipping will not occur.

As a matter of experiment, the author finds that with such a smooth hub and an arc of contact of half a turn slipping takes place in riding up steep hills only when the spokes are initially slacker than is usual in ordinary tangent wheels.

Arc of Contact between Spokes and Hub.—The pair of spokes (fig. 347) is shown having an arc of contact with the hub of nearly two right angles. The arc of contact may be varied. For example, keeping the end a_1 fixed, the other end of the spoke may be moved from a^1_1 to a^1_2, or even further, so that the arc of contact may be as shown in figure 350. In this case there are five spoke ends left between the ends of one pair. In general, $4n + 1$ spokes must be left between the ends of the same pair, n being an integer.

FIG. 350.

In this wheel, should one of the spokes break, a whole loop of wire must be removed. Of course the tendency to break is, as already shown, far less than in direct or tangent spokes of the usual type. If the arc of contact, however, is as shown in figure 347, and a pair of spokes are removed from the wheel, a great additional tension will be thrown on the spoke between the two vacant spaces. If the angle of contact shown in figure 350 be adopted, there will still remain five spokes between the two vacant spaces, so that the additional tension thrown on any single spoke will not be abnormally great.

Grooved Hubs.—The hub surface in contact with the spokes may be left quite smooth, with merely a small flange to preserve the spread of the spokes. The parts of the spokes wrapped round the hub will lie in contact side by side (fig. 348) Should one break and be removed from the wheel, the remaining spokes in contact with the hub will close up the space vacated by the broken one. In putting in a new spoke they will have to be again separated. Spiral grooves may be cut round the hub, so that each spoke may lie in its own special groove, and if one breaks, the space will be left quite clear for the new spoke to replace it.

The grooves may be made so as to considerably increase the frictional grip on the nave. Figure 351 shows the section of a

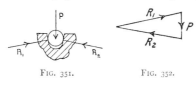

FIG. 351.   FIG. 352.

spoke in a groove, the spoke touching the sides, but not the bottom of the groove. It is pressed to the hub by a radial force, P, and the reactions R_1 and R_2 are at right angles to the side of the grooves. Figure 352 shows the corresponding force-triangle. The sum of the forces R_1 and R_2, between the spoke and the hub, may be increased to any desired multiple of P by making the angle between the sides of the groove sufficiently small, and the frictional grip will be correspondingly increased. If the angle of the sides of the grooves is such that $R_1 + R_2 = n\,P$, $n\,\mu$ must be used instead of μ in equations (4) and (5).

Example II.—If the spokes in the wheel in the above example lie in grooves, the sides of which are inclined 60°; find the driving effort that can be transmitted without slipping.

In this example the force-triangle (fig. 352) becomes an equilateral triangle, and $R_1 + R_2 = 2\,P$. Taking $\mu = \cdot 15$ and $\theta = \pi$ as before, $n\,\mu = \cdot 3$, and

$$log\,\frac{t_1}{t_2} = \cdot 4343 \times \cdot 3 \times 3\cdot 141 = \cdot 4093,$$

from which, consulting a table of logarithms,

$$\frac{t_1}{t_2} = 2\cdot 566.$$

But $t_1 + t_2 = 120$ lbs. Solving these two simultaneous simple equations, we get

$$t_1 = 86\cdot3$$
$$t_2 = 33\cdot7,$$

the driving effort is $t_1 - t_2 = 52\cdot6$ lbs.

Thus, the effect of the grooves inclined 60° is to nearly double the driving effort that can be transmitted.

252. **Spread of Spokes.**—If the spokes of a tension wheel all lay in the same plane, then, considering the rim fixed, any couple tending to move the spindle would distort the wheel, as shown in figure 353. The distortion would go on until the moment of the pull of the spokes on the hub was equal to the moment applied to the shaft. If the spindle remains fixed in position, any lateral force applied to the rim causes a deviation of its plane, the relative motion of the rim and spindle being the same as before ; the wheel, in fact, wobbles. If the spokes are spread out at the hub (fig. 354), the rim being fixed and the same bending-moment being applied at the spindle, the tension on the

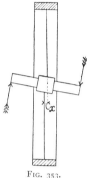

FIG. 353.

spokes A at the bottom right-hand side, and on the spokes B at the top left-hand side, is decreased, and that on the spokes C at the left-hand bottom side, and on the spokes D at the right-hand top side is increased. This increase and diminution of tension takes place with a practically inappreciable alteration of length of the spokes, and therefore the wheel is practically rigid.

The lateral spreading of the spokes of a cycle wheel should be looked upon as *a means of connecting the hub rigidly to the rim*, rather than of giving the rim lateral stability relative to the hub. The rim must be of a form possessing initially sufficient lateral stability, otherwise it cannot be built up into a good wheel. The lateral components of the pulls of the spokes on the rim, instead of preserving the lateral stability of the rim, rather tend to destroy it.

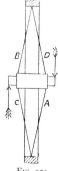

FIG. 354.

They form a system of equal and parallel forces, but alternately in opposite directions (fig. 355), and thus cause bending of the rim at right angles to its plane. If the rim be very narrow in the direction of the axis of the wheel, it may be distorted by the pull of the spokes into the shape shown exaggerated in figure 355.

FIG. 355.

The 'Westwood' rim (fig. 373), on account of its tubular edges, is very strong laterally.

253. **Disc Wheels.**—Instead of wire spokes to connect the rim and hub, two conical discs of very thin steel plate have been used, the discs being subjected to an initial tension. It was claimed—and there seems nothing improbable in the claim—that the air resistance of

FIG. 356.

these wheels was less than that of wheels with wire spokes. Later, the Disc Wheel Company (Limited) made the front wheel of a Safety with four arms, as shown in figure 356.

Nipples.—The nipples used for fastening the ends of the spokes to the rim are usually of steel or gun-metal. Perhaps, on the whole, gun-metal nipples are to be preferred to steel, since they do not corrode, and being of softer metal than the spokes, they cannot cut into and destroy the screw threads on the spoke ends. Figure 357 is a section of an ordinary form of nipple which can be used for both solid and hollow rims, and figure 358 is an external view of the same nipple, showing its hexagonal external surface for screwing up. The hole in the nipple is not tapped throughout its whole length, but the ends towards the centre of the wheel are drilled the full diameter of the spoke, so that the few extra screw threads left on the spoke to provide for the necessary adjustment are protected by the nipple. Figure 359 shows a square-bodied nipple, otherwise the same as that in figure 358.

Fig. 357.   Fig. 358.   Fig. 359.

When solid rims are used, the nipple heads must be flush with the rim surface, so as not to damage the tyre ; but when hollow rims are used, the nipple usually bears on the inner surface of the rim, and is therefore quite clear of the tyre. Figures 360 and 361 show forms of nipples for use with hollow rims, the screw thread of the spoke being protected by the latter nipple.

Fig. 360.   Fig. 361

In rims of light section, such as the hollow rims in general use, the greatest stress is the *local* stress due to the screwing up of the spokes. With a very thin rim, which otherwise might be strong enough to resist the forces on it, the bearing surfaces of the nipples shown above are so small that the nipple would be actually pulled through the rim by the pull due to tightening the spoke. To distribute the pressure over a larger surface of the rim, small washers (fig. 362) may be used with advantage.

Fig. 362.

With wood rims, washers should be used below the nipples, otherwise the wood may be crushed as the tension comes on the spokes.

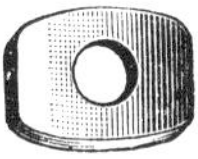

Figure 363 shows the form of steel nipple to be used with Westwood's rim when the spokes are attached, not at the middle, but at the sides of the rim. Figures 357–363 are taken from the catalogue of the Abingdon Works Company (Limited), Birmingham.

254. **Rims.**—We have already seen that the rim is subjected to a force of compression due to the initial pull on the spokes. Let us consider more minutely the stresses on the rim when the wheel is not supporting any external load. Let figure 364 be the elevation of a wheel with centre C, $A B$ being the chord between the ends of two adjacent spokes. Then

Fig. 363.

the stress-diagram (see figs. 334 and 335) of the structure will be a similar regular polygon, the pull on each spoke being repre-

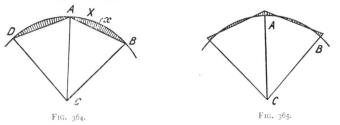

FIG. 364.   FIG. 365.

sented by the side and the compression on the rim by the radius of the polygon.

If the rim were polygonal, the axes of the rim and the compressive force on it would coincide, and the compressive stress would be equally distributed over the section. But since the rim is circular, its axis will differ from the axis of the compression, and there will be a bending-moment introduced. Since at any point X this bending-moment is equal to the product of the compression P into the distance x between the axis of the rim and the line of action of P, the bending-moment on the rim will be proportional to the intercept between the rim and the chord $A B$, formed by joining the ends of two adjacent spokes, provided that the bending-moment on the rim at the points where the spokes are fastened is zero. The shaded area (fig. 364) would thus form a bending-moment diagram. But if the rim initially had no bending stress on it, it is likely that at the points A and B the pull of the spokes will tend to straighten the rim, and therefore a bending-moment, m, of some magnitude will exist at these points. The bending-moment at any point X will be diminished by the amount m, and the diagram will be as shown in figure 365, the bending-moments being of opposite signs at the ends of, and midway between, the spokes. From an inspection of figure 365, it is clear that in a wheel with 32 to 40 spokes, the bending-moment on the rim due to the compression will be negligibly small in comparison with the latter.

When the wheel supports a load the distribution of stress on the rim is much more complex, and a satisfactory treatment of the subject is beyond the scope of the present work. The simplest treatment—which, however, the author does not think will give even rough approximations to the truth—will be to assume that the segments of the rim are *jointed* together at the points of attachment of the spokes. With this assumption, if the wheel supports a weight W, when the lowest spoke is vertical, the force-triangle at A, the point of contact with the ground, will be made up of the two compressions along the adjacent segments of the rim, and the pull on the vertical spoke plus the upward reaction of the ground, W. The rest of the stress-diagram will be as in the former case ; consequently, if the pull on the vertical spoke is zero, that on the other spokes will be W; if the pull on the vertical spoke is t, that on the other spokes will be $(W + t)$.

When the two bottom spokes are equally inclined to the vertical, the lower rim segment is in the condition of a beam supported at the ends and carrying in the middle a load, W; therefore the bending-moment is $\dfrac{Wl}{4}$, l being the length of the rim segment.

The assumption made above does not agree, even approximately, with the actual condition of things in the continuous rim

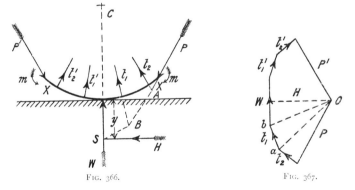

FIG. 366.   FIG. 367.

of a bicycle wheel. A general idea of the nature of the forces acting may be obtained from figure 366, which represents a small portion, XX, of the rim near the ground. This is acted on by the known force W, the upward reaction of the ground ; by the unknown forces t_1, t_2, . . . the pulls on the spokes directed towards the centre, C, of the wheel ; by forces of compression, P, on the rim, unknown both in direction and magnitude ; and by unknown bending-moments, m, at the section X. The portion of the rim considered is, therefore, somewhat in the condition of an inverted arch. If the forces P, t_1, t_2, . . . and the bending-moments, m, were known, the straining action at any point on the rim could be determined as follows : Figure 367 shows the force-polygon, on the assumption that the forces considered are symmetrically situated with regard to the vertical centre line. The horizontal thrust on the rim at its point of contact with the ground is H, the resultant of the forces P, t_1, t_2, . . . on one side of the vertical. This, however, acts at a point S, at a vertical distance y below the rim, determined as follows : Produce the lines of action of P and t_2 to meet at A ; their resultant, which is parallel to $O\,a$ (fig. 367), passes through the point A. Draw, therefore, $A\,B$ parallel to $O\,a$, cutting the line of action of t_1 at B. Through B draw a line parallel to $O\,b$, giving the resultant of P, t_2, and t_1, and cutting the vertical through the point of contact at S. The rim at its point of contact with the ground is thus subjected to a compression H, and a bending-moment $m + H\,y$. To make the solution complete, the unknown forces P, t_1 and t_2 should be determined ; this can be done by aid of the theory of elasticity.

Steel Rims.—Figure 368 shows a section of a rim for a solid tyre, figure 369 for a cushion tyre. The edges of the latter are

Fig. 368. Fig. 369.

slightly bent over, so that the tyre when it bulges out on touching the ground will not be cut by the rim edge. Figure 370 shows a section of Warwick's hollow rim, which is rolled from one strip of steel bent to the required section, its edges scarfed, and brazed together. The part of the rim of smallest radius is thickened, so

that the local stresses due to the screwing-up of the spokes may be
better resisted. Figure 371 shows the 'Invincible' rim which was

FIG. 370. FIG. 371.

made by the Surrey Machinists Company, rolled from two distinct
strips, the inner being usually much thicker than the outer. The strips

FIG. 372.

were brazed together right round the circumference. Figure 372
shows the Nottingham Machinists' hollow rim. In this the local
strength for the attachment of the
nipple is provided by folding over the
plate from which the rim is made, so
that four thicknesses are obtained.
Figure 373 shows the 'Westwood' rim,

FIG. 373.

which is formed from one plate bent round at each edge to form
a complete circle. The spokes can be attached at the edges of the
rim as indicated, or at the middle of the rim in the usual way.

All the above rims are rolled to different sections to fit
the different forms of pneumatic tyres. They are all made
from straight strips of steel, and have, therefore, one joint in the
circumference, the ends being brazed together. This joint, however carefully made, is always weaker than the rest of the rim,
and adds to the difficulty of building the wheel true. The
Jointless Rim Company roll each rim from a weldless steel
ring, in somewhat the same way as railway tyres are rolled.

This rim, though perhaps more costly, is therefore much stronger weight for weight than a rim with a brazed joint.

Wood Rims.—The fact that the principal stress on the rim of a bicycle wheel is compression, and that, therefore, the material must be so distributed as to resist buckling or collapse, and not concentrated as in a steel wire, suggests the use of wood as a suitable material. Hickory, elm, ash, and maple are used. Two types are in use : in one the rim is made from a single piece of wood, the two ends being united by a convenient joint. Figure 374 shows the 'Plymouth' joint. The other type is a built-up

FIG. 374.

rim composed of several layers of wood. Figures 375 and 376 show the 'Fairbank' laminated rim, for a solutioned tyre and for the Dunlop tyre respectively, the grain of each layer of wood

FIG. 375.　　　　　　FIG. 376.

running in an opposite direction to that next it. Each layer or ring is made with a scarfed joint, and the various rings are fastened together with marine glue under hydraulic pressure. The built-up rim is then covered with a waterproof linen fabric, and varnished.

255. Hubs.—Figure 404 shows a section of the ordinary form of hub for a direct spoke-wheel, and figure 377 an external view of a driving hub. The hub proper in this is made as short as possible, and the spindle, with its adjusting cones, projects considerably beyond the hub, so as to allow the wheel to clear the frame of the machine.

Figure 378 shows a driving hub, in which the hub proper is extended considerably beyond the spoke flanges, and the ball-races kept as far apart as possible. This hub is intended for tangent-spokes, the flanges being thinner than in figure 377.

FIG. 377.

Hubs for direct-spokes are made either of gun-metal or steel ; tangent-spoke hubs should be invariably of steel, as the local

FIG. 378.

stress due to the pull of the spoke cannot be resisted by the softer metal.

Figure 379 shows a pair of semi-tangent hubs, as made by Messrs. W. A. Lloyd & Co., the flanges for the attachment of the

FIG. 379

hub in this case forming cylindrical drums instead of flat discs, as in figure 378. The spokes may leave the circumference of the drum at any angle between the radius and the tangent, hence the name *semi-tangent*.

In all the hubs above described the adjusting cones are screwed on the spindle, and the hard steel cups are rigidly fixed to the hub. In the 'Elswick' hub (fig. 380), the adjusting cone

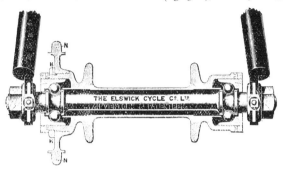

FIG. 380.

is screwed to the hub and the ball-races on the spindle are rigidly fixed. One important advantage of this form of hub is that the clear space which must always be preserved between the fixed spindle and the rotating hub is of much smaller radius than in the others. The area by which dust and grit may enter the bearing

FIG. 381.

is smaller, the bearing should therefore be more dust-proof than the others. Another important feature is the fact that the hub is oil-retaining, and the balls may have oil-bath lubrication at the lowest point of their path. Figure 381 shows the 'Centaur' hub, also possessing dust-proof oil-retaining properties.

In recent years 'barrel' hubs of large diameter have been used, whereas the earlier hubs were made just large enough to clear the spindle inside. The 'Centaur' is an example of a barrel hub.

The best hubs are turned out of solid steel bar, the diameter of which must be as great as that of the flanges for the attachment of the spokes. To avoid this excessive amount of turning, the 'Yost' hub is made of two end pieces and a middle tube.

The hubs of Sharp's tangent wheel may, with advantage, be made of aluminium, since the pull of the spokes has not to be transmitted by flanges.

The 'Gem' hub, made by the Warwick and Stockton Company, has the hard steel cup screwed to the end of the hub. The balls lie between the cup and an inner projecting lip of the hub, so that they remain in place when the spindle is removed.

256. **Fixing Chain-wheel to Hub.**—The chain-wheel should not be fixed by a key or pin, as this will usually throw it slightly eccentric to the hub. In testing the resistance of the chain gearing of a Safety it is often noticed that the chain runs quite slack in some places and tight in others. This can only mean that the centres of the pitch-polygons of the chain-wheels do not coincide with the axes of rotation. The chain-wheel and the corresponding surface on the hub, being turned to an accurate fit, are often fastened by simply soldering. The temperature at which the solder melts is sufficiently low to prevent injury to the temper of the ball-races of the hub. Another method is to screw the chain-wheel, N, on the hub ; the screw should then be arranged that the driving effort in pedalling ahead tends to screw the chain-wheel up against the projecting hub flange. This is done in the 'Elswick' hub (fig. 380). If the chain is at the right-hand side of the machine looking forwards, the screw on the chain-wheel should be right-handed. During back-pedalling the driving effort will tend to unscrew the chain-wheel. This is counteracted by having a nut, K, with left-handed screw, screwed up hard against the chain-wheel. If the chain-wheel, N, tends to unscrew during back-pedalling, it will take with it the nut K, which will then be screwed more tightly against the wheel, and its further unscrewing prevented.

A method adopted by the Abingdon Company a few years ago was to have the chain-wheel and hub machined out to a polygonal surface of ten sides, and the wheel then soldered on.

257. **Spindle.**—The spindle, strictly speaking, is a part of the frame, and serves to transfer the weight of the machine and rider to the wheel. Let the spindle be connected to the frame at *A* and *B* (fig. 382), *C* and *D* be the points at which it rests on the

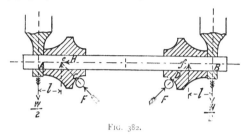

Fig. 382.

balls of the bearing, and *W* be the total load on the wheel. Then the spindle may be considered as a beam loaded at *A* and *B* with equal weights $\frac{W}{2}$, and supported at the points *C* and *D* ; the direction of the forces of reaction *F*, at *C* and *D*, coinciding with the radii of the balls to their points of contact with their paths. Let *e* and *f* be the points at which the forces *F* cut the axis of the spindle ; then *F* can be resolved into vertical and horizontal forces, $\frac{W}{2}$ and *H* respectively, acting through *e*. The horizontal forces, *H*, produce a tension on the part *e f* of the spindle, the remaining forces produce bending stresses. The spindle may thus be considered as a beam supported at *e* and *f* and loaded at *A* and *B* with equal weights, $\frac{W}{2}$. The bending-moment on any section between *e* and *f* is $\frac{W l}{2}$, *l* being the distance *A e*. It is evident that this bending-moment will be zero if the points *A* and *e* coincide, and will be greater the greater the distance *A e* ; hence the spindle in figure 378 is subjected to a far smaller bending stress than that in figure 377.

Example I.—In a bearing the distance $A\ e$ (fig. 382) is $\frac{7}{8}$ in., and the total weight on the wheel is 120 lbs., what is the necessary size of spindle, the maximum stress allowed being 10 tons per sq. in.?

The bending-moment on the spindle will be

$$\frac{120}{2} \times \frac{7}{8} = 52\cdot5 \text{ inch-lbs.}$$

Substituting in the formula $M = \frac{d'^3}{10} f$ (sec. 94), we get

$$52\cdot5 = \frac{d^3}{10} \times 10 \times 2240,$$

that is

$$d^3 = \cdot0234, \text{ and } d = \cdot286 \text{ in.}$$

This gives the least permissible diameter of the spindle, that is, the diameter at the bottom of the screw threads.

Step.—The most convenient step for mounting a Safety bicycle is formed either by prolonging the spindle itself, or by forming a long tube on the outer nut that serves to fasten the spindle to the frame and lock the adjusting cone in position. If the length of this step be $1\frac{1}{2}$ in., W the weight of the rider, and if the rider in mounting the machine press on its outer edge, the bending-moment produced on the spindle will be $1\frac{1}{2} W$ inch-lbs.

Example II.—If $W = 150$ lbs., $M = 225$ inch-lbs.; substituting in the formula $M = Z f$, we get

$$225 = \frac{d^3}{10} \times 10 \times 2240,$$

from which

$$d^3 = \cdot100 \text{ and } d = \cdot464 \text{ in.}$$

A common diameter for the spindle is $\frac{3}{8}$ in.; if the $\frac{3}{8}$ in. spindle resist the whole of the above bending-moment, the maximum stress on it will be much greater than 10 tons per sq. in.; it will be

$$\frac{\cdot464^3}{\cdot375^3} \times 10 = 18\cdot9 \text{ tons per sq. in.}$$

The tube from saddle-pin to driving-wheel spindle may take up some of the bending due to the weight on the step, in which case the maximum stress on the spindle may be lower than given above.

258. **Spring Wheels.**—Different attempts have been made to make the wheels elastic, so that vibration and bumping due to the unevenness of the road may not be communicated to the frame. One of the earliest successful attempts in this direction was the corrugated spokes used in the 'Otto' dicycle. These spokes, instead of being straight, were made wavy or corrugated, and of a harder quality of steel than used in the ordinary straight spokes. Their elastic extension was great enough to render the machine provided with them much more comfortable than one with the ordinary straight spokes.

A spring wheel has the advantage over a spring frame, that it intercepts vibration sooner, so that practically only the wheel rim partakes of the jolting due to the roughness of the road. On the other hand, the springs of a wheel extend and contract once every revolution, and as this cannot be done without the expenditure of energy, a spring wheel must require more power than a rigid wheel to propel it over a good road. The springs of a frame remain quiescent under a steady load while running over a smooth road, only extending or shortening when the wheel passes over a hollow or lump in the road.

In the 'Everett' spring wheel the spokes, instead of being connected directly to the hub, are connected to short spiral springs, thus giving an elastic connection between the hub and the rim, so that the rim may run over an obstacle on the road without communicating much shock to the frame. One objection to a wheel with spring spokes is the want of lateral stiffness of the rim, it being quite easy to deflect the rim sideways by a lateral pressure. The author is inclined to think that this objection may be over-rated, since in a bicycle the pressure on the rim of a wheel must be in, or nearly in, the plane of the wheel. The 'Everett' wheel is satisfactory in this respect. In the 'Persil' spring wheel two rims are used, the springs being introduced between them. The introduction of such a mass of material near the periphery of the wheel will make the bicycle provided with 'Persil' wheels slower in starting than one with ordinary wheels (see sec. 68).

In the 'Deburgo' spring wheel the springs are introduced at the hub, which is much larger than that of an ordinary wheel. Figure 383 shows a section of the 'Deburgo' hub, and figure 384 an end elevation with the outer dust cover removed, so as to show the springs. The outer hub or frame *1*, to which the spokes are attached, is suspended from the inner hub or axle-box, *5*, by spiral springs, *11* and *12*. Frames *2* and *4*, forming rectangular guides at right angles to each other, are fixed respectively to the outer and inner hubs ; an intermediate slide, *3*, is formed with corresponding guides, the combination com-

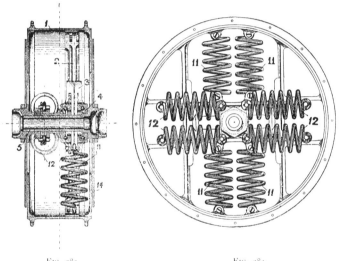

FIG. 383. FIG. 384.

pelling the outer to turn with the inner hub, while retaining their axes always parallel to each other, and allowing their respective centres perfect freedom of linear motion. To diminish friction a number of balls are introduced between the slides. Dust-caps, *14*, fixed to the inner hub enclose the springs and guides.

This spring wheel is quite rigid laterally, the only possible relative motion of the outer and inner hubs being at right angles to the direction of their axes.

259. **Definition.**—A *bearing* is the surface of contact of two pieces of mechanism having relative motion. In a machine the frame is the structure which supports the moving pieces, which are divided into *primary* and *secondary*, the former being those carried direct by the frame, the latter those carried by other moving pieces. In a more popular sense the bearing is generally spoken of as the portions of the frame and of the moving piece in the immediate neighbourhood of the surface of contact. In this sense the word 'bearing' will be used in this chapter. The bearings of a piece which has a motion of translation in a straight line must have cylindrical or prismatic surfaces, the straight lines of the cylinder or prism being parallel to the direction of motion. The bearings of pieces having rotary motion about a fixed axis must be surfaces of revolution. A part of a mechanism may have a helical motion—that is, a motion of rotation together with a motion of translation in the direction of the axis of rotation ; in this case the bearings must be formed to an exact screw.

The three forms of bearing above mentioned correspond to the three lower pairs in kinematics of machinery, viz. the sliding pair, the turning pair, and the screw pair. In each of these three cases the two parts having relative motion may have contact with each other over a surface.

260. **Journal, Pivot, and Collar Bearings.**— Figure 385 shows the simplest form of *journal* bearing for a rotating shaft, the section of the shaft and journal being circular. In this bearing no provision is made to prevent motion of the shaft in the direction of its axis. A bearing in which provision is made

to prevent the longitudinal motion of the shaft is called a *pivot* or *collar* bearing. Figure 386 shows the simplest form of pivot bearing, figure 387 a combined journal and pivot bearing, the end of the shaft being pressed against its bearing by a force in the direction of the axis. Figure 388 shows a simple form of collar bearing in

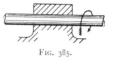

FIG. 385.

which the same object is attained. A rotating shaft provided with journal bearings may be constrained longitudinally, either by fixing a pivot bearing at each end, or by having a double collar bearing at some point along the shaft. This double collar bearing is usually combined with one of the journals, as at *A* (fig. 389), a collar being formed at each end of the cylindrical bearing. In a long shaft supported by a number of journals it is only necessary to have one double collar bearing; the other bearings should be quite free lon-

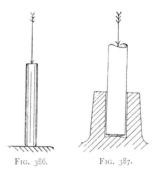

FIG. 386. FIG. 387.

gitudinally. Thus, in a tricycle axle with four bearings, the best result will be got by having the longitudinal motion of the axle

FIG. 388. FIG. 389.

controlled at only one of the bearings; if more collars, or their equivalents, are placed on the axle, the only effect is to increase the pressure of the collars on their bearings, and so increase the frictional resistance.

From the point of view of the constraint of the motion it would be quite sufficient for a journal bearing to have contact with the shaft at three points (fig. 390), but as there is usually a considerable pressure on the bearings they would soon be worn. The area of the surfaces of contact should be such that the pressure per square inch does not exceed a certain limit, depending on the material used and the speed of rubbing.

The bearings of the wheel of an 'Ordinary bicycle were originally made as at *A* (fig. 389), the bearing at each side of

FIG. 390.

the wheel being provided with collars, since the lateral flexibility of the forks was so great that otherwise the bearings would have sprung apart. It was impossible to keep the lubrication of the bearings constantly perfect, and with no film of oil between the surfaces the coefficient of friction rose rapidly and the resistance became serious.

Journal Friction.—In a well-designed journal the diameter of the surface of the fixed bearing should be a little greater than that of the rotating shaft (fig. 391). The direction of the motion

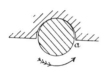

FIG. 391.

being then as indicated by the arrow, if the pressure is not too great, the lubricant at *a* is carried by the rotating shaft, and held by capillary attraction between the metal surfaces, so that the shaft is not in actual contact with its bearing, but is separated from it by a thin film of oil. From the experiments carried out by the Institution of Mechanical Engineers it appears that the friction of a perfectly lubricated shaft is very small, the coefficient being in some cases as low as ·001. This compares favourably with the friction of a ball-bearing.

Pivot Friction.—With a pivot or collar bearing the case is quite different. The rubbing surface of the shaft is continually in contact with the bearing, and cannot periodically get a fresh supply of oil (as in fig. 391) to keep between the two surfaces. The consequence is that, with the best form of collar bearing, the coefficient of friction is much higher. From the experiments of the Institution of Mechanical Engineers it appears that ·03 to ·06 may be taken as an average value of μ for a well lubricated collar bearing.

261. **Conical Bearings.**—In machinery subjected to much friction and wear, after running some time a shaft will run loose in its bearing. When the slackness exceeds a certain amount the

bearing must be readjusted. One of the simplest means for providing for this adjustment is shown in the *conical bearing* often used for the back wheel of an 'Ordinary' (fig. 392). The hub, *H*, ran loose on the spindle, *S*, which was fastened to the fork ends, F_1 and F_2. The surfaces of contact of the hub and spindle were conical, a *loose* cone, *C*, being screwed on near one

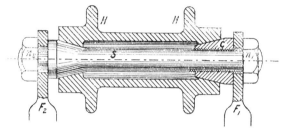

FIG. 392.

end of the spindle. If the bearing had worn loose, the cone *C* was screwed one or two turns further on the spindle until the shake was taken up. The cone was then locked in position by the nut n_1, which also fastened the end of the spindle to the fork. During this adjustment the other end of the spindle was held rigidly to the fork end F_2, by the nut n_2.

262. **Roller-bearings.**—The first improvement on the plain cylindrical bearing was the *roller-bearing*. Figure 393 is a

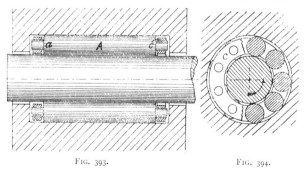

FIG. 393.   FIG. 394.

longitudinal, and figure 394 an end section of a roller-bearing. In this a number of cylindrical rollers, *A*, are interposed between the cylindrical shaft and the bearing-case, the axes of the rollers, *A*, being parallel to that of the shaft. These rollers were sometimes quite loose in the bearing-case, in which case as many rollers as could be placed in position round the shaft were used. More often, however, the ends of the rollers were turned down, forming small cylindrical journals, supported in cages *c*, one at each end of the roller. This cage served the purpose of keeping the distance between the rollers always the same, so that each roller revolved free of the others ; whereas, without the cage, two adjacent rollers would often touch, and a rubbing action would occur at the point of contact.

The chief advantage of a roller-bearing over a plain cylindrical bearing is that the lubrication need not be so perfect. While a plain bearing, if allowed to run dry, will very soon get hot ; a roller-bearing will run dry with little more friction than when lubricated.

A plain collar bearing must be used in conjunction with a roller-bearing, to prevent the motion of the shaft endways.

263. **Ball-bearings.**—Instead of cylindrical rollers, a number of balls, *B* (fig. 395), might be used. The principal difference in

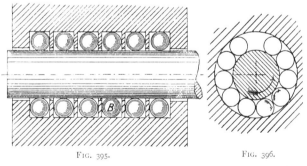

Fig. 395.   Fig. 396.

this case would be that each ball would have contact with the shaft and the bearing-case at a *point*, while each cylindrical roller had contact along a *line*. As a matter of fact, the surface of contact in the case of the ball-bearing would be a circle of very small diameter (point contact), while in the case of the roller-bearing it would be a very small, narrow rectangle of length equal to that of the roller (line contact). Other things being equal, the roller-bearing should carry safely a much greater load than the ball-bearing before crushing took place.

The motion of the balls in the bearing shown in figure 395 loaded at right angles to the axis, is one of pure rolling, the axis of rotation of the ball being always parallel to that of the axes of the rolling surfaces of the shaft and bearing-case.

264. **Thrust Bearings with Rollers.**—If a ball- or roller-bearing be required to resist pressure along the shaft, as in figures 386 and 387, the arrangement must be quite different. Two conical surfaces, *a v a* and *b v b*, formed on the frame and the rotating spindle respectively (fig. 397), having a common vertex at *v*, and a common axis coincident with the axis of the spindle, with conical rollers, *a v b*, having the same vertex, *v*, will satisfy the condition of pure rolling. If the axis, *v c*, of the conical roller be supposed fixed, and the spindle be driven, the cone *b v b*

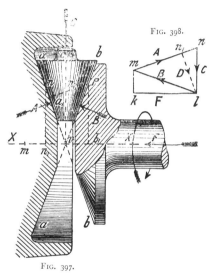

FIG. 398.

FIG. 397.

will drive the roller by friction contact, and it in turn will drive the cone *a v a*. If the cone *a v a* be fixed, and the spindle be driven, the relative motion of the three conical surfaces will remain the same ; but in this case the axis of the roller, *v c*, will also rotate about the axis *X X*. With perfectly smooth surfaces, the direction of the pressure is at right angles to the surface of contact, and very nearly so with well lubricated surfaces. On the conical roller, *a v b*, there will therefore be two forces, *A* and *B*, acting at right angles to its sides, *v a* and *v b*, respectively. These have a resultant along the axis *v c*, and unless a third force, *C*, be applied to the conical roller, it will be forced outwards during the motion.

The magnitudes of the forces *A*, *B*, and *C* can easily be found if the force, *F*, along the axis is given. In figure 398 draw *l k* equal to *F* and parallel to the axis *X X*, draw *l m* at right angles to *v b*, and *k m* at right angles to *X X* ; *l m* will give the magnitude of the force *B* ; draw *m n* and *l n* respectively at right angles to *v a* and *a b*, meeting at *n* ; *m n* and *n l* will be the magnitudes of the forces *A* and *C* respectively.

In figure 397 the conical roller is shown with a prolongation on its axis rubbing against the bearing-case, so that its further

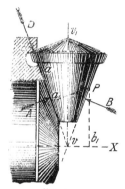

outward motion is prevented. With this arrangement there will be considerable rubbing friction between the end of the roller and the bearing-case. In Purdon & Walters' thrust bearing for marine engines the resultant outward pressure on the roller is balanced by letting its edge bear against a part of the bearing-case (fig. 399). The generating line, *v a*, of the roller is produced to a point *d* ; *d v₁* is drawn perpendicular to *v d*, and forms the generating line of a second conical surface coaxial with the first. A small portion on each side of *d* is the only part of this surface that presses against the bearing-case. The instantaneous axis of rotation being *v d*, there is no rubbing of the roller on the case, but only a relative spinning motion at *d*. In this case, the force-triangle (fig. 398) will have to be modified by drawing *l n₁* at right angles to *m n* ; *m n₁* will then be equal to the force *A*, and *l n₁* to the pressure *D* at *d*.

Relative Speeds of Roller and Spindle.— Let *P* (figs. 397 and 399) be any point in the line of contact of the conical roller with the spindle ; draw *P a₁* and *P b₁* at right angles to *v a* and *v X* respectively, and let *V* be the linear speed of the point *P* at any instant. Since *a₁* is a point on the instantaneous axis of rotation of the roller, and *b₁* a point on the fixed axis of rotation of the spindle, the angular speeds ω_2 and ω_1 of the roller and spindle are respectively

$$\omega_2 = \frac{V}{P a_1} \text{ and } \omega_1 = \frac{V}{P b_1} . \quad . \quad . \quad . \quad . \quad (1)$$

Therefore

$$\frac{\omega_2}{\omega_1} = \frac{P b_1}{P a_1} \quad . \quad . \quad . \quad . \quad . \quad . \quad . \quad (2)$$

Comparing figures 397, 399, and 398, the triangles $P v b_1$ and $l m k$ are similar; the triangles $P v a_1$ and $l m n_1$ are similar; so also are the four-sided figures $l k m n_1$ and $P b_1 v a_1$. Therefore,

$$\frac{\omega_1}{\omega_2} = \frac{P a_1}{P b_1} = \frac{l n_1}{l k} = \frac{D}{F} . \quad . \quad . \quad . \quad (3)$$

or,

$$F \omega_1 = D \omega_2 \quad . \quad . \quad . \quad . \quad . \quad . \quad . \quad (4)$$

That is, if only one roller be used, the angular speeds of the roller and spindle are inversely proportional to the pressures along their instantaneous axes of rotation.

If $v m$ (fig. 397) be set off along the axis of the spindle equal to $P a_1$, and $v n$ along $v a$ equal to $P b_1$, the vectors $v m$ and $v n$ will represent the rotations of the spindle and roller respectively, both in magnitude and direction. $v n$, the rotation of the roller, can be resolved into the rotations $v n_1$ and $v n_2$ about the axes of the shaft and roller respectively. It can easily be shown, from the geometry of the figure, that $v n_1 = \frac{1}{2} v m$; therefore the axis of the roller turns about the axis of the shaft at half the speed of the shaft.

The rotation $\overline{v n_1} = \frac{1}{2} \omega_1$, is equivalent to an equal rotation about a parallel axis through c (fig. 397), together with a translation $\frac{1}{2} \omega \times \overline{v c}$. This translation and rotation constitute a rubbing of the roller on the bearing at c. Thus, finally, the relative motion at c consists of a rubbing with speed $\overline{v c} \times \frac{1}{2} \omega$ and a spinning with speed $\omega_3 = \overline{v n_2}$.

From figures 397 and 398, $\dfrac{\omega_3}{\omega_1} = \dfrac{\overline{v n_2}}{2 \overline{v n_1}} = \dfrac{F}{C}$ (fig. 398).

Therefore, $\qquad F \omega_1 = C \omega_3,$ . . . . (5)

and if n rollers be used, with the total thrust W along the shaft,

$$W \omega_1 = n C \omega_3 \qquad . . . . (6)$$

If a number of conical rollers are interposed between the two conical surfaces on the shaft and bearing respectively, as in figure 397, the radial thrust, C, on the rollers may be provided for by a steel live-ring against which the ends of the rollers bear. This live-ring will rotate at half the speed of the shaft, and there will be no rubbing of the roller ends relative to it. But it should be noted that the speed of rotation ω_3 of each roller relative to the live-ring will be as a rule greater than the speed of rotation of the shaft, and therefore with a heavy end thrust on the shaft, the risk of abrasion of the outer ends of the rollers will be great. In a

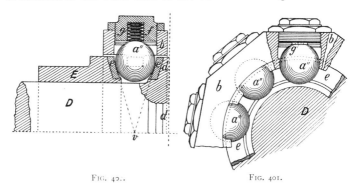

FIG. 400.   FIG. 401.

thrust bearing for marine engines, designed by the author, a number of lens-shaped steel discs were introduced between the outer end of each roller and the live-ring, so that the average relative spinning motion of two surfaces in contact is made equal to the relative speed between the roller and the live-ring, divided by the number of pairs of surfaces in contact. Figures 400 and 401 show a modification of this design, in which the conical rollers are replaced by balls, a'', rolling between hard steel rings, e, fixed on the shaft and the pedestal respectively. The small portions of these rings and of the balls in contact may be considered as conical surfaces with a common vertex, v. Anti-friction discs, g, are carried in a nut, f, which is screwed into and can be locked in position on the live-ring, b. This design (figs. 400 and 401) is arranged so that if one ball breaks it can be removed and replaced without disturbing any other part of the bearing. In this thrust-block a plain cylindrical bearing is used to support the shaft.

This bearing may be simplified by the omission of the anti-friction discs, and allowing the balls to run freely in the space

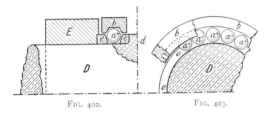

FIG. 402.   FIG. 403.

enclosed by the two steel rings, e, and the live-ring, b. Figure 402 is a part longitudinal section of such a simplified thrust bearing, and figure 403 a part cross section.

In a journal bearing the work lost in friction is proportional to the product of the pressure and the speed of rubbing, provided the coefficient of friction remains constant for all loads. In the same way, in a pivot bearing, the work lost in friction—other things being equal—is proportional to the product of the pressure and the angular speed. Equations (4) and (6), therefore, assert that it is impossible by any arrangement of balls or rollers to diminish the friction of a pivot bearing below a certain amount. If a shaft subjected to a longitudinal force can be supported by a plain pivot bearing (fig. 386), the work lost in friction will be a minimum. If, however, the circumstances of the case necessitate a collar bearing (fig. 388), an arrangement of balls or conical rollers may serve to get rid of the friction due to the rubbing of the collar on its bearings. In other words, the effective arm at which the frictional resistance acts may be reduced by a properly designed ball- or roller-bearing to a minimum, so that it may be equivalent to that illustrated in figures 400–1. The pressure on the pivot may sometimes be so great as to make it undesirable to support it by a bearing of the type shown in figure 386 ; the use of a bearing of either of the types shown in figures 400–1 and 402–3 with a number of balls or conical rollers, is equivalent to the subdivision of the total pressure into as many parts as there are rollers in the bearing.

265. **Adjustable Ball-bearing for Cycles.**—Figure 404 shows diagrammatically one of the forms of ball-bearing used almost universally for cycles. The external load on the bearings of a bicycle or a tricycle is always, with the exception of the ball steering-head, at right-angles to its axis ; any force parallel to the axis being simply due to the reaction of the bearing necessary to keep the spindle in its place. Figure 404 represents the section of the hub of a bicycle wheel ; the spindle, $S\,S$, is fixed to the

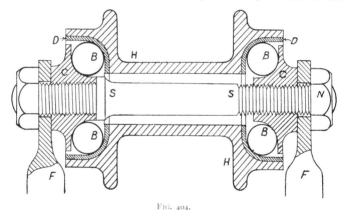

Fig. 404.

fork ; hardened steel 'cones,' $C\,C$, are screwed on its ends, and hardened steel cups, D, are fixed into the ends of the hub, H, which is of softer metal. The balls, B, run freely between the cone C and the cup D. One of the cones C is screwed up tight against a shoulder of the spindle S, the other is screwed up until the wheel runs freely on the spindle without undue shake, it is then locked in position by a lock-nut N, which usually also serves to fasten the spindle to the fork end, F.

266. **Motion of Ball in Bearing.**—Consider now the equilibrium of the ball B. It is acted on by two forces, f_1 and f_2 (fig. 405), the pressure of the wheel and the reaction of the

spindle respectively. Since the ball is in equilibrium, these two forces must be equal and opposite ; therefore the points of contact, *a* and *b*, of the ball with the cup and cone must be at the extremities of a diameter. During the actual motion in the bicycle the cone *C* is at rest, the ball *B* rolls round it, and the cup *D* rolls on the balls. The relative motion will be the same,

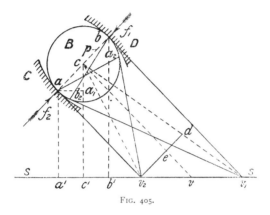

FIG. 405.

however, if a motion of rotation about the axis of the spindle, *S S*, be impressed on the whole system, equal in amount but opposite in direction to that of the centre of the balls round the axis, *S S.* The centre *c* of the ball *B* may thus be considered to be at rest, the ball to turn about an axis through its centre, the cup *D* and cone *C* to rotate in opposite directions about their common axis, *S S.*

Draw $b\,v_1$ at right angles to $a\,b$, cutting the spindle *S S* in v_1 (fig. 405, which is part of fig. 404 to a larger scale) ; from v_1 draw a tangent $v_1\,b_2$ to the circle *B*, of which $a\,b$ is the diameter. If the relative motion of the ball and cup at *b* be one of pure rolling, the portion of the ball in contact at *b* may be considered as a small piece of a cone $b\,v_1\,b_2$, and the portion of the ball-race at *b* part of a cone coaxial with *S S*, both cones having the common vertex v_1. The axis of rotation of the ball will pass through v_1 and the centre *c* of the ball *B*.

Now draw $a\,v_2$ at right angles to $b\,a$, cutting *S S* at v_2. If the relative motion of the ball and cone at a be one of pure rolling, the portions of the ball and cone surfaces in contact may be considered portions of cones having a common vertex v_2; the axis of rotation of the ball will thus be $c v_2$. But the ball cannot be rotating at the same instant about two separate axes $v_1 c$ and $v_2 c$, so that motions of pure rolling cannot exist at a and b simultaneously. If the surfaces of the cone and the cup be not equally smooth, it is possible that pure rolling may exist at the point of contact of the ball with the rougher surface. Suppose the rougher surface is that of the cup, the axis of rotation of the ball would then be $v_1 c$, and the motion at a would be rolling combined with a spinning about the axis $a c$ at right angles to the surface of contact. Draw $v_2 d$ parallel to $a c$, cutting $v_1 c$ at d. Then, if $c d$ represent the actual angular velocity of the ball about its axis of rotation $v_1 c$, $v_2 d$ will represent the angular velocity of the ball about the axis $a c$; since the rotation $c d$ is the resultant of a rotation $c v_2$ about the axis $c v_2$, and a rotation $v_2 d$ about the axis $c a : d c v_2$ is, in fact, the triangle of rotations about the three axes intersecting at c.

If the surfaces of the cone and cup be equally smooth the axis of rotation of the ball will be $c v$, v being somewhere between v_1 and v_2. If the angular speeds of the spinning motions at a and b be equal, $c v$ will bisect $d v_2$. If e be this point of intersection, $e c v_2$ and $e c d$ will be the triangles of rotation at the points a and b respectively.

The above investigation clearly shows that a grinding action is continually going on in all ball-bearings at present used in cycle construction. The grooves formed in the cone and cup after

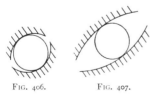

FIG. 406.   FIG. 407.

running some time are thus accounted for, while the popular notion that all but rolling friction is eliminated in a well-designed ball-bearing is shown to be erroneous. The effect of this grinding action will depend on the closeness with which the balls fit the cone and cup. If the radii of curvature of the ball, cone, and cup be nearly the same (fig. 406), friction due to the spinning will be great ; while, if they are perceptibly different (fig. 407), the friction of the bearing will be much less. On the other hand, a ball in the bearing (fig. 406) will be able to withstand greater pressure than a ball in the bearing (fig. 407), the surface of contact with a given load being so very much less in figure 407 than in figure 406.

267. **Magnitudes of the Rolling and Spinning of the Balls on their Paths.**—From a, c, and b (fig. 405) draw perpendiculars to the axis SS, and let ω be the actual angular speed of the wheel on its spindle, T the sum of the angular speeds of the spinning motions of the ball on its two bearing surfaces, r the radius ca of the ball, and R the radius cc^1 of the circle in which the ball centres run. From a draw aa_1 perpendicular to cc^1. Considering the motion relative to a plane passing through the spindle SS and the line vc—that is, considering the point c to be at rest, as described in section 266—let ω_1 be the angular speed of rotation of the ball about the axis vc, which may be assumed at right angles to ab. The linear speeds of the points a and b of the ball will be $\omega_1 r$. The angular speeds of the spindle and the wheel will be respectively

$$-\frac{\omega_1 r}{a\,a^1} \text{ and } \frac{\omega_1 r}{b\,b^1}.$$

But the spindle is actually at rest ; so, if the angular speed $\frac{\omega_1 r}{a\,a^1}$ about the axis SS be now added to the whole system, the actual angular speed of the wheel will be

$$\omega = \left(\frac{1}{a\,a^1} + \frac{1}{b\,b^1} \right) \omega_1 r \quad . \quad . \quad . \quad . \quad . \quad (7)$$

Denoting the length ca_1 by q,

$$a\,a^1 = R - q, \text{ and } b\,b^1 = R + q ;$$

equation (7) may therefore be written

$$\omega_1 = \frac{R^2 - q^2}{2\,R\,r} \omega \quad . \quad . \quad . \quad . \quad . \quad (8)$$

But by section 266

$$T = \frac{v_2\,d}{e\,c} \boldsymbol{\omega}_1 \quad . \quad . \quad . \quad . \quad . \quad . \quad (9)$$

Combining (8) and (9)

$$T = \frac{v_2\,d}{e\,c} \cdot \frac{R^2 - q^2}{2\,R\,r}\,\omega \quad . \quad . \quad . \quad . \quad . \quad (10)$$

An inspection of the diagram (fig. 405) will show that the fractions $\frac{v_2\,d}{e\,c}$ and $\frac{R^2 - q^2}{2\,R\,r}$ are smaller the nearer the diameter $a\,b$ of contact of the ball with its bearings is to a perpendicular to the spindle $S\,S$. Also, the distance $e\,c$ depends on the position of the actual axis of rotation, $c\,v$, of the ball; but it does not vary greatly, its maximum value being when it coincides with $c\,v_2$, its minimum when it is perpendicular to $a\,b$.

The above considerations show that a ball-bearing arranged as in the full lines (fig. 408) will be much better than the one

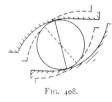

FIG. 408.

arranged as shown by the dotted lines. The end thrust in bicycle bearings is always small, so that the line of contact $a\,b$ need not be inclined 45° to the axis, but be placed nearer a perpendicular to the axis.

The rolling of the balls on the bearings will be much less prejudicial than the spinning; it may be calculated as follows:

The linear speed of the point b of the ball (fig. 405) is $\omega \times b\,b^1 = (R + q)\,\omega$. The angular speed of rolling of the ball about the axis $a\,v_2$ is therefore $\frac{(R + q)}{2\,r}\,\omega$. Consider now the outer path D to be fixed, and the inner path C to revolve with the angular speed $-\omega$; the relative motion will be, of course, the same as before. The linear speed of the point a of the ball is $\omega \times \overline{a\,a^1} = (R - q)\,\omega$, and the angular speed of rolling of the ball about the axis $b\,v_1$ is therefore $\frac{(R - q)}{2\,r}\,\omega$. The sum of the rolling speeds of the ball at a and b is therefore

$$\frac{R}{r}\,\omega, \quad . \quad . \quad . \quad . \quad . \quad . \quad . \quad . \quad (11)$$

a result independent of the angle that the diameter of contact $a\,b$

makes with the axis of the bearing. The pressure on the ball, however, and therefore also the rolling friction depends on this angle.

Example.—In the bearing of the driving-wheels of a Safety bicycle the balls are $\frac{1}{4}$ in. diameter, the ball circle—that is, the circle in which the centres of the balls lie—is ·8 in. diameter, and the line of contact of the ball is inclined 45°; find the angular speed of the spinning of the balls on their bearing. Figure 409 is the diagram for this case drawn to scale, from which $v_2\,d =$ ·21 in., $e\,c =$ ·44 in., and $q =$ ·09. Substituting these values in (10)

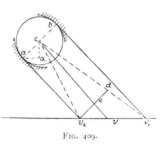

Fig. 409.

$$T = \frac{·21\,(·16 - ·0081)}{·44 \times 2 \times ·4 \times ·125}\,\omega = ·72\,\omega.$$

That is, for every revolution of the hub, the total spinning of each ball relative to the bearings is nearly three-fourths of a revolution.

The pressure on each ball in this case is $\sqrt{2}$ times the *vertical* load on it. Hence the resistance due to spinning friction of the balls will be $·72\sqrt{2}, = 1·018$ times that of a simple pivot-bearing formed by placing a single ball between the end of the pivot and its seat, the total load being the same in each case.

The sum of the speeds of rolling of the ball is, by (11),

$$\frac{·8}{·25}\,\omega = 3·2\,\omega.$$

268. **Ideal Ball-bearing.**—The external load on the ball-bearing of a cycle is usually at right angles to the axis, but from the arrangement of the bearing (fig. 404) the pressure on the balls has a component parallel to the axis. This component has to be resisted by the bearing acting practically as a collar bearing,

as described in section 260. Thus not only is the actual pressure on the balls increased, but instead of having a motion of pure rolling, a considerable amount of spinning motion under considerable pressure is introduced. The actual force in the direction of the axis necessary to keep the wheel hub in place is very small compared with the total external load ; a ball-bearing in which the load is carried by one set of balls, arranged as in figure 395, and the end thrust taken up by another set, might therefore be expected to offer less frictional resistance than those in use at present. Such a bearing is shown in figure 410. The main balls, *B* (fig. 410), transmitting the load from the wheel to the

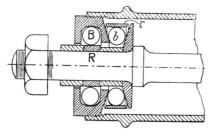

spindle run between coaxial cylindrical surfaces on the spindle and hub respectively ; the motion of the balls, *B*, relative to both surfaces, is thus one of pure rolling. The space in which the balls run is a little longer, parallel to the axis of the spindle,

FIG. 410.

than their diameter, so that they do not bear sideways. The wheel is kept in position along the spindle by a set of balls, *b*, running between two conical surfaces on the spindle and hub respectively, having a common vertex, and kept radially in place by a live-ring, *r*. One of these cones is fixed to the spindle, the other forms part of the main ball cup. This bearing is therefore a combination of the ball-bearing (fig. 395) and the thrust bearing (fig. 402–3). The motion of the main balls, *B*, being pure rolling, the necessity of providing means of adjustment will not be so great as with the usual form ; in fact, the bearing being properly made by the manufacturer may be sent out without adjustment. A play of a hundredth part of an inch might be allowed in the two main rows of balls, *B*, and a longitudinal play of one-twentieth of an inch for the secondary rows, *b*. If the main row of balls ultimately run loose, a new hard steel ring, *R*, can be easily slipped on the spindle.

If adjustments for wear are required in this type of bearing, they can be provided by making the hard steel ball ring, *R*, slightly tapered (fig. 411), and screwing it on the spindle. It would be locked in position by the nut fixing the spindle to the frame. There would be an adjustment at each end.

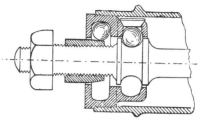

FIG. 411.

These bearings may be somewhat simplified in construction, though the frictional resistance under an end thrust will be theoretically increased, by omitting the live-ring confining the secondary balls, and merging it in either the cup or the conical disc (fig. 412). If this be done a single ball will probably be sufficient for each row of secondary balls, *b*. If a double collar be formed near one end of the spindle, one row of secondary balls, *b*, would be sufficient for the longitudinal constraint. They could be put in place through a hole in the ball cup (fig. 411), or by screwing an inner ring on the cup

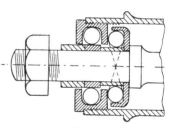

FIG. 412.

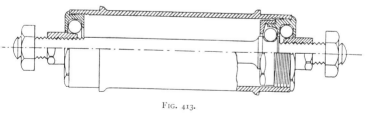

FIG. 413.

(fig. 413). The other end of the bearing will have only the main row of balls.

269. **Mutual Rubbing of Balls in the Bearing.**—Figure 414
may be taken to represent a section of a ball-bearing by a plane
at right angles to the axis, the central spindle being fixed and
the outer case revolving in the direction of the arrow a. The
balls will therefore roll on the fixed spindle in the direction
indicated. If two adjacent balls, B_1 and B_2, touch each other
there will be rubbing at the point of contact, and of course the
friction resistance of the bearing will be increased. Now, in

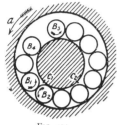

FIG. 414.

a ball-bearing properly adjusted the adjusting cone is not screwed up quite tight,
but is left in such a position that the
balls are not all held at the same
moment between the cones and cups ;
in other words, there is a little play left
in the bearing. Figure 414 shows such
a bearing sustaining a vertical load, as in
the case of the steering-wheel of a bicycle,
with the play greatly exaggerated for
the sake of clearness of illustration. The cone on the wheel
spindle will rest on the balls near the lowest part of the bearing,
and the balls at the top part of the bearing will rest on the cone,
but be clear of the cup of the wheel. Thus, a ball in its course
round the bearing will only be *pressed* between the two surfaces
while in contact at any point of an arc, c_1 c_2, and will run loose
the rest of the revolution. The balls should never be jammed
tightly round the bearing, or the mutual rubbing friction will be
abnormally great. The ascending balls will all be in contact, the
mutual pressure being due merely to their own weight. A ball,
B_3, having reached the top of the bearing will roll slightly forward
and downward, until stopped by the ball in front of it, B_4. The
descending balls will all be in contact, the mutual pressure being
again due to their own weight. On coming into action at the arc
c_1 c_2, the pressure on the balls tends to flatten them slightly in the
direction of the pressure, and to extend them slightly in all directions at right angles. The mutual pressure between the balls may
thus be slightly increased, but it is probable that it cannot be
much greater than that due to the weight of the descending balls.
As this only amounts to a very small fraction of an ounce, in com-

parison with the spinning friction above described under a total load of perhaps 100 lbs., the friction due to the balls rubbing on each other is probably negligibly small.

Figure 414 represents the actions in the bearings of non-driving wheels of bicycles and tricycles, and in the driving-wheels of chain-driven Safety bicycles ; also, supposing the outer case fixed and the inner spindle to revolve, it represents the action in the crank-bracket of a rear-driving Safety.

In the bearings of the front wheel of an 'Ordinary,' or the front-driving Safety, the action is different, and is represented in figure 415. In these cases the balls near the upper part of the bearing transmit the pressure, the lower balls being idle. The motion being in the direction shown by the arrow a, the ball B_1 is just about to roll out of the arc of action, and will drop on the top of the ball B_2. The ball B_3, ascending upwards, will move into the arc of action c_1 c_2, and will be carried round, while the ball behind it, B_4, will lag slightly behind. In this way, it is

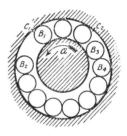

possible that there may be no actual contact between the balls transmitting the pressure.

It would be interesting to experiment on the coefficient of friction of the same ball-bearing under the two different conditions illustrated in figures 414 and 415. In some of the earlier ball-bearings the balls were placed in cages, so as to prevent their mutual rubbing. Figures 416 and 417 show the 'Premier' bearing with ball-cage. It does not appear that the rubbing of the balls on the sides of the cage is less prejudicial than their mutual rubbing ; and as, with a cage, a less number of balls could be put into a bearing, cages were soon abandoned.

Effect of Variation in Size of Balls.—If one ball be slightly larger than the others used in the bearing, it will, of course, be subjected to a greater pressure than the others ; in fact, the whole load of the bearings may at times be transmitted by it, and there will be a probability of it breaking and consequent damage to the surface of the cone and cup. Let V be the linear speed of the point of the cup in contact with the ball (fig. 414), R the radius of the ball centre, and r the radius of the ball ; the linear speed of the ball centre is $\dfrac{V}{2}$, and its angular speed round the axis of

FIG. 416.

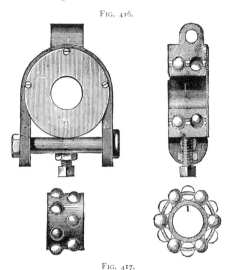

FIG. 417.

the spindle is $\dfrac{V}{2\,R}$. The radius R is the sum of the radii of a ball and of the circle of contact with the cone ; consequently the angular speed round the centre of the spindle of a ball slightly larger than the others will be less than that of the others, the large ball will tend to lag behind and press against the following ball.

If P be the bearing pressure on the large ball, the mutual pressure, F, between it and the following ball may amount to $\mu\,P$, and the frictional resistance of the bearing will be largely increased.

The mutual rubbing of the balls may be entirely eliminated by having the balls which transmit the pressure alternating with others slightly smaller in diameter. The latter will be subjected only to the mutual pressure between them and the main balls, and will rotate in the opposite direction. They may rub on the bearing-case or spindle, but, since the pressure at these points approaches zero, there will be very little resistance. This device may be used satisfactorily in a ball-thrust bearing, but in a bicycle ball-bearing the number of balls in action at any moment may be too small to permit of this.

270. **The Meneely Tubular Bearing.**—In the Meneely tubular bearing, made by Messrs. Siemens Brothers (fig. 418), the mutual rubbing of the rollers is entirely eliminated by an ingenious arrangement. "The bearing is composed of steel tubes, uniform in section, which are grouped closely, although not in contact with each other, around and in alignment with the

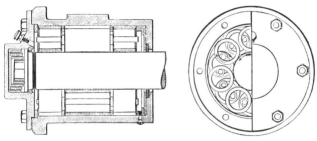

Fig. 418.

journal; these rollers are enclosed within a steel-lined cylindrical housing. They are arranged in three series, the centre series being double the length of the outer series. Each short tube is in axial alignment with the corresponding tube of the opposite end series, while exactly intermediate to these end lines are arranged the axes of the centre series, thus making the lines of bearing equal. Each end tube overlaps two centre tubes, as shown in figure 418. To keep the long and short tubes in proper relative position, there are threaded through their insides round steel rods. These rods both lock the rollers together and hold them apart in their proper relative position, collars on the rods also serving to aid in maintaining the endwise positions. These connecting-rods share in the general motion, rolling without friction in contact with the tubes. They intermesh the long and short tubes, and keep them rigidly in line with the axis." For a journal 3 in. diameter, the external and internal diameters of the rollers are 2 in. and $1\frac{1}{2}$ in. respectively.

271. **Ball-bearing for Tricycle Axle.**—Figure 419 represents a form of ball-bearing often used for supporting a rotating axle,

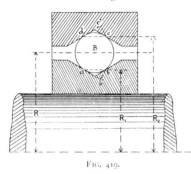

FIG. 419.

as the front axle of an 'Ordinary,' tricycle axles, &c. This bearing supports the load at right angles to the axle and at the same time resists end-way motion. A ball has contact with the ball-races at four points, *a, b, c, d,* which for the best arrangement should be in pairs parallel to the axis ; the motion of the ball will then be one of rotation, the instantaneous axis being *c d,* its line of contact with the bearing case. The motion of the ball relative to any point of the surface it touches will, however, be one of rolling combined with spinning about an axis perpendicular to the surface of contact.

Figure 420 shows this form as made adjustable by Mr. W. Bown. The outer ball-cup is screwed into the bearing case, and

B.

FIG. 420.

when properly adjusted is fixed in position by a plate and set screw. If this bearing be attached to the frame or fork by a bolt having its axis at right angles to the rotating spindle, it will automatically adjust itself to any deflection of the frame or spindle ; the axes of the spindle and bearing case always remaining coincident.

Let ω be the angular speed of the axle, *r* the radius of the ball, R_1 *R,* and R_2 the distances of the points *a,* *B,* and *d* from the axis *S S.* The linear speed of the point *a* common to the ball and the axle will be ω R_1. The angular speed of the ball about its instantaneous axis of rotation *d c* will be

$$- \frac{\omega\,R_1}{a\,d} = - \frac{\omega\,R_1}{R_2 - R_1} \qquad \cdot \quad \cdot \quad \cdot \quad \cdot \quad \cdot \quad (12)$$

The relative angular speed of the ball and axle about their instantaneous axis $a\,b$ will be

$$\omega + \frac{\omega\,R_1}{R_2 - R_1} = \frac{\omega\,R_2}{R_2 - R_1} \qquad \cdots \qquad (13)$$

Draw $c\,c^1$ at right angles to the tangent to the ball at d; then at the point d the actual rotation of the ball about the axis $c\,d$ can be resolved into a rolling about the axis $d\,c^1$ and a spinning about an axis $d\,B$ at right angles; $d\,c\,c^1$ will be the triangle of rotations at the point d. If the angular speed of spinning of the ball at d is T_d, we have

$$-\frac{T_d}{\dfrac{\omega\,R_1}{R_2 - R_1}} = \frac{c\,c^1}{c\,d}, \quad i.e., \quad T_d = -\frac{c\,c^1}{c\,d}\cdot\frac{R_1}{(R_2 - R_1)}\omega \quad . \quad (14)$$

Draw $b\,b^1$ perpendicular to the tangent at a; then, in the same way, it may be shown that the angular speed of the relative spinning at the point a is

$$T_a = \frac{b\,b^1}{b\,a}\cdot\frac{R_2}{(R_2 - R_1)}\omega \quad . \quad \cdots \quad (15)$$

From (14) and (15) the speeds of spinning at a and d are inversely proportional to the radii; the circumferences of the bearings at a and b are also proportional to the radii. If the wear of the bearing be proportional to the relative spinning speed of the ball, and inversely proportional to the circumference—both of which assumptions seem reasonable—the wear of the inner and outer cases at a and d will be inversely proportional to the squares of their radii. If the bearing surfaces at a, b, c, and d are all equally inclined to the axis, $\dfrac{c\,c^1}{c\,d} = \dfrac{b\,b^1}{b\,a}$; then adding (14) and (15), the sum of the angular speeds of spinning at a, b, c, and d will be

$$T = 2\frac{c\,c^1}{c\,d}\cdot\frac{R_2 + R_1}{R_2 - R_1}\omega$$

$$= 4\frac{c\,c^1}{c\,d}\cdot\frac{R}{a\,d}\omega \quad . \quad \cdots \quad (16)$$

If θ be the angle that the tangent $d\ c^1$ makes with the axis of the bearing, $\dfrac{c\ c^1}{c\ d} = sin\ \theta,\ a\ d = 2\ r\ cos\ \theta$, and (16) may be written,

$$T = \frac{2\ R\ tan\ \theta}{r}\omega \quad . \quad . \quad . \quad . \quad . \quad . \quad (17)$$

Equation (17), therefore, shows that the spinning motion in this form of bearing is proportional to the radius of the ball circle, inversely proportional to the radius of the ball, and directly proportional to the tangent of the angle the bearing surfaces make with the axis.

Example.—Let the four bearing surfaces be each inclined 45° to the axis ; then $tan\ \theta = 1$, and (17) becomes

$$T = \frac{2\ R}{r}\omega \quad . \quad . \quad . \quad . \quad . \quad . \quad (18)$$

If the diameter of the ball is $\dfrac{1''}{4}$, $r = \frac{1}{8}$, and if R is $\dfrac{1''}{2}$;

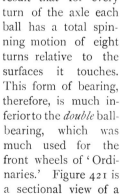

substituting in (18),

$$T = 8\omega.$$

This gives the startling result that for every turn of the axle each ball has a total spinning motion of eight turns relative to the surfaces it touches. This form of bearing, therefore, is much inferior to the *double* ball-bearing, which was much used for the front wheels of 'Ordinaries.' Figure 421 is a sectional view of a

FIG. 421.

double ball-bearing as used by Messrs. Singer & Co. The motion of the balls in this bearing is the same as that analysed in section 266.

272. **Ordinary Ball Thrust Bearing.**—Figure 422 is a section of a form of ball thrust bearing which is sometimes used in light drilling and milling machines. The lower row in the ball-head of a cycle also forms such a bearing.

The arrangement of the ball and its grooves, shown in figure 422, is almost as bad as it could possibly be. Let a, b, c,

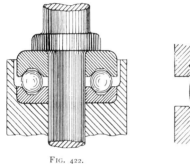

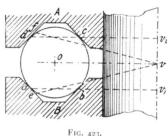

FIG. 422.   FIG. 423.

and d (fig. 423) be the points of contact of the ball with the sides of the groove, and o the centre of the ball. If no rubbing takes place at the points a and b, the instantaneous axis of rotation of the ball relative to the groove B must be the line $a\,b$; that is, the motion of the ball is the same as that of a cone, with vertex v_1 and semi-angle $o\,v_1\,a$, rolling on the disc of which the line $a\,v_1$ is a section. Suppose now that there is no rubbing at the point d, and let ω be the angular speed of the spindle A. Drop a perpendicular $d\,v_2$ on to the axis. Then V_d, the linear speed of the point D, will be

$$\omega \times \overline{v_2\,d,}$$

and V_c, the linear speed of the point c of the spindle, will be

$$\omega \times \overline{v_2\,c.}$$

The angular speed of the ball is

$$\frac{V_d}{d\,a} = \omega \times \frac{v_2\,d}{d\,a}.$$

The linear speed of the point c on the ball must be equal to the speed of the point d on the ball, since these two points are at the same distance from the instantaneous axis of rotation $a\,v_1$. Therefore the speed of rubbing at the point c is

$$\omega \times (\overline{v_2\,d} - \overline{v_2\,c})$$

$$= \omega \times \overline{c\,d}.$$

If the grooves are equally smooth it seems probable that the actual vertex, v, of the rolling cone will be about midway between v_1 and v_2, and the rolling cone, equivalent to the ball, will be $e\,v\,f$; the points a and d will lie inside this cone, the points b and c outside, and the rubbing will be equally distributed between the points a, b, c, and d.

In comparison with the rubbing, the rolling and spinning frictions will be small. A much better arrangement would be to have only one groove, the other ball-race being a flat disc.

273. **Dust-proof Bearings**.—If the ball-bearing (fig. 404) be examined it will be noticed that there is a small space left between the fixed cone C, and the cup D, fastened to the rotating hub. This is an essential condition to be attended to in the design of ball-bearings. If actual contact took place between the cone and the cup, the rubbing friction introduced would require a greater expenditure of power on the part of the rider. Now, for a ball-bearing to work satisfactorily, the adjusting cone should not be screwed up quite tight, but a perceptible play should be left between the hub and the spindle; the clearance between the cup C and cone D should therefore be a little greater than this.

In running along dusty roads it is possible that some may enter through this space, and get ground up amongst the balls. In so-called dust-proof bearings, efforts are made to keep this opening down to a minimum, but no ball-bearing can be absolutely dust-proof unless there is actual rubbing contact between the rotating hub and a washer, or its equivalent, fastened to the spindle. Approximately dust-proof bearings can be made by arranging that there shall be no corners in which dust may easily find a lodgment. Again, it will be noticed that the diameter of the annular opening for the ingress of dust is smaller in the bearing figure 413 than in the bearing figure 404 ; the former bearing should, therefore, be more nearly dust-proof than the latter.

The small back wheels of ' Ordinaries' often gave trouble from dust getting into the bearings, such dust coming, not only from the road direct, but also being thrown off from the driving-wheel. When a bearing has to run in a very dusty position a thin washer of leather may be fixed to the spindle and press lightly on the rotating hub, or *vice versâ*. The frictional resistance thus introduced is very small, and does not increase with an increase of load on the bearing.

274. **Oil-retaining Bearings.**—Any oil supplied through a hole at the middle of the hub in the bearing shown in figure 404 will sooner or later get to the balls, and then ooze out between the cup and cone. In the bearing shown in figure 413, on the other hand, the diameter of the opening between the spindle and hub being much less than the diameter of the outer ball-race, oil will be retained, and each ball at the lowest part of its course be immersed in the lubricant.

Figure 380 is a driving hub and spindle, with oil-retaining bearings, made by Messrs. W. Newton & Co., Newcastle-on-Tyne. Figure 381 is a hub, also with oil-bath lubrication, by the Centaur Cycle Company.

275. **Crushing Pressure on Balls.**—In a row of eight or nine balls, all exactly of the same diameter and perfectly spherical, running between properly formed races, it seems probable that the load will be distributed over two or three balls. If one ball is a trifle larger than the others in the bearing, it will have, at intervals, to sustain all the weight. In a ball thrust bearing with balls of uniform size, the total load is distributed amongst all the balls. The following table of crushing loads on steel balls is given by the Auto-Machinery Company (Limited), Coventry, from which it would appear that if P be the crushing load in lbs., and d the diameter of the ball in inches,

$$P = 82400 \, d^2 \quad . \quad . \quad . \quad . \quad . \quad . \quad (19)$$

TABLE XIII.—WEIGHTS, APPROXIMATE CRUSHING LOADS, AND SAFE WORKING LOADS OF DIAMOND CAST STEEL BALLS.

Diameter of ball	Weight per gross	Crushing load	Working load
in.	lbs.	lbs.	lbs.
$\frac{1}{8}$	·0415	1,288	160
$\frac{3}{16}$	·1401	2,900	360
$\frac{1}{4}$	·3322	5,150	640
$\frac{5}{16}$	·6488	8,050	1,000
$\frac{3}{8}$	1·1213	11,600	1,450
$\frac{1}{2}$	2·6576	20,600	2,570

276. **Wear of Ball-bearings.**—It is found that the races in ball-bearings are grooved after being some time in use. This grooving may be due partly to an actual removal of material owing to the grinding motion of the balls, and partly to the balls gradually pressing into the surfaces, the balls possibly being slightly harder than the cups and cones.

Professor Boys has found that the wear of balls in a bearing is practically negligible ('Proc. Inst. Mech. Eng.,' 1885, p. 510).

277. **Spherical Ball-races.**—If by any accident the central spindle in a ball-bearing gets bent, the axes of the two ball-races will not coincide, and the bearing may work badly. Messrs. Fichtel & Sachs, Schweinfurt, Germany, get over this difficulty

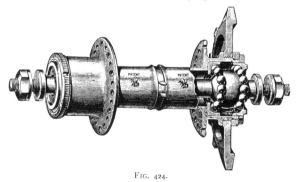

FIG. 424.

by making the inner ball-race spherical (fig. 424), so that however the spindle be bent the ball-race surface will remain unaltered.

278. **Universal Ball-bearing**. — Figure 425 shows a ball-bearing designed by the author, in which either the spindle or the hub may be considerably bent without affecting its smooth running. The cup and cone between which the balls run, instead of being rigidly fixed to the hub and spindle respectively, rest on concentric spherical surfaces. One of the spindle spherical surfaces is made on the adjusting

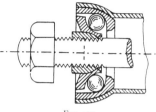

FIG. 425.

nut. This bearing, automatically adjusting itself, requires no care to be taken in putting it together. The working parts, being loose, can be renewed, if the necessity arises, by an unskilled person.

In the case of a bicycle falling, the pedal-pin runs a great chance of being bent, a bearing like either of the two above described seems therefore desirable for, and specially applicable to, pedals.

CHAPTER XXVI

279. **Transmission of Power by Flexible Bands.**—A flexible steel band passing over two pulleys was used in the 'Otto' dicycle to transmit power from the crank-axle to the driving-wheels. The effort transmitted is the difference of the tensions of the tight and slack sides of the band ; the maximum effort that can be transmitted is therefore dependent on the initial tightness. Like belts or smooth bands, chains are flexible transmitters. If the speed of the flexible transmitter be low, the tension necessary to transmit a certain amount of power is relatively high. In such cases the available friction of a belt on a smooth pulley is too low, and gearing chains must be used. Projecting teeth are formed on the drums or wheels, and fit into corresponding recesses in the links of the chain.

A chain has the advantage over a band, that there is, or should be, no tension on its slack side, so that the total pressure on the bearing due to the power transmitted is just equal to the tension on the driving side.

For chain gearing to work satisfactorily, the pitch of the chain should be equal to that of the teeth of the chain-wheels over which it runs. Unfortunately, gearing chains subjected to hard work gradually stretch, and when the stretching has exceeded a certain amount they work very badly.

Gear.—The total effect of the gearing of a cycle is usually expressed by giving the diameter of the driving-wheel of an 'Ordinary' which would be propelled the same distance per turn of the pedals. Thus, if a chain-driven Safety has a 28-in. driving-wheel which makes two revolutions to one of the crank-axle, the machine is said to be geared to 56 in. Let N_1 and N_2 be the numbers of teeth on the chain-wheels on driving-wheel hub and crank-axle respectively, and d the diameter of the driving-wheel in inches, then the machine is geared to

$$\frac{N_2}{N_1}\, d \text{ inches} \qquad\qquad\qquad (1)$$

The distance travelled by the machine and rider per turn of the crank-axle is of course

$$\frac{N_2}{N_1}\, \pi\, d \text{ inches} \qquad\qquad\qquad (2)$$

The following table of gearing may be found useful for reference :

TABLE XIV.—CHAIN GEARING.

Gear to which Cycle is Speeded.

Number of teeth on		Diameter of driving wheel											
Crank-axle	Hub	22	24	26	28	30	32	34	36	38	40	42	44
16	7	$50\frac{2}{7}$	$54\frac{6}{7}$	$59\frac{3}{7}$	64	$68\frac{4}{7}$	$73\frac{1}{7}$	$77\frac{5}{7}$	$82\frac{2}{7}$	$86\frac{6}{7}$	$91\frac{3}{7}$	96	$100\frac{4}{7}$
16	8	44	48	52	56	60	64	68	72	76	80	84	88
16	9	$39\frac{1}{9}$	$42\frac{2}{3}$	$46\frac{2}{9}$	$49\frac{7}{9}$	$53\frac{1}{3}$	$56\frac{8}{9}$	$60\frac{4}{9}$	64	$67\frac{5}{9}$	$71\frac{1}{9}$	$74\frac{2}{3}$	$78\frac{2}{9}$
16	10	$35\frac{1}{5}$	$38\frac{2}{5}$	$41\frac{3}{5}$	$44\frac{4}{5}$	48	$51\frac{1}{5}$	$54\frac{2}{5}$	$57\frac{3}{5}$	$60\frac{4}{5}$	64	$67\frac{1}{5}$	$70\frac{2}{5}$
17	7	$53\frac{3}{7}$	$58\frac{2}{7}$	$63\frac{1}{7}$	68	$72\frac{6}{7}$	$77\frac{5}{7}$	$82\frac{4}{7}$	$87\frac{3}{7}$	$92\frac{2}{7}$	$97\frac{1}{7}$	102	$106\frac{6}{7}$
17	8	$46\frac{3}{4}$	51	$55\frac{1}{4}$	$59\frac{1}{2}$	$63\frac{3}{4}$	68	$72\frac{1}{4}$	$76\frac{1}{2}$	$80\frac{3}{4}$	85	$89\frac{1}{4}$	$93\frac{1}{2}$
17	9	$41\frac{5}{9}$	$45\frac{1}{3}$	$49\frac{1}{9}$	$52\frac{8}{9}$	$56\frac{2}{3}$	$60\frac{4}{9}$	$64\frac{2}{9}$	68	$71\frac{7}{9}$	$75\frac{5}{9}$	$79\frac{1}{3}$	$83\frac{1}{9}$
17	10	$37\frac{2}{5}$	$40\frac{4}{5}$	$44\frac{1}{5}$	$47\frac{3}{5}$	51	$54\frac{2}{5}$	$57\frac{4}{5}$	$61\frac{1}{5}$	$64\frac{3}{5}$	68	$71\frac{2}{5}$	$74\frac{4}{5}$
18	7	$56\frac{4}{7}$	$61\frac{5}{7}$	$66\frac{6}{7}$	72	$77\frac{1}{7}$	$82\frac{2}{7}$	$87\frac{3}{7}$	$92\frac{4}{7}$	$97\frac{5}{7}$	$102\frac{6}{7}$	108	$113\frac{1}{7}$
18	8	$49\frac{1}{2}$	54	$58\frac{1}{2}$	63	$67\frac{1}{2}$	72	$76\frac{1}{2}$	81	$85\frac{1}{2}$	90	$94\frac{1}{2}$	99
18	9	44	48	52	56	60	64	68	72	76	80	84	88
18	10	$39\frac{3}{5}$	$43\frac{1}{5}$	$46\frac{4}{5}$	$50\frac{2}{5}$	54	$57\frac{3}{5}$	$61\frac{1}{5}$	$64\frac{4}{5}$	$68\frac{2}{5}$	72	$75\frac{3}{5}$	$79\frac{1}{5}$
19	8	$52\frac{1}{4}$	57	$61\frac{3}{4}$	$66\frac{1}{2}$	$71\frac{1}{4}$	76	$80\frac{3}{4}$	$85\frac{1}{2}$	$90\frac{1}{4}$	95	$99\frac{3}{4}$	$104\frac{1}{2}$
19	9	$46\frac{4}{9}$	$50\frac{2}{3}$	$54\frac{8}{9}$	$59\frac{1}{9}$	$63\frac{1}{3}$	$67\frac{5}{9}$	$71\frac{7}{9}$	76	$80\frac{2}{9}$	$84\frac{4}{9}$	$88\frac{2}{3}$	$92\frac{8}{9}$
19	10	$41\frac{4}{5}$	$45\frac{3}{5}$	$49\frac{2}{5}$	$53\frac{1}{5}$	57	$60\frac{4}{5}$	$64\frac{3}{5}$	$68\frac{2}{5}$	$72\frac{1}{5}$	76	$79\frac{4}{5}$	$83\frac{3}{5}$
20	8	55	60	65	70	75	80	85	90	95	100	105	110
20	9	$48\frac{8}{9}$	$53\frac{1}{3}$	$57\frac{7}{9}$	$62\frac{2}{9}$	$66\frac{2}{3}$	$71\frac{1}{9}$	$75\frac{5}{9}$	80	$84\frac{4}{9}$	$88\frac{8}{9}$	$93\frac{1}{3}$	$97\frac{7}{9}$
20	10	44	48	52	56	60	64	68	72	76	80	84	88

280. **Early Tricycle Chain.**—Figure 426 illustrates the 'Morgan' chain, used in some of the early tricycles, which was

composed of links made from round steel wire alternating with
tubular steel rollers. There being only line contact between

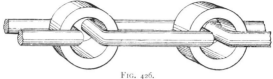

FIG. 426.

adjacent links and rollers, the wear was great, and this form of
chain was soon abandoned.

281. **Humber Chain.**—Figure 427 shows the 'Humber' chain,
formed by a number of hard steel blocks (fig. 428) alternating

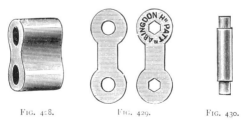

FIG. 427.

with side-plates (fig. 429). The side-plates are riveted together
by a pin (fig. 430), which passes through the hole in the block.
The rivet-pin is provided with shoulders at each end, so that the
distance between the side-plates is preserved a trifle greater than

FIG. 428.   FIG. 429.   FIG. 430.

the width of the block. In the 'Abingdon-Humber' chain the
holes in one of the side-plates are hexagonal, so that the pair of
rivet-pins, together with the pair of side-plates they unite, form
one rigid structure, and the pins are prevented from turning in the
side-plates.

Figure 431 shows a 'Humber' pattern chain, made by Messrs. Perry & Co. The improvement in this consists principally in the addition of a pen-steel bush surrounding the rivet. The ends of the bush are serrated, and its total length between the points is a trifle greater than the distance between the shoulders of the

FIG. 431.

rivet-pin. The act of riveting thus rigidly fixes the bush to the side-plate, and prevents the rivet-pins turning in the side-plates. The hard pen-steel bush bears on the hard steel block, and there is, therefore, less wear than with a softer metal rubbing on the block.

Messrs. Brampton & Co. make a 'self-lubricating' chain of the 'Humber' type (fig. 432). The block is hollow, and made in

FIG. 432.

two pieces; the interior is filled with lubricant—a specially prepared form of graphite—sufficient for several years.

282. **Roller Chain.**— Figure 433 shows a *roller* or *long-link chain*, as made by the Abingdon Works Co., the middle block of the 'Humber' chain being dispensed with, and the number of rivet-pins required being only one-half. Each chain-link is formed by two side-plates, symmetrically situated on each side of the centre line, and each rivet thus passes through four plates. The two outer plates are riveted together, forming one chain-link; while the two inner plates, forming the adjacent chain-link, can rotate

on the rivet-pin as a bearing. If the inner plates were left as narrow as the outer plates, the bearing surface on the rivet would

FIG. 433.

be very small, and wear would take place rapidly. Figure 434 shows the inner plate provided with bosses, so that the bearing

surface is enlarged ; and figure 435 shows the plates riveted together. The rivet, shown separately (fig. 436), thus bears along the whole

FIG. 434.

width of the inner chain-link. Loose rollers surround the bosses ; these are not shown in figure 435, but are shown in figure 433.

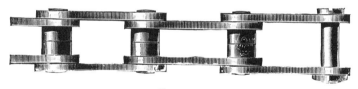

FIG. 435.

Single-link Chain.—The chain illustrated in figure 433 is a two-link chain ; that is, its length must be increased or diminished by two links at a time. Thus, if the chain stretches and becomes too long for the cycle, it can only be shortened by two inches at a time. Figure 437 shows a *single-link chain* ; that is, one which can be shortened by removing one link at a time. The side-plates in this case are not straight, but one pair of ends are brought closer together than the other ; the details of boss, rivets, and rollers are

FIG. 436.

the same as in the double-link chain.

The width of the space between the side-plates of figure 433

is different for two consecutive links. If the narrow link fit the
side of the chain-wheel, the side of the wide link will be quite

FIG. 437.

clear ; in other words, the chain will be guided sideways on to
the chain-wheel only at every alternate link. The single-link chain

FIG. 438.

is in this respect superior to the double-link chain. In the ' R. F.
Hall ' corrugated-link chain (fig. 438) the alternate side-plates were
depressed, so that the inside width was the same for all links.

283. **Pivot-chain.**—In the pivot-chain (fig. 439), made by
the Cycle Components Manufacturing Company, (Limited) the
pins and bushes of the ' Humber' or long-link chain are replaced
by hard steel knife edges. The relative motion of the parts is

FIG. 439.

smaller, and therefore the work lost in friction may also be expected
to be smaller than in the ' Humber ' chain, though it remains
to be seen whether the bearing surfaces will be able to stand for
a few years the great intensity of pressure to which they are subjected in ordinary running.

284. **Roller-chain Chain-wheel.**—The pitch-line of a long-
link chain-wheel must be a regular polygon of as many sides as

there are teeth in the wheel. Let a, b, c . . . (fig. 440) be consecutive angles of the polygon. When the chain is wrapped round the wheel the centres of the chain rivets will occupy the positions a, b, c . . . The relative motion of the chain and wheel will be the same, if the wheel be considered fixed and the chain to be wound on and off. If the wheel be turning in the direction of the arrow, as the rivet a leaves contact with the wheel, it will move relative to the wheel in the circular arc $a\,a_1$, having b as centre, a_1 lying in the line $c\,b$ produced. Assuming that the chain is tight, the links $a\,b$ and $b\,c$ will now be in the same straight line, and the rivet a will move, relative to the chain-wheel, in circular arc $a_1\,a_2$, with centre c ; a_2 lying in the straight line $d\,c$

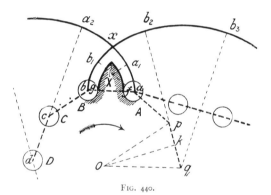

F<small>IG</small>. 440.

produced. Thus, the relative path of the centre of the rivet A as it leaves the wheel is a series of circular arcs, having centres b, c, d . . . respectively. It may be noticed that this path is approximately an involute of a circle, the approximation being closer the larger the number of teeth in the wheel. In the same way, the relative path of the centre of rivet b as it moves into contact with the wheel is an exactly similar curve $b\,b_1$ b_2 . . ., which intersects the curve $a\,a_1\,a_2$ . . . at the point x. If the rivets and rollers of the chain could be made indefinitely small, the largest possible tooth would have the outline $a\,a_1\,x\,b_1\,b$. Taking account of the rollers actually used, the outline of the largest possible tooth will be a pair of parallel curves $A\,X$ and

$B\,X$ intersecting at X, and lying inside $a\,a_1$ . . . and $b\,b_1$ . . .,
a distance equal to the radius of the rollers.

Kinematically there is no necessity for the teeth of a chain-wheel projecting beyond the pitch-line, as is absolutely essential in spur-wheel gearing.   If the
pitches of the chain and wheel
could be made exactly equal,
and the distance between the
two chain-wheels so accurately
adjusted that the slack of the
chain could be reduced to zero,
and the motion take place without side-swaying of the chain,

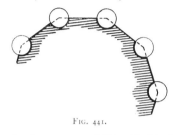

FIG. 441.

the chain-wheel might be made as in figure 441.   With this ideal
wheel there would be no rubbing of the chain-links on it as they
moved into and out of gear.

But, owing to gradual stretching, the pitch of the chain is
seldom exactly identical with that of the wheel ; this, combined
with slackness and swaying of the chain, makes it desirable, and
in fact necessary, to make the cogs project from the pitch-line.
If the cogs be made to the outline $A\,X\,B$ (fig. 440), each link of
the chain will rub on the corresponding cog along its whole length
as it moves into and out of gear ; or rather, the roller may *roll* on
the cog, and rub with its inner surface on the bosses of the inner
plates of the link.   To eliminate this rubbing the outline of the
cog should therefore be drawn as follows : Let a and b (fig.
442) be two adjacent corners of
the pitch-polygon, and let the
rollers, with a and b as centres, cut
$a\,b$ at f and g respectively.   The
centres m and n of the arcs of outline through f and g respectively
should lie on $a\,b$, but closer together than a and b ; in fact, f may

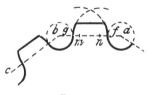

FIG. 442.

conveniently be taken for the centre of the arc through g, and
vice versâ.   The addendum-circle may be conveniently drawn
touching the straight line which touches, and lies entirely outside
of, two adjacent rollers.

The '*Simpson*' *Lever-chain* has triangular links, the inner
corners, A, B, C . . . (fig. 443), are pin-jointed and gear in the
ordinary way with the chain-wheel on the crank-axle. Rollers
project from the outer corners, a, b, . . . and engage with the
chain-wheel on the driving-hub. As the chain winds off the
chain-wheel the relative path, p p_1 p_2 . . . of one of the inner
corners is, as in figure 440, a smooth curve made up of circular
arcs, while that of an outer corner has cusps, a_1, a_2, . . . corre-

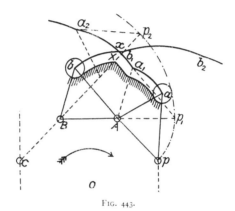

Fig. 443.

sponding to the sudden changes of the relative centre of rotation
from A to B, from B to C, ... As the chain is wound on
to the wheel, the relative path of an adjacent corner, b, is a
curve, b b_1 b_2 . . . of the same general character, but not of
exactly the same shape, since the triangular links are not equal-
sided. These two curves intersect at x, and the largest possible
tooth outline is a curve parallel to a a_1 x b. If the actual tooth
outline lie a little inside this curve, as described in figure 442,
the rubbing of the rollers on their pins will be reduced to a
minimum, and the frictional resistance will not be greater than
that of an ordinary roller chain. Thus there is no necessity for
the cusp on the chain-wheel; the latter may therefore be made
with a smaller addendum-circle.

Let a, p, and q (fig. 440) be three consecutive corners of the pitch-polygon of a long-link chain-wheel, one-inch pitch. The circumscribing circle of the pitch-polygon may, for convenience of reference, be called the *pitch-circle*. Let R be the radius of the pitch-circle, and N the number of teeth on the wheel. From O, the centre, draw $O\,k$ perpendicular to $p\,q$. The angle $p\,O\,q$ is evidently $\dfrac{360}{N}$ degrees, and the angle $p\,O\,k$ therefore $\dfrac{180}{N}$ degrees. And

$$R = O\,p = \frac{p\,k}{\sin p\,O\,k} = \frac{0\cdot5}{\sin \dfrac{180°}{N}} \text{ inches} \quad . \quad . \quad . \quad (3)$$

TABLE XV.—CHAIN-WHEELS, 1-IN. PITCH.

N Number of teeth in chain-wheel	R Radius of circumscribing circle of pitch-polygon		Radius of circle whose circumference is N inches
	Long-link chain	Humber chain	
	Inches	Inches	Inches
6	1·000	·967	·955
7	1·153	1·125	1·114
8	1·307	1·283	1·274
9	1·462	1·441	1·433
10	1·618	1·599	1·592
11	1·775	1·758	1·751
12	1·932	1·916	1·910
13	2·089	2·074	2·069
14	2·247	2·233	2·228
15	2·405	2·392	2·387
16	2·563	2·551	2·546
17	2·721	2·710	2·705
18	2·880	2·870	2·865
19	3·039	3·029	3·024
20	3·197	3·188	3·183
21	3·356	3·347	3·342
22	3·514	3·505	3·501
23	3·672	3·664	3·660
24	3·831	3·824	3·820
25	3·990	3·983	3·979
26	4·148	4·142	4·138
27	4·307	4·301	4·297
28	4·466	4·460	4·456
29	4·626	4·620	4·616
30	4·785	4·779	4·775

The values of R for wheels of various numbers of teeth are given in Table XV.

285. **Humber Chain-wheel.**—The method of designing the form of the teeth of a 'Humber' chain-wheel is, in general, the same as for a long-link chain, the radius of the end of the hardened block being substituted for the radius of the roller; but the distance between the pair of holes in the block is different from that between the pair of holes in the side-plates, these distances

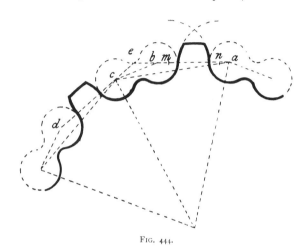

being approximately ·4 in. and ·6 in. respectively. The pitch-line of a 'Humber' chain-wheel will therefore be a polygon with its corners all lying on the circumscribing circle, but with its sides ·4 in. and ·6 in. long alternately. Figure 444 shows the method of drawing the tooth, the reference letters corresponding to those in figures 440 and 442, so that the instructions need not be repeated.

Let a, b, c, d (fig. 444) be four consecutive corners of the pitch-polygon of a 'Humber' chain-wheel. Produce the sides $a\,b$ and $d\,c$ to meet at e. Then, since a, b, c, and d lie on a circle, it is evident, from symmetry, that the angles $e\,b\,c$ and $e\,c\,b$ are equal. If N be the number of teeth in the wheel, there are $2\,N$

sides of the pitch-polygon, and the external angle $e\,b\,c$ will be $\dfrac{360}{2\,N} = \dfrac{180}{N}$ degrees.

Let $\quad a\,c = D$, then $R = \dfrac{D}{2\,sin\,\dfrac{180°}{N}}$.

But $\qquad D^2 = \overline{a\,b}^2 + \overline{b\,c}^2 + 2\,\overline{a\,b}\,.\,\overline{b\,c}\,cos\,\dfrac{180°}{N}$

$$= \cdot 36 + \cdot 16 + \cdot 48\,cos\,\dfrac{180°}{N}$$

$$\therefore R = \dfrac{\sqrt{\cdot 52 + \cdot 48\,cos\,\dfrac{180°}{N}}}{2\,sin\,\dfrac{180°}{N}} \quad . \quad . \quad . \quad . \quad . \quad (4)$$

The radii of the pitch-circles of wheels having different numbers of teeth are given in Table XV.

286. **Side-clearance and Stretching of Chain.**—With chain-wheels designed as in sections 284–5, with the pitch of the teeth exactly the same as the pitch of the chain, there is no rubbing of the chain links on the wheel-teeth, the driving arc of action is the same as the arc of contact of the chain with the wheel, and all the links in contact with the wheel have a share in transmitting the effort. But when the pitch of the chain is slightly different from that of the wheel-teeth the action is quite different, and the chain-wheels should be designed so as to allow for a slight variation in the pitch of the chain by stretching, without injurious rubbing action taking place. The thickness of the teeth of the long-link chain-wheel (fig. 440) is so great that it can be considerably reduced without impairing the strength. Figure 445 shows a wheel in which the thickness of the teeth has been reduced. If the pitch of the chain be the same as that of the wheel, each tooth in the arc of contact will be in contact with a roller of the chain, and there will be a clearance space x between each roller and tooth. Let N be the number of teeth in the wheel ; then the number of teeth in action will be in

general not more than $\dfrac{N}{2} + 1$. The original pitch of the chain

may be made $\dfrac{x}{\dfrac{N}{2} + 1}$ *less* than the pitch of the wheel-teeth, the

wheel and chain will gear perfectly together. Figure 445 illustrates the wheel and chain in this case. After a certain amount of wear and stretching, the pitch of the chain will become exactly the same as that of the teeth, and each tooth will have a roller in contact with it. The stretching may still continue until the pitch

of the chain is $\dfrac{x}{\dfrac{N}{2} + 1}$ *greater* than that of the wheel-teeth, with-

out any injurious action taking place.

The mutual action of the chain and wheel having different pitches must now be considered. First, let the pitch of the chain be a little *less* than that of the teeth (fig. 445), and suppose the

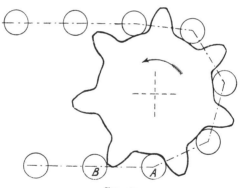

FIG. 445.

wheel *driven* in the direction of the arrow. One roller, *A*, just passing the lowest point of the wheel will be driving the tooth in front of it, and the following roller, *B*, will sooner or later come in contact with a tooth. Figure 445 shows the roller *B* just coming into contact with its tooth, though it has not yet reached the pitch-line of the wheel. The motion of the chain and wheel continuing, the roller *B* rolls or rubs on the tooth, and the roller *A* gradually recedes from the tooth it had been driving. Thus the total effort is transmitted to the wheel by one tooth, or at most two, during the short period one roller is receding from, and another coming into, contact.

If the pitch of the chain be a little too great, and the wheel be *driven* in the same direction, the position of the acting teeth is at the top of the wheel (fig. 446). The roller, *C*, is shown driving the tooth in front of it, but as it moves outwards along the tooth surface the following roller, *D*, will gradually move up to, and drive, the tooth in front of it.

The action between the chain and the *driving*-wheel is also explained on the same general principles ; if the direction of the arrow be reversed, figures 445 and 446 will illustrate the action.

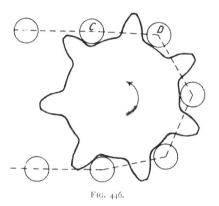

FIG. 446.

In a chain-wheel made with side-clearance, assuming the pitches of chain and teeth equal, there will be two pitch-polygons for the two directions of driving. Let *a* and *b* (fig. 447) be two consecutive corners of one of the pitch-polygons, and let the roller

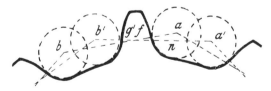

FIG. 447.

with centre *a* cut *a b* at *f.* The centre *m* of the arc of tooth outline through *f* lies on *a b*. Let *a a'* = *b b'* be the side-clearance measured along the circumscribing circle ; *a'* and *b'* will therefore be consecutive corners of the other pitch-polygon. Let the roller with centre b' cut a' b' in g ; the centre n of the arc of tooth outline through f lies on a' b'. The bottom of the tooth space should be a circular arc, which may be called the root-circle, concentric with the pitch-polygons, and touching the circles of the rollers a and a'.

287. **Rubbing and Wear of Chain and Teeth.** If the outline of the teeth be made exactly to the curve $f X$ (fig. 440), the roller A will knock on the top of the tooth, and will then roll or rub along its whole length. If the tooth be made to a curve lying inside $f X$, the roller will come in contact with the tooth at a point p (fig. 448), such that the distance of p from the curve $f X$ is equal to the difference of the pitches of the teeth and chain ; $p f$ will be the arc of the tooth over which contact takes place. The length of this arc will evidently be smaller (and therefore also the less will be the work lost in friction), the smaller the radius of the tooth outline.

In the 'Humber' chain the block comes in contact with the teeth, and there is relative rubbing over the arc $p f$ (fig. 448).

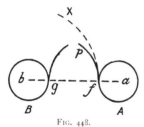

FIG. 448.

The same point of the block always comes in contact with the teeth, so that after a time the wear of the blocks of the chain and the teeth of the wheel becomes serious, especially if the wheel-teeth be made rather full.

The chief advantage of a roller-chain lies in the fact that the roller being free to turn on the rivet, different points of the roller come successively in contact with the wheel-teeth. If the chain be perfectly lubricated the roller will actually roll over its arc of contact, $f p$, with the tooth, and will rub on its rivet-pin. The rubbing is thus transferred from a higher pair to a lower pair, and the friction and wear of the parts, other things being equal, will be much less than in the 'Humber' chain. Even with imperfect lubrication, so that the roller may be rather stiff on its rivet-pin, and with rubbing taking place over the arc $f p$, the roller will at least be slightly disturbed in its position relative to its rivet-pin, and a fresh portion of it will next come in contact with the wheel-teeth. Thus, even under the most unfavourable conditions, the wear of the chain is distributed over the cylindrical surface of the roller, consequently the alteration of form will be much less than in a 'Humber' chain under the same conditions.

It must be clearly understood that the function performed by rollers in a chain is quite different from that in a roller-bearing. In the latter case rubbing friction is eliminated, but not in the former.

288. **Common Faults in Design of Chain-wheels.**—The portions of the teeth lying outside the pitch-polygon are often made far too full, so that a part of the tooth lies beyond the circular arc *f X* (fig. 440); the roller strikes the corner of the tooth as it comes into gear, and the rubbing on the tooth is excessive. This faulty tooth is illustrated in figure 449.

In long-link chain-wheels the only convex portion is very often merely a small circular arc rounding off the

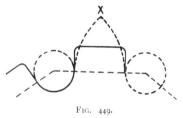

FIG. 449.

side of the tooth into the addendum-circle of the wheel. This rounding off of the corner is very frequently associated with the faulty design above mentioned. If the tooth outline be made to *f p*, a curve lying well within the circular arc *f X* (fig. 448), this rounding off of the corners of the teeth is quite unnecessary.

Another common fault in long-link chain-wheels is that the bottom of the tooth space is made one circular arc of a little larger radius than the roller. There is in this case no clearly defined circle in which the centres of the rollers are compelled to lie, unless the ends of the link lie on the cylindrical rim from which the teeth project. In back-hub chain-wheels this cylindrical rim is often omitted. Care

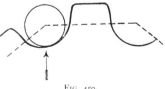

FIG. 450.

should then be taken that the tooth space has a small portion made to a circle concentric with the pitch-circle of the wheel. Again, in this case, the direction of the mutual force between the roller and wheel is not along the circumference of the pitch-polygon; there is therefore a radial component tending to force the rollers out of the tooth spaces, that is, there is a tendency of the chain to mount the wheel (fig. 450).

In 'Humber' pattern chain-wheels the teeth are often quite straight (fig. 451). This tooth-form is radically wrong. If the

FIG. 451.

teeth are so narrow at the top as to lie inside the curve $f\,X$, the force acting on the block of the chain will have an outward component, and the chain will tend to mount the wheel. This faulty design is sometimes carried to an extreme by having the teeth *concave* right to the addendum-circle.

Either of the two faults above discussed gives the chain a tendency to mount the wheel, and this tendency will be greater the more perfect the lubrication of the chain and wheel.

289. **Summary of conditions determining the proper form of Chain-wheels.**—1. Provision should be made for the gradual stretching of the chain. This necessitates the gap between two adjacent teeth being larger than the roller or block of the chain.

2. The centres of the rollers in a long-link chain, or the blocks in a 'Humber' chain, must lie on a perfectly defined circle concentric with the chain-wheel. When the wheel has no distinct cylindrical rim, the bottom of the tooth space must therefore be a circular arc concentric with the pitch-polygon.

3. In order that there should be no tendency of the chain to be forced away from the wheel, the point of contact of a tooth and the roller or block of the chain should lie on the side of the pitch-polygon, and the surface of the tooth at this point should be at right angles to the side of the pitch-polygon. The centre of the circular arc of the tooth outline must therefore lie on the side of the pitch-polygon.

4. The blocks or rollers when coming into gear must not strike the corners of the teeth. The rubbing of the roller or block on the tooth should be reduced to a minimum. Both these conditions determine that the radius of the tooth outline should be less than 'length of side-plate of chain, minus radius of roller or block.'

The following method of drawing the teeth is a *résumé* of the results of sections 285–8, and gives a tooth form which satisfies the above conditions : Having given the type of chain, pitch, and number of teeth in wheel, find R, the radius of the pitch-circle $c\,c$, by calculation or from Table XV. On the pitch-circle $c\,c$ (fig. 452), mark off adjacent corners a and b of the pitch-polygon. With centres a and b, and radius equal to the radius of the roller (or the radius of the end of the block in a ' Humber' chain), draw circles, that from a as centre cutting $a\,b$ at f. Through f draw a circular arc, $f\,h$, with centre m on $a\,b$, $m\,f$ being less than $b\,f$. Mark off,

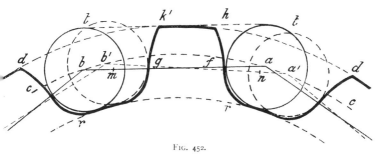

FIG. 452.

along the circle $c\,c$, $a\,a^1 = b\,b^1 =$ side clearance required, and with centres a^1 and b^1, and the same radius as the rollers, draw circles, that from centre b^1 cutting the straight line $a^1\,b^1$ at g. With centre n^1 lying on $a^1\,b^1$, and radius equal to $m\,f$, draw a circular arc $g\,k^1$. Draw the root-circle $r\,r$ touching, and lying inside, the roller circles. The sides $f\,h$ and $g\,k^1$ of the tooth should be joined to the root-circle $r\,r$ by fillets of slightly smaller radius than the rollers. Draw a common tangent $t\,t$ to the roller circles a and b, and lying outside them ; the addendum-circle may be drawn touching $t\,t$.

It should be noticed that this tooth form is the same whatever be the number of teeth in the wheel, provided the side-clearance be the same for all. The form of the spaces will, however, vary with the number of teeth in the wheel. A single milling-cutter to cut the two sides of the same tooth might therefore serve for all sizes of wheels ; whereas when the milling-cutter cuts out the space between two adjacent teeth, a separate cutter is required for each size of wheel.

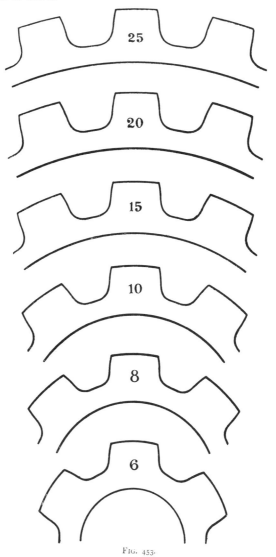

FIG. 453.

Figure 453 shows the outlines of wheels for inch-pitch long-links made consistent with these conditions, the diameter of the roller being taken $\frac{3}{8}$ in.   The radius of the side of the tooth is in

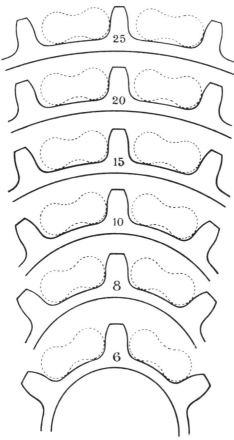

FIG. 454.

each case $\frac{3}{8}$ in. (it may with advantage be taken less), and the radius of the fillet at root of tooth $\frac{1}{8}$ in.   The width of the roller space measured on the pitch-polygon is ($\cdot375 + \cdot005\ N$) in. ; N being the number of teeth in the wheel.

Figure 454 shows the outlines of wheels for use with the 'Humber' chain, the pitch of the rivet-pins in the side-plates being ·6 in. and in the blocks ·4 in., and the ends of the blocks being circular, ·35 in. diameter.

290. **Section of Wheel Blanks.**—If the chain sways sideways, the side-plates may strike the tops of the teeth as they come into gear, and cause the chain to mount the wheel, unless each

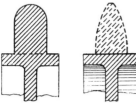

FIG. 454A.   FIG. 454B.

link is properly guided sideways on to the wheel. The cross section of the teeth is sometimes made as in figure 454A, the sides being parallel and the top corners rounded off. A much better form of section, which will allow of a considerable amount of swaying without danger, is that shown in figure 454B. The thickness of the tooth at the root is a trifle less than the width of the space between the side-plates of the link. The thickness at the point is very small—say, $\frac{1}{32}$ in. to $\frac{1}{16}$ in.—and the tooth section is a wedge with curved sides.

If the side-plates of the chain be bevelled, as in Brampton's bevelled chain (fig. 432), an additional security against the chain coming off the wheel through side swaying will be obtained.

291. **Design of Side-plates of Chain.**—The side-plates of a well-designed chain should be subjected to simple tension. If P be the total pull on the chain, and A the least sectional area of the two side-plates, the tensile stress is $\dfrac{P}{A}$. Such is the case with side-plates of the form shown in figure 429.

Example I.—The section of the side-plates (fig. 429) is ·2 in deep and ·09 in. thick. The total sectional area is thus

$$2 \times ·2 \times ·09 = ·036 \text{ sq. in.}$$

The proof load is 9 cwt. = 1,008 lbs. The tensile stress is, therefore,

$$f = \frac{1,008}{·036} = 28,000 \text{ lbs. per sq. in.}$$

$$= 12·5 \text{ tons per sq. in.}$$

A considerable number of chains are being made with the side-plates recessed on one side, and not on the other (fig. 455). These side-plates are subjected to combined tensile and bending stresses. Let b be the width of the plate, t its thickness, b_1 be the depth of the recess, and let

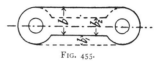

FIG. 455.

$b_2 = b - b_1$; that is, b_2 would be the width of the plate if recessed the same amount on both sides. The distance of the centre of the section from the centre line joining the rivets is $\frac{b_1}{2}$.

The bending-moment M on the link is $\frac{P b_1}{2}$. The modulus of the section, Z, is $\frac{2\,t\,b^2}{6}$. The maximum tensile stress on the section is (sec. 101)

$$f = \frac{P}{A} + \frac{M}{Z} = \frac{P}{2\,b\,t} + \frac{6\,P\,b_1}{2.\,2\,t b^2}$$

$$= \frac{P}{2\,b\,t}\left\{ 1 + \frac{3\,b_1}{b} \right\} \quad . \quad . \quad . \quad . \quad . \quad (5)$$

The stress on the side-plate if recessed on both sides would be

$$f = \frac{P}{2\,(b - b_1)\,t} \quad . \quad . \quad . \quad . \quad . \quad . \quad (6)$$

The stress f calculated from equation (6) is always, within practical limits, less than the stress calculated from equation (5); and, therefore, the recessed side-plates can be actually strengthened by cutting away material. This can easily be proved by an elementary application of the differential calculus to equation (5).

Example II.—Taking a side-plate in which $t = \cdot 09$ in., $b = \cdot 3$ in., $b_1 = \cdot 1$ in., and therefore $b_2 = \cdot 2$ in., we get

$$A = 2 \times \cdot 09 \times \cdot 3 = \cdot 054 \text{ sq. in.,}$$

and

$$Z = \frac{2 \times \cdot 09 \times \cdot 3^2}{6}$$

$$= \cdot 0027 \text{ in.}^3$$

The distance of the centre of the section from the centre-line of the side-plate is ·05 in., and the bending-moment under a proof load of 9 cwt. is

$$M = 9 \times 112 \times ·05 = 50·4 \text{ inch-lbs.}$$

The maximum stress on the section is

$$f = \frac{1008}{·054} + \frac{50·4}{·0027} = 37,320 \text{ lbs. per sq. in.}$$
$$= 16·7 \text{ tons per sq. in.}$$

Thus this link, though having 50 per cent. more sectional area, is much weaker than a link of the form shown in figure 429.

If the plate be recessed on both sides, $A = ·036$ sq. in., and

$$f = \frac{1008}{·036} = 28,000 \text{ lbs. per sq. in.}$$
$$= 12·5 \text{ tons per sq. in.}$$

Thus the side-plate is strengthened, even though 33 per cent. of its section has been removed.

From the above high stresses that come on the side-plates of a chain during its test, and from the fact that these stresses may occasionally be reached or even exceeded in actual work when grit gets between the chain and wheel, it might seem advisable to make the side-plates of steel bar, which has had its elastic limit artificially raised considerably above the stresses that will come on the links under the proof load.

The Inner Side-plates of a Roller Chain, made as in figure 434, are also subjected to combined tension and bending in ordinary working. Assuming that the pressure between the rivet and inner link is uniformly distributed, the side-plate of the latter will be subjected to a bending-moment $M = \frac{P}{2} \times l$, l being the distance measured parallel to the axis of the rivet, between the centres of the side-plate and its boss, respectively.

Example III.—Taking $l = ·08$ in., and the rest of the data as in the previous example, the maximum additional stress on the side-plate due to bending is

$$\frac{M}{Z} = \frac{504 \times ·08 \times 6}{·2 \times ·0081} = 149,000 \text{ lbs. per sq. in.}$$
$$= 66·7 \text{ tons per sq. in.},$$

which, added to the 12·5 tons per sq. in. due to the direct pull, gives a total stress of 79·2 tons per sq. in. Needless to say, the material cannot endure such a stress ; what actually happens during the test is, the side-plates slightly bend when the elastic limit is reached, the pressure on the inner edge of the boss is reduced, so that the resultant pressure between the rivet and side-plate acts nearly in line with the latter. Thus the extra bearing surface for the rivet, supposed to be provided by the bosses, is practically got rid of the first time a heavy pull comes on the chain.

A much better method of providing sufficient bearing surface for the rivet-pins is to use a tubular rivet to unite the inner side-plates (fig. 456), inside which the rivet-pin uniting the outer side-plates bears, and on the outside of which the roller turns.

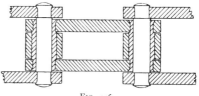

FIG. 456.

This is the method adopted by Mr. Hans Renold for large gearing chains.

Side-plates of Single-link Chain.—In the same way, it will be readily seen that the maximum stress on the side-plates of the chain shown in figure 437 is much greater than on a straight plate with the same load. If the direction of the pull on the plate be parallel to the centre line of the chain, each plate will be subjected to a bending action. The transverse distance between the centres of the sections of the two ends is t, the bending-moment on the section will therefore be Pt. A more favourable assumption will be that the pull on each plate will be in a line joining the middle points of its ends. The greatest distance between this line and the middle of section will be then nearly $\frac{t}{2}$, and $M = \frac{Pt}{2}$. The bending in this case is in a plane at right angles to the direction of the bending in the recessed side-plate (fig. 455). The modulus of the section Z is $\frac{bt^2}{6}$.

Example IV.—Taking the same data as in the former examples, the load on the chain is 9 cwt., $b = \cdot 3$ in., and $t = \cdot 09$ in., the pull on each plate is 504 lbs., the bending-moment is $\dfrac{504 \times \cdot 09}{2}$

$= 22 \cdot 7$ inch-lbs., $A = \cdot 027$ sq. in., $Z = \dfrac{\cdot 3 \times \cdot 09^2}{6} = \cdot 000405$ in.^{3}

The maximum stress on the section is therefore

$$f = \frac{P}{A} + \frac{M}{Z}$$

$$a = \frac{504}{\cdot 027} + \frac{22 \cdot 7}{\cdot 000405}$$

$$= 18,670 + 56,050 = 74,720 \text{ lbs. per sq. in.}$$
$$= 33 \cdot 3 \text{ tons per sq. in.}$$

292. **Rivets.**—The pins fastening together the side-plates must be of ductile material, so that their ends may be riveted over without injury. A soft ductile steel has comparatively low tensile and shearing resistances. The ends of these pins are subjected to shearing stress due to half the load on the chain. If d be the diameter of the rivet, its area is $\dfrac{\pi d^2}{4}$, and the shearing stress on it will be

$$\frac{2P}{\pi d^2} \quad . \quad . \quad . \quad . \quad . \quad . \quad . \quad (7)$$

Example I.—If the diameter of the holes in the side-plate (fig. 429) be $\cdot 15$ in., under a proof load of 9 cwt. the shearing stress will be

$$\frac{1008}{2 \times \cdot 01767} = 28,500 \text{ lbs. per sq. in.}$$

$$= 12 \cdot 78 \text{ tons per sq. in.}$$

The end of the rivet is also subjected to a bending-moment $\dfrac{P}{2} \times \dfrac{t}{2}$. The modulus of the circular section is approximately $\dfrac{d^3}{10}$, the stress due to bending will therefore be

$$f = \frac{10\, P t}{4\, d^3} \quad . \quad . \quad . \quad . \quad . \quad . \quad (8)$$

Example II.—Taking the dimensions in Example I. of section 291, and substituting in (8), the stress due to bending is

$$f = \frac{10 \times 1008 \times \cdot 09}{4 \times \cdot 15^3} = 67,200 \text{ lbs. per sq. in.}$$

$$= 30 \text{ tons per sq. in.}$$

The rivet is thus subjected to very severe stresses, which cause its ends to bend over (fig. 457).

The stretching of a chain is probably always due more to the yielding of the rivets than to actual stretching of the side-plates, if the latter are properly designed. A material that is soft enough to be riveted cold has not a very high tensile or shearing resistance. It would seem advisable, therefore, to make the pins of

FIG. 457.

hard steel with a very high elastic limit, their ends being turned down with slight recesses (fig. 458), into which the side plates, made of a softer steel, could be forced by pressure. The Cleveland Cycle Company, and the Warwick and Stockton Company, manufacture chains on this system.

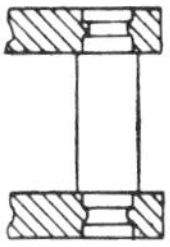

293. **Width of Chain and Bearing Pressure on Rivets.**—In the above investigations it will be noticed that the width of the chain does not enter into con-

FIG. 458.

sideration at all. The only effect the width of the chain has is on the amount of bearing surface of the pins on the block. If l be the width of the block, and d the diameter of the pin, the projected bearing area is $l\,d$, and the intensity of pressure is $\dfrac{P}{d\,l}$ If the diameter of the pin (fig. 430) be ·17 in., and the width of the block be $\frac{5}{16}$ in. = ·3125 in., the bearing pressure under the proof load will be

$$\frac{1008}{\cdot 17 \times \cdot 3125} = 18,980 \text{ lbs. per sq. in.}$$

This pressure is very much greater than occurs in any other example of engineering design. Professor Unwin, in a table of 'Pressures on Bearings and Slides,' gives 3,000 lbs. per sq. in. as the maximum value for bearings on which the load is intermittent and the speed slow. Of course, in a cycle chain the period of relative motion of the pin on its bearing is small compared to that during which it is at rest, so that the lubricant, if an oil-tight gear-case be used, gets time to find its way in between the surfaces.

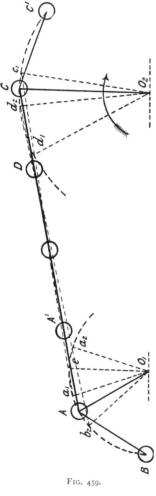

FIG. 459.

294. **Speed-ratio of Two Shafts Connected by Chain Gearing.**—The *average* speeds of two shafts connected by chain gearing are inversely proportional to the numbers of teeth in the chain wheels; but the speed-ratio is not *constant*, as in the case of two shafts connected together by a belt or by toothed-wheels. Let O_1 and O_2 (fig. 459) be the centres of the two shafts, let the wheel O_2 be the driver, the motion being as indicated by the arrow, and let $A C$ be the straight portion of the chain between the wheels at any instant. The instantaneous angular speed-ratio of the wheels is the same as that of two cranks $O_1 A$ and $O_2 C$ connected by the coupling-rod $A C$. Let B and D be the rivets consecutive to A and C respectively; then, as the motion of the wheels continues, the rivet D will ultimately touch the chain-wheel at the point $d_1 - a_1$, d_1 and c_1 being in the same straight line— and the angular speed-ratio of the wheels will be the same as that of the two cranks $O_1 A$ and $O_2 D$ connected by the straight coup-

ling-rod $A D$, shorter by one link than the coupling-rod $A C$. The motion continuing, the rivet A leaves contact with the chain-wheel at a_2, and the virtual coupling-rod becomes $B D$; the points b_2, a_2, and d_2 lying in one straight line. The angular speed-ratio of the wheels is now the same as that of the two cranks $O_1 B$ and $O_2 D$ connected by the coupling-rod $B D$, of the same length as $A C$.

Thus, with a long-link chain, the wheels are connected by a virtual coupling-rod whose length changes twice while the chain moves through a distance equal to the length of one of its links. The small chain-wheel, being rigidly connected to the driving-wheel of the bicycle, will rotate with practically uniform speed; since the whole mass of the machine and rider acts as an accumulator of energy (or fly-wheel), keeping the motion steady. The chain-wheel on the crank-axle will therefore rotate with variable speed. The speed-ratio in any position, say $O_1 A C O_2$, can be found, after the method of section 32, by drawing $O_1 e$ parallel to $O_2 C$, meeting $C A$ (produced if necessary) at e; the intercept $O_1 e$ is proportional to the angular speed of the crank-axle. If this length be set off along $O_1 A$, and the construction be repeated, a polar curve of angular speed of the crank-axle will be obtained. It will be noticed that in figure 459 the angular speed of the crank-axle decreases gradually shortly after passing the position $O_2 d_1$ until the position $O_2 d_2$ is reached, and the rivet A attains the position a_2. The length of the coupling-rod being now increased by one link, the angular speed of the crank-axle increases gradually until the rivet C attains the position c_1. Here the length of the coupling-rod is decreased by one link, and the virtual crank of the wheel changing suddenly from $O_1 c_1$ to $O_2 d_1$, the length of the intercept $O e$ also changes suddenly, corresponding to a sudden change in the angular speed of the crank-axle.

With a 'Humber' chain the speed will have four maximum and minimum values while the chain moves over a distance equal to one link.

The magnitude of the variation of the angular speed of the crank-axle depends principally, as an inspection of figure 459 will show, on the number of teeth in the smaller wheel. If the crank-axle be a considerable distance from the centre of the driving-wheel, and if the number of teeth of the wheel on the crank-axle be great, the longest intercept, $O\,e$, will be approximately equal to the radius of the pitch-circle, and the smallest intercept to the radius of the inscribed circle of the pitch-polygon. The variation of the angular speed of the crank can then be calculated as follows :

Let N_1 and N_2 be the numbers of teeth in the chain-wheels on the driving-hub and crank-axle respectively, R_1 and R_2 the radii of the pitch-circles, r_1 and r_2 the radii of the *inscribed* circles of the pitch-polygon. Then for a long-link chain the average speed-ratio $= \dfrac{N_1}{N_2}$. Assuming the maximum intercept $O_1\,e$ (fig. 459) to be equal to R_1, then from (3), the

$$\text{maximum speed-ratio} = \frac{R_1}{R_2} = \frac{sin\ \dfrac{\pi}{N_2}}{sin\ \dfrac{\pi}{N_1}}\ \text{approx.} \quad . \quad (9)$$

Assuming the minimum intercept $O\,e$ (fig. 459) to be equal to r_1, then from (3)

$$\text{minimum speed-ratio} = \frac{r_1}{R_2} = \frac{sin\ \dfrac{\pi}{N_2}}{tan\ \dfrac{\pi}{N_1}}\ \text{approx.} \quad . \quad (10)$$

Then, for the crank-axle,

$$\frac{\text{maximum speed}}{\text{mean speed}} = \frac{N_2\ sin\ \dfrac{\pi}{N_2}}{N_1\ sin\ \dfrac{\pi}{N_1}}\ \text{approx.} \quad . \quad . \quad . \quad (11)$$

$$\frac{\text{minimum speed}}{\text{mean speed}} = \frac{N_2\ sin\ \dfrac{\pi}{N_2}}{N_1\ tan\ \dfrac{\pi}{N_1}}\ \text{approx.} \quad . \quad . \quad . \quad (12)$$

$$\frac{\text{maximum speed}}{\text{minimum speed}} = \frac{O\,p}{O\,k}\ (\text{fig. 440}) = \frac{1}{cos\ \dfrac{\pi}{N_1}}\ \text{approx.} \quad (13)$$

In the same way, we get for a 'Humber' chain,

$$\frac{\text{maximum speed}}{\text{minimum speed}} = \frac{1}{\cos a\,O\,b} \text{ (fig. 444)} = \frac{R_1}{\sqrt{R_1{}^2 - 0\cdot3^2}}. \quad (14)$$

Table XVI. is calculated from formulæ (13) and (14).

TABLE XVI.—VARIATION OF SPEED OF CRANK-AXLE.

Assuming that centres are far apart, and that the number of teeth on chain-wheel of crank-axle is great.

Number of Teeth on Hub	6	7	8	9	10	11	12
Ratio of maximum to minimum speed { Humber .	1·052	1·038	1·028	1·022	1·018	1·015	1·012
Long-link.	1·155	1·110	1·082	1·064	1·051	1·042	1·035

Discarding the assumptions made above, when N_2 is much greater than N_1, the maximum intercept $O\,e$ (fig. 459) may be appreciably greater than R_1, while the minimum intercept *may* not be appreciably greater than r_1. The variation of the speed-ratio *may* therefore be appreciably greater than the values given in Table XVI.

An important case of chain-gearing is that in which the two chain-wheels are equal, as occurs in tandems, triplets, quadruplets, &c. Drawing figure 459 for this case, it will be noticed that if the distance between the wheel centres be an exact multiple of the pitch of the chain, the lines $O_1\,A$ and $O_2\,C$ are always parallel, the intercept $O_1\,e$ always coincides with $O_1\,A$, and therefore the speed-ratio is constant. If, however, the distance between

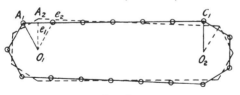

FIG. 460.

the wheel centres be $(k + \frac{1}{2})$ times the pitch, k being any whole number, the variation may be considerable.

In this case, the minimum intercept $O_1\,e_1$ (fig. 460), the radius $O_1 A_1$ and the chain line $A_1 C_1$ form a triangle $O_1 A_1 e_1$, which is very nearly right-angled at e_1. Therefore for a long-link chain,

$$\frac{\text{minimum speed}}{\text{mean speed}} = \frac{O_1 e_1}{O_1 A_1} = cos \frac{\pi}{N} \text{ approx.} \quad . \quad . \quad . (15)$$

The corresponding triangle $O_1 A_2 e_2$ formed by the maximum intercept is very nearly right-angled at A_2. Therefore,

$$\frac{\text{maximum speed}}{\text{mean speed}} = \frac{O_1 e_2}{O_1 A_2} = \frac{1}{cos \dfrac{\pi}{N}} \text{ approx.} \quad . \quad . \quad . (16)$$

$$\frac{\text{maximum speed}}{\text{minimum speed}} = \frac{1}{cos^2 \dfrac{\pi}{N}} \text{ approx.} \quad . \quad . \quad . . \quad (17)$$

For a 'Humber' chain,

$$\frac{\text{maximum speed}}{\text{minimum speed}} = \frac{R^2}{R^2 - 0\cdot3^2} \text{ approx.} \quad . \quad . \quad . . \quad (18)$$

Table XVII. is calculated from formulæ (17) and (18).

TABLE XVII.—GREATEST POSSIBLE VARIATION OF SPEED-RATIO OF TWO SHAFTS GEARED LEVEL.

Number of Teeth		8	10	12	16	20	30
Ratio of maxi-mum to mini-mum speed-ratio .	Humber chain .	1·058	1·036	1·025	1·014	1·009	1·004
	Long-link chain .	1·190	1·105	1·072	1·039	1·025	1·011

The figures in Table XVI. show that the variation of speed, when a small chain-wheel is used on the driving-hub, is not small enough to be entirely lost sight of. The 'Humber' chain is better in this respect than the long-link chain.

Again, in tandems the speed-ratio of the front crank-axle and the driving-wheel hub is the product of two ratios. The ratio of the maximum to the minimum speed of the front axle *may* be as great as given by the product of the two suitable numbers from Tables XVI. and XVII.

Example.—With nine teeth on the driving-hub, and the two

axles geared by chain-wheels having twelve teeth each, the maximum speed of the front axle *may* be

$$1\cdot064 \times 1\cdot072 = 1\cdot14 \text{ times its minimum speed,}$$

with long-link chains ; and

$$1\cdot022 \times 1\cdot025 = 1\cdot047 \text{ times}$$

with ' Humber ' chains.

With triplets and quadruplets the variation *may* be still greater ; and it is open to discussion whether the crank-axles should not be fixed, without chain-tightening gear, at a distance apart equal to some exact multiple of the pitch.

If a hypothetical point be supposed to move with a uniform speed exactly equal to the average speed of a corresponding point actually on the pitch-line of the crank-axle chain-wheel, the distance at any instant between the two is never very great. Suppose the *maximum* speed of the actual point be maintained for a travel of half the pitch, and that it then travels the same distance with its minimum speed. For a speed variation of one per cent. the hypothetical point will be alternately $\frac{1}{800}$th of an inch before and behind the actual point during each inch of travel. This small displacement, occurring so frequently, is of the nature of a vibratory motion, superimposed on the uniform circular motion.

295. **Size of Chain-wheels**.—The preceding section shows that the motion of the crank-axle is more nearly uniform the greater the number of teeth in the chain-wheels. Also, if the ratio of the numbers of teeth in the two wheels be constant, the larger the chain-wheel the smaller will be the pull on the chain. Instead of having seven or eight teeth on the back-hub chain-wheel it would be much better, from all points of view, to have at least nine or ten, especially in tandem machines.

296. **Spring Chain-wheel**.—Any sudden alteration of speed, that is, jerkiness of motion, is directly a waste of energy, since bodies of sensible masses cannot have their speeds increased by a finite amount in a very short interval of time without the application of a comparatively large force. The chain-wheel on the crank-axle revolving with variable speed, if the crank be rigidly connected the pedals will also rotate with variable speed. In the cycle spring chain-wheel (fig. 461) a spring is interposed between the wheel and

the cranks. If, as its inventors and several well-known bicycle manufacturers claim, the wheel gives better results than the ordinary construction, it may be possibly due to the fact that the

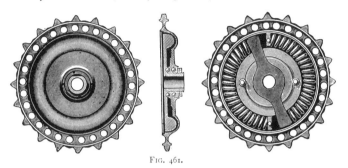

FIG. 461.

spring absorbs as soon as possible the variations of speed due to the chain-driving mechanism, and does not allow it to be transmitted to the pedals and the rider's feet.

If direct spokes are used for the driving-wheel they act as a flexible connection between the hub and rim, allowing the former to run with variable, the latter with uniform speed.

297. **Elliptical Chain-wheel.**—An elliptical chain-wheel has been used on the crank-axle, the object aimed at being an increased speed to the pedals when passing their top and bottom positions, and a diminution of the speed when the cranks are passing their horizontal positions. The pitch-polygon of the chain-wheel in this case is inscribed in an ellipse, the minor axis of which is in line with the cranks (fig. 462).

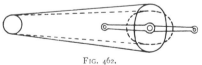

FIG. 462.

The angular speed of the driving-wheel of the cycle being constant and equal to ω, that of the crank-axle is approximately

$$\frac{r_1\,\omega}{r_2}$$

where r_1 and r_2 are the radii from the wheel centres to the ends of the straight portion of the chain. r_1 and ω being constant, the angular speed of the crank is therefore inversely proportional to the radius from the centre of the crank-axle to the point at which the driving side of the chain touches the chain-wheel. The speed of the pedals will therefore be least when the cranks are horizontal, and greatest when the cranks are vertical, as indicated by the dotted lines (fig. 462).

If both sides of the chain connecting the two wheels be straight, the total length of the chain as indicated by the full lines (fig. 462) is greater than that indicated by the dotted lines, the difference being due to the difference of the obliquities of the straight portions when the cranks are vertical and horizontal respectively. This difference is very small, and may be practically left out of account. If the wheel centres are very far apart, so that the top and bottom sides of the chain may be considered parallel, the length in contact with the elliptical chain-wheel in any position is evidently equal to half the circumference of the ellipse ; similarly, the length in contact with the chain-wheel on the hub is half its circumference, and the length of the straight portions is approximately equal to twice the distance between the wheel centres. Thus the total length is approximately the same, whatever be the position of the chain-wheel.

A pair of elliptical toothed-wheels are sometimes used to connect two parallel shafts. The teeth of these wheels are all of different shapes ; there can be at most four teeth in each wheel of exactly the same outline. It has therefore been rather hastily assumed that the teeth of an elliptical chain-wheel must all be of different shapes ; but a consideration of the method of designing the chain-wheel (sec. 289) will show that this is not necessarily the case. The investigation there given is applicable to elliptical chain-wheels, and therefore all the teeth may be made from a single milling-cutter, though the form of the spaces will vary from tooth to tooth.

298. **Friction of Chain Gearing**.—There is loss by friction due to the rubbing of the links on the teeth, as they move into, and out of, contact with the chain-wheel. We have seen (sec. 286) that the extent of this rubbing depends on the difference of

the pitches of the chain and wheel ; if these pitches be exactly equal, and the tooth form be properly designed, theoretically there is no rubbing.   If, however, the tooth outline fall at or near the curve $f X$ (fig. 440), the rubbing on the teeth may be the largest item in the frictional resistance of the gearing.

As a link moves into, and out of, contact with the chain-wheel, it turns through a small angle relative to the adjacent link, there is therefore rubbing of the rivet-pin on its bush.   While the pin A (fig. 459) moves from the point b_2 to the point a_2 the link $A B$ turns through an angle of $\dfrac{360°}{N_1}$, and the link $A A^1$ moves practically parallel to itself.   The relative angular motion of the adjacent links, $B A$ and $A A^1$, and therefore also the angle of rubbing of the pin A on its bush, is the same as that turned through by the wheel O_1.   In the same way, while the pin D moves from d_1 to c_1 the relative angular motion of the adjacent links $D C$ and $C C^1$ is the same as the angle turned through by the wheel O_2, viz. $\dfrac{360°}{N_2}$.   The pressure on the pins A and C during the motion is equal to P, the pull of the chain.   On the slack side of the chain the motion is exactly similar, but takes place with no pressure between the pins and their bushes.   Therefore, the frictional resistance due to the rubbing of the pins on their bushes is the sum of that of two shafts each of the same diameter as the rivet-pins, turning at the same speeds as the crank-axle and driving-hub respectively, when subjected to no load and to a load equal to the pull of the chain.

299. **Gear-Case**.—From the above discussion it will be seen that the chain of a cycle is subjected to very severe stresses, and in order that it may work satisfactorily and wear fairly well it must be kept in good condition.   The tremendous bearing pressure on the rivets necessitates, for the efficient working of the chain, constant lubrication.   Again, the bending and shearing stresses on the rivets, sufficiently great during the normal working of the chain, will be greatly increased should any grit get between the chain and the teeth of the chain-wheel, and stretching of the chain will be produced.   Any method of keeping the chain constantly lubricated and preserving it from dust and grit should add

to the general efficiency of the machine. The gear-case introduced by Mr. Harrison Carter fulfils these requirements. The
Carter gear-case is oil-tight, and the chain at its lowest point dips
into a small pool of oil, so that the lubrication of the chain is
always perfect. The stretching of the chain is not so great with,
as without a gear-case ; in fact, some makers go the length of
saying that with a Carter gear-case, and the chain properly adjusted
initially, there is no necessity for a chain-tightening gear. A great
variety of gear-cases have recently been put on the market ; they
may be subdivided into two classes : (1) Oil-tight gear-cases, in
which the chain works in a bath of oil ; and (2) Gear-cases which
are not oil-tight, and which therefore serve merely as a protection
against grit and mud. A gear-case of the second type is probably
much better than none at all, as the chain, being kept comparatively free from grit, will probably not be stretched so much as
would be the case if no gear-case were used.

300. **Comparison of Different Forms of Chain**. — The ' Roller '
has the advantage over the ' Humber,' or block chain, that its
rubbing surface is very much larger, and that the shape of the
rubbing surface—the roller—is maintained even after excessive
wear. The ' Roller,' or long-pitched chain, on the other hand, gives
a larger variation of speed-ratio than the ' Humber,' or short-
pitched chain, the number of teeth in the chain-wheels being the
same in both cases ; but a more serious defect of the ' Roller ' chain
is the imperfect design of the side-plates (sec. 291). If, however,
the side-plates of a ' Roller ' chain be properly designed, there should
be no difficulty in making them sufficiently strong to maintain
their shape under the ordinary working stresses. Undoubtedly
the weakest part of a cycle chain, as hitherto made, is the rivet,
the bending of the rivets probably accounting for most of the
stretch of an otherwise well-designed chain. A slight increase in
the diameter of the rivets would enormously increase their strength,
and slightly increase their bearing surface.

In the ' Humber ' there are twice as many rivets as in a ' Roller '
chain of the same length. It would probably be improved by
increasing the length of the block, until the distance between the
centres of the holes was the same as between the holes in the
side-plates. This would increase the pitch to 1·2 inches, without

in any way increasing the variation of the speed-ratio. In order
to still further reduce the number of rivets, the pitch might be

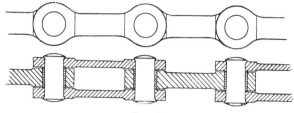

FIG. 463.

increased to one inch, giving a total pitch of two inches for the
chain. If the side-plates be made to the same outline as the
middle block (fig. 463), they may also be used to come in contact
with the teeth of the chain-wheel. The chain-wheel would then

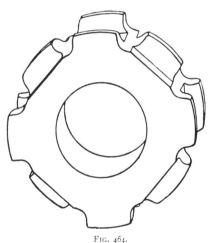

FIG. 464.

have the form shown in
figure 464, in which the
alternate teeth are in
duplicate at the edges of
the rim. For tandems,
triplets, &c., a still greater
pitch, say $1\frac{1}{2}$ inches, may
be used with advantage ;
the back-hub chain-
wheel, with six teeth of
this pitch, would have
the same average radius
of pitch-polygon as a
chain-wheel with nine
teeth of 1 inch pitch ;
the chain would have
only one-third the num-
ber of rivets in an ordinary 'Humber' chain of the same
length, and if the rivets were made slightly larger than usual,
stretching of the chain might be reduced to zero.

301. **Chain-tightening Gear.**—The usual method of providing
for the chain adjustment is to have the back-hub spindle fastened

to a slot in the frame, the length of slot being at least equal to half the pitch of the chain. In the swinging seat-strut adjustment, the slot is made in the lower back fork, and the lower ends of the seat-struts are provided with circular holes through which the spindle passes. These have been described in the chapter on Frames.

The 'eccentric' adjustment is almost invariably used for the front chain of a tandem bicycle. The front crank-axle is carried on a block, the outer surface of which is cylindrical and eccentric to the centre of the axle. The adjustment is effected by turning

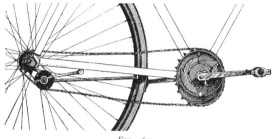

FIG. 465.

the block in the bottom-bracket, and clamping it in the desired position.

A loose pulley carried at the end of a rod controlled by a spring (fig. 465) is used in conjunction with Linley & Biggs' expanding chain-wheel.

FIG. 466.

Figure 466 illustrates a method used at one time by Messrs. Hobart, Bird & Co. When the chain required to be tightened, the loose chain-wheel was placed nearer the hub chain-wheel.

CHAPTER XXVII

TOOTHED-WHEEL GEARING

302. Transmission by Smooth Rollers.—Before beginning the study of the motion of toothed-wheels, it will be convenient to take that of wheels rolling together with frictional contact; since a properly designed toothed-wheel is kinematically equivalent to a smooth roller.

Parallel Shafts.—Let two cylindrical rollers be keyed to the

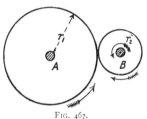

FIG. 467.

shafts A and B (fig. 467); if one shaft revolves it will drive the other, provided the frictional resistance at the point of contact of the rollers is great enough to prevent slipping. When there is no slipping, the linear speeds of two points, one on the circumference of each roller, must be the same. Let ω_1 and ω_2 be the angular speeds of the shafts, r_1 and r_2 the radii of the rollers; then the above condition gives

$$\omega_1 r_1 = - \omega_2 r_2,$$

or

$$\frac{\omega_1}{\omega_2} = - \frac{r_2}{r_1} \quad . \quad . \quad . \quad . \quad . \quad . \quad (1)$$

the negative sign indicating that the shafts turn in opposite directions. Thus the angular speeds are inversely proportional to the radii (or diameters) of the rollers.

If the smaller roller lie inside the larger, they are said to have internal contact, and the shafts revolve in the same direction.

Intersecting Shafts.—Two shafts, the axes of which intersect, may be geared together by conical rollers, the vertices of the two cones coinciding with the point of intersection of the shafts. Figure 468 shows diagrammatically two shafts at right angles, geared together by rollers forming short frusta of cones. If there be no slipping at the point of contact, the linear speeds of two points, situated one on each wheel, which touch each other during contact, must be equal. Equation (1) will hold in this case, r_1 and r_2 being the radii of the bases of the cones.

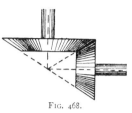

FIG. 468.

Two shafts whose axes are not parallel and do not intersect may be connected by rollers, the surfaces of which are hyperboloids of revolution. The relative motion will, however, not be pure rolling, but there will be a sliding motion along the line of contact of the rollers, which will be a generating straight line of each of the hyperboloids. This form of gear, or its equivalent hyperboloid skew-bevel gear, has not been used to any great extent in cycle construction, and will therefore not be discussed in the present work.

303. **Friction Gearing.**—If two smooth rollers of the form above described be pressed together there will be a certain frictional resistance to the slipping of one on the other, and hence if one shaft is a driver the other may be driven, provided its resistance to motion is less than the frictional resistance at the surface of the roller. Friction rollers are used in cases where small driving efforts have to be transmitted, but when the driving effort is large, the necessary pressure between the rollers would be so great as to be very inconvenient. In ' wedge gearing,' the surfaces are made so that a projection of wedge section on one roller fits into a corresponding groove on the other ; the frictional resistance, for a given pressure, being thereby greatly increased.

304. **Toothed-wheels.**—When the effort to be transmitted is too large for friction gearing to be used, projections are made on one wheel and spaces on the other ; a pair of toothed-wheels are thus obtained.

Toothed-wheels should have their teeth formed in such a manner that the relative motion is the same as that of a pair of toothless rollers. The surfaces of the equivalent toothless rollers are called the *pitch surfaces* of the wheels. By the radius or diameter of a toothed-wheel is usually meant that of its pitch surface; equation (1) will therefore be true for toothed-wheels. The distance between the middle points of two consecutive teeth measured round the pitch surface is called the *pitch* or the *circular pitch* of the teeth. The pitch must evidently be the same for two wheels in gear. Let p be the pitch, N_1 and N_2 the numbers of teeth in the two wheels, and n_1 and n_2 the numbers of revolutions made per minute; then the spaces described by two points, one on each pitch surface, in one minute are equal; therefore

$$2 \pi n_1 r_1 = 2 \pi n_2 r_2.$$

Since

$$2 \pi r_1 = N_1 p, \text{ and } 2 \pi r_2 = N_2 p, \text{ we get}$$

$$N_1 n_1 p = N_2 n_2 p$$

or,

$$\frac{N_1}{N_2} = \frac{n_2}{n_1} \quad . \quad . \quad . \quad . \quad . \quad . \quad (2)$$

That is, the angular speeds of the toothed-wheels in contact are inversely proportional to the numbers of teeth.

If the pitch diameter be a whole number, the circular pitch will be an incommensurable number. The *diametral pitch* is defined by Professor Unwin as "A length which is the same

FIG. 469.

fraction of the diameter as the circular pitch is of the circumference." The American gear-wheel makers define the diametral pitch as "The number of teeth in the gear divided by the pitch diameter of the gear." The latter may be called the *pitch-number*.

It is much more convenient to use the pitch-number than the circular pitch to express the size of wheel-teeth. Figure 469 shows the actual sizes of a few teeth, with pitch-numbers suitable for use in cycle-making. If p be the circular pitch, s the diametral pitch, and P the pitch-number,

$$s = \frac{p}{\pi}$$

$$P = \frac{1}{s} = \frac{\pi}{p} \quad . \quad . \quad . \quad . \quad . \quad . \quad (3)$$

305. Train of Wheels.—If the speed-ratio of two shafts to be geared together by wheels be large, to connect them by a single pair of wheels will be in most cases inconvenient ; one wheel of the pair will be very large and the other very small. In such a case one or more intermediate shafts are introduced, so that the speed-ratio of any pair of wheels in contact is not very great. The whole system is then called a *train* of wheels. For example, in a watch the minute hand makes one complete revolution in one hour, the seconds hand in one minute ; the speed-ratio of the two spindles is 60 to 1 ; here intermediate spindles are necessary.

If the two shafts to be connected are coaxial, it is *kinematically necessary*, not merely convenient, to employ a train of wheels. This is the case of a wheel or pulley rotating loosely on a shaft, the two being geared to have different speeds. Figure 470 shows the simplest form of gearing of this description, universally used to form the slow gearing of lathes, and which has been extensively used to form gears for front-driving Safeties. *A* is the shaft to which is rigidly fixed the wheel *D*, gearing with the wheel *E* on the intermediate

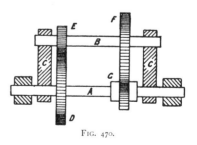

FIG. 470.

shaft *B*. The bearings of the shafts *A* and *B* are carried by the frame *C*. On the shaft *B* is fixed another wheel *F*, gearing with the wheel *G*, rotating loosely on the shaft *A*.

Denoting the number of teeth in a wheel by the corresponding small letter, the speed-ratio of the shafts B and A will be $-\dfrac{d}{e}$; the negative sign indicating that the shafts turn in opposite directions. The speed-ratio of the wheels G and F will be $-\dfrac{f}{g}$, and the speed-ratio of the wheels G and D will be the product

$$\frac{d}{e} \cdot \frac{f}{g} \qquad \ldots \quad \ldots \quad \ldots \quad \ldots \quad (4)$$

The wheels D and G (fig. 470) revolve in the same direction, the four wheels in the gear all having external contact. If one of the pairs of wheels has internal contact, the wheels A and G will revolve in opposite directions. The speed-ratio will then be

$$-\frac{d}{e} \cdot \frac{f}{g} \qquad \ldots \quad \ldots \quad \ldots \quad \ldots \quad (4)$$

306. **Epicyclic Train.**—The mechanism (fig. 470) may be inverted by fixing one of the wheels D or G and letting the frame-link C revolve ; such an arrangement is called an *epicyclic train*. The speed-ratio of the wheels D and G relative to C will still be expressed by (4). Suppose D the wheel fixed, also let its angular speed relative to the frame-link C be denoted by unity, and that of G by n. When the frame-link C is at rest its angular speed about the centre A is zero. The angular speeds of D, C, and G are then proportional to 1, 0, and n. Let an angular speed -1 be added to the whole system ; the angular speeds of D, C, and G will then be respectively

$$0, -1, \text{ and } n-1 \quad . \quad \ldots \quad \ldots \quad \ldots \quad (5)$$

If one pair of wheels has internal contact, the angular speeds of D, C, and G will be represented by -1, 0, and n ; adding a speed $+1$ to the system, the speeds will become respectively

$$0, 1, \text{ and } n+1 \quad . \quad \ldots \quad \ldots \quad \ldots \quad (6)$$

An epicyclic train can be formed with four bevel-wheels (fig. 471) ; also, instead of two wheels, E and F (fig. 470), only one may be used which will touch A externally and G internally (fig.

472) ; this is the kinematic arrangement of the well-known 'Crypto'
gear for Front-drivers. Again, in a bevel-wheel epicyclic train the
two wheels on the interme-
diate shaft B may be merged
into one ; this is the kinematic arrangement of Starley's
balance gear for tricycle axles
(fig. 219).

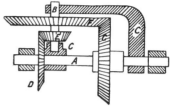

FIG. 471.

In a Crypto gear, let N_1
and N_2 be the numbers of
teeth on the hub wheel and
the fixed wheel respectively, then $n = \dfrac{N_2}{N_1}$; and from (6), the
speed-ratio of the hub and crank is

$$\frac{N_2}{N_1} + 1 = \frac{N_1 + N_2}{N_1}. \quad . \quad . \quad . \quad . \quad (7)$$

From (7) it is evident that if a speed-ratio greater than 2 be
desired, N_2 must be greater than N_1, and the annular wheel must
therefore be fixed to the frame and the inner wheel be fixed to
the hub.

Example.—The fixed wheel D of a Crypto gear has 14 teeth,
the wheel E mounted on the arm C has 12 teeth ; the number of

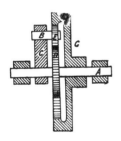 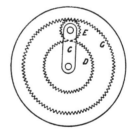

FIG. 472.

teeth in the wheel G fixed to the hub of the driving-wheel must
then be $12 + 12 + 14 = 38$. The driving-wheel of the bicycle
is 46 in. diameter ; what is it geared to ? Substituting in (7),
the speed-ratio of the hub and the crank is $\dfrac{52}{38}$, and the bicycle is
geared to $\dfrac{52 \times 46}{38} = 62\cdot95$ inches.

307. Teeth of Wheels.—The projection of the pitch-surface of a toothed-wheel on a plane at right angles to its axis is called the *pitch-circle*; a concentric circle passing through the points of the teeth is called the *addendum-circle*; and a circle passing through the bases of the teeth is called the *root-circle* (fig. 473). The part of the tooth surface *b c* outside the pitch-line is called the *face*, and the part *a b* inside the pitch-circle the *flank* of

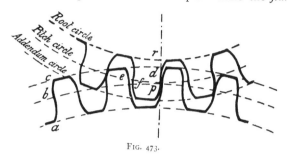

FIG. 473.

the tooth. The portion of the tooth outside the pitch-circle is called the *point*; and the portion inside, the *root*. The line joining the wheel centres is called the *line of centres*. The *top and bottom clearance* is the distance *r d* measured on the line of centres, between the addendum-circle of one wheel and the root-circle of the other. The *side-clearance* is the difference *e f* between the pitch and the sum of the thicknesses of the teeth of the two wheels, measured on the pitch-circle.

For the successful working of toothed-wheels forming part of the driving mechanism of cycles it is absolutely necessary not only that the tooth forms should be properly designed, but also that they be accurately formed to the required shape. This can only be done by cutting the teeth in a special wheel-cutting machine. In these machines, the milling-cutter being made initially of the proper form, all the teeth of a wheel are cut to exactly the same shape, and the distances measured along the pitch-line between con-secutive teeth are exactly equal. In slowly running gear teeth, as in the bevel-wheels in the balance gear of a tricycle axle, the necessity for accurate workmanship is not so great, and the teeth of the wheels may be cast.

308. **Relative Motion of Toothed-wheels.**—Let $a\,F\,a$ and
$b\,F\,b$ (fig. 474) be the outlines of the teeth of wheels, F being the
point of contact of the two teeth.   Let D be the centre of curvature of the portion of the curve $a\,F\,a$ which lies very close to the
point F; that is, D is the centre of a cir-
cular arc approximating very closely to a
short portion of the curve $a\,F\,a$ in the neighbourhood of the point F.   Similarly, let C
be the centre of curvature of the portion of
the curve $b\,F\,b$ near the point F.   Whatever
be the tooth-forms $a\,F\,a$ and $b\,F\,b$, it will
in general be possible to find the points D and
C, but the positions of C and D on the respective wheels change as the wheels rotate
and the point of contact F of the teeth
changes.   While the wheels A and B rotate
through a small angle near the position
shown, their motion is exactly the same as
if the points C and D on the wheels were
connected by a link $C\,D$.   The instantaneous motion of the
two wheels is thus reduced to that of the levers $A\,C$ and $B\,D$
connected by the coupler $C\,D$.

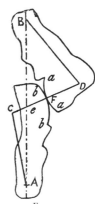

Fig. 474.

In figure 21 (sec. 32) let $B\,A$ and $C\,D$ be produced to meet
at J; then

$$\frac{D\,e}{C\,B} = \frac{D\,J}{C\,J}, \ \text{ or } \ D\,e = \frac{D\,J}{C\,J}\,.\,C\,B.$$

And it has been already shown that the speed-ratio of the two
cranks is $\dfrac{D\,e}{D\,A}$.   Therefore the speed-ratio may be written equal

to $\dfrac{C\,B}{D\,A}\,.\,\dfrac{D\,J}{C\,J}$.

Therefore, since $C\,B$ and $D\,A$ are constant whatever be the
position of the mechanism (fig. 21), the angular speeds of the two
cranks in a four-link mechanism are inversely proportional to the
segments into which the line of centres is divided by the centre-
line of the coupling-link.   Therefore if the straight line $C\,D$ cut
$A\,B$ at e (fig. 474) the speed-ratio of the wheels A and B is

$$\frac{B\,e}{A\,e} \quad . \quad . \quad . \quad . \quad . \quad . \quad (8)$$

For toothed-wheels to work smoothly together the angular speed-ratio should remain constant ; (8) is therefore equivalent to the following condition : *The common normal to a pair of teeth at their point of contact must always pass through a fixed point on the line of centres.* This fixed point is called the *pitch-point*, and is evidently the point at which the pitch-circles cut the line of centres.

If the form of the teeth of one wheel be given, that of the teeth of the other wheel can in general be found, so that the above condition is satisfied. This problem occurs in actual designing when one wheel of a pair has been much worn and has to be replaced. But in designing new wheels it is of course most convenient to have the tooth forms of both wheels of the same general character. The only curves satisfying this condition are those of the trochoid family, of which the cycloid and involute are most commonly used.

309. **Involute Teeth.**—Suppose two smooth wheels to rotate about the centres A and B (fig. 475), the sum of the radii being

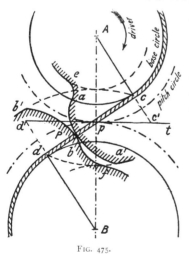

FIG. 475.

less than the distance between the centres. Let a very thin cord be partially wrapped round one wheel, led on to the second wheel, and partially wrapped round it. Let c and d respectively be the points at which the cord leaves A and touches the wheel B. Let a pencil, P, be fixed to the cord, and imagine a sheet of paper fixed to each wheel. Then the cord not being allowed to slip round either wheel, while the point P of the string moves from c to d, the wheels A and B will be driven, and the pencil will trace out on the paper fixed to A an arc of an involute $a\,a^1$, and on the paper fixed to B an involute arc $b^1\,b$. If teeth-outlines be made to these curves they must touch each other at some point on the line $c\,d$, which is their common normal at their point of contact ; and since $c\,d$ intersects the line of centres at a fixed point, p, the tooth-outlines satisfy the condition for constant speed-ratio.

The circles round which the cord was supposed to be wrapped are called the *base-circles* of the involute teeth ; the line $c\,d$ is called the *path of the point of contact*, or simply the *path of contact*. The longest possible involute teeth are got by taking the addendum-circles of wheels A and B through d and c respectively ; for though the involutes may be carried on indefinitely outwards from the base-circles, no portion can lie inside the base-circle. Except in wheels having small numbers of teeth, the arcs of the involutes used to form the tooth outlines are much smaller than shown in figure 475 ; the path of contact being only a portion of the common tangent $c\,d$ to the base-circles.

The *angle of obliquity* of action is the angle between the normal to the teeth at their point of contact, and the common tangent at p to the pitch-circles. The angle of obliquity of involute teeth is constant, and usually should not be more than $15°$.

No portion of a tooth lying inside the base-circle has working contact with the teeth of the other wheel, but in order that the points of the teeth of the wheels may get past the line of centres, the space between two adjacent teeth must be continued inside the base-circle. If the teeth be made with no clearance the continuation of the tooth outline $b^1\,P\,b$, between the base- and root-circles, is an arc of an epitrochoid $b\,f$, described on the wheel B by the point a^1 of the wheel A. The continuation of the tooth outline $a^1\,P\,a$ between the base- and root-circles is an arc of an epitrochoid $a\,e$, described on the wheel A by the point b^1 of the wheel B. This part of the tooth outline lying between the root-circle and the working portion of the tooth outline is sometimes called the *fillet*. The flanks are sometimes continued radially to the root-circle ; but where the strength of the teeth is of importance, the fillet should be properly designed as above. The *fillet-circle* is a circle at which the fillets end and the working portions of the teeth begin. When involute teeth are made as long as possible, the base- and fillet-circles coincide. In any case, the fillet-circle of one wheel and the addendum-circle of the other pass through the same point, at the end of the path of contact.

Let the centres A and B of the wheels (fig. 475) be moved farther apart ; the teeth will not engage so deeply, and the line $d\,c$ will make a larger angle with the tangent to the pitch-circles at p. The form of the involutes traced out by the pencil will, however, be exactly the same though a longer portion will be drawn. Therefore the teeth of the wheels will still satisfy the condition of constant speed-ratio. Wheels with involute teeth have therefore the valuable property that the distance between

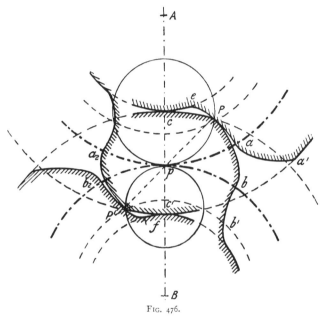

FIG. 476.

their centres may be slightly varied without prejudicially affecting the motion.

310. **Cycloidal Teeth.**—Let A and B (fig. 476) be the centres of two wheels, and let p be the pitch-point. Let a third circle with centre C lying inside the pitch-circle of A roll in contact with the two pitch-circles at the pitch-point. Suppose a pencil P fixed to the circumference of the rolling-circle. If the three circles roll so that p is always their common point of contact, the

pencil will trace out an epicycloid on wheel B, and an hypocycloid on A. Let P be any position of the pencil, then the relative motion of the two circles A and C is evidently the same as if A were fixed and C rolled round inside it ; p is therefore the instantaneous centre of rotation of the circle C, and the direction of motion of P relative to the wheel A must be at right angles to the line $p\,P$; that is, $p\,P$ is the normal at P to the hypocycloid $P\,a$.

In the same way it can be shown that the line $p\,P$ is the normal to the epicycloid $P\,b$. If tooth outlines be made to these curves they will evidently satisfy the condition for constant speed-ratio.

The tooth $P\,a$ is all flank, and the tooth $P\,b$ all face. Another rolling-circle C^1 may be taken inside the pitch-circle of wheel B, a tracing-point P^1 on it will describe an epicycloid on wheel A and a hypocycloid on wheel B. The tooth outline $P^1\,a_2$ is all face, and the tooth outline $P^1\,b_2$ is all flank. They may be combined with the former curves, so that the tooth outlines $P\,a\,a^1$ and $P\,b\,b^1$ may be used.

The path of contact $P^1\,p\,P$ in this case is evidently made up of arcs of the two rolling-circles. If the diameter of the rolling-circle be equal to the radius of the pitch-circle, the hypocycloid described reduces to a straight line a diameter of the pitch-circle. The flanks of cycloidal teeth may therefore be made radial.

If contact begins and ends at the points P and P^1 respectively, the addendum-circles of B and A pass through these points. If the teeth are made without clearance, the fillet will be, as in involute teeth, an arc of an epitrochoid, $P\,e$, described on the wheel A by the point P of the wheel B. Similarly, the fillet of wheel B between P^1 and the root-circle is an arc of an epitrochoid $P^1\,f$ described on wheel B by the point P^1 of the wheel A.

If a set of wheels with cycloidal teeth are required, one wheel of the set to gear with any other, the same rolling-circle must be taken for the faces and flanks of all.

An important case of cycloidal teeth is that in which the rolling-circle is equal to the pitch-circle of one of the wheels of the pair. The teeth of one wheel become points ; those of the other, epicycloids described by one pitch-circle rolling on the other.

If two tooth outlines gear properly together with constant

speed-ratio, tooth outlines formed by parallel curves will in general also gear together properly. In the above case the *point* teeth of one wheel may be replaced by round *pins*, the epicycloid teeth of the other wheel by a parallel curve at a distance equal to the radius of the pin. Loose rollers are sometimes put round the pins so that the wear is distributed over a larger surface.

An example of pin-gearing is found in the early patterns of the ' Collier' two-speed gear (sec. 319).

311. **Arcs of Approach and Recess.**—The *arc of approach* is the arc through which a point on the pitch-circle moves from the time that a pair of teeth come first into contact until they are in contact at the pitch-point. The *arc of recess* is the arc through which a point on the pitch-circle moves from the time a pair of teeth are in contact at the pitch-point until they go out of contact. The *arc of contact* is, of course, the sum of the arcs of approach and recess.

With cycloidal teeth (fig. 476), if P and P^1 be the points of contact when the teeth are just beginning and just leaving contact respectively, $a\,p$ or $b\,p$ will be the arc of approach, and $p\,a_2$ or $p\,b_2$ the arc of recess, provided the wheel A is the driver, in watch-hand direction. From the mode of generation of the epicycloid and the hypocycloid it is evident that the arc of the rolling circle, $P\,p$, is equal to the arc of approach, and $p\,P^1$ to the arc of recess.

With involute teeth (fig. 475) the path of contact is the straight line $c\,d$. The arc of contact, measured along either of the base-circles, is equal to $c\,d$. The arc of contact, measured along the pitch-circle, is equal to $\overline{c\,d}$ multiplied by $\dfrac{A\,p}{A\,c}$, the ratio of the radii of the pitch- and base-circles. Draw the tangent $p\,t$ at p to the pitch-circles, and produce $A\,c$ and $B\,d$ to meet $p\,t$ at c^1 and d^1 respectively. $c^1\,p$ and $p\,d^1$ are the lengths of the arcs of approach and recess respectively, measured along the pitch-circle. For from similar triangles,

$$\frac{p\,c^1}{p\,c} = \frac{A\,p}{A\,c}, \text{ or } p\,c^1 = \frac{A\,p}{A\,c} \cdot p\,c.$$

Similarly $\qquad p\,d^1 = \dfrac{A\,p}{A\,c} \cdot p\,d.$

312. **Friction of Toothed-wheels.**—There is a widespread impression, even among engineers, that, if the form of wheel-teeth be correctly designed, the relative motion of the teeth is one of pure rolling. Probably the use of the term *rolling-circles* in connection with cycloidal teeth has given rise to this impression ; but a very slight inspection of figures 475 and 476 will show that the teeth rub as well as roll on each other. In figure 476 a pair of teeth are shown in contact at P. While the teeth are passing the pitch-point, the points a and b touch each other at p. Now the length Pa of one tooth is much less than the length Pb of the other. The teeth must therefore rub on each other a distance equal to the difference between these two arcs. The same thing is apparent from figure 475.

The speed of rubbing at any point can be easily expressed as follows : Let a pair of wheels rotate about the centres A and B (fig. 477) ; let their pitch-lines touch at p ; let $a' p a''$ be the path of contact ; let r_1 and r_2 be the radii of the pitch-circles, and V be their common linear speed, the angular speeds will be $\dfrac{V}{r_1}$ and $-\dfrac{V}{r_2}$ respectively. Suppose a pair of teeth to be in contact at a ; the relative motion of the two wheels will be the same if the whole system be given a rotation $\dfrac{V}{r_2}$ about B in the direction opposite to the rotation of the wheel B. Wheel B will now be at rest, and

FIG. 477.

the pitch-line of wheel A will roll on the pitch-line of B. The angular speed of wheel A is now $\dfrac{V}{r_1} + \dfrac{V}{r_2}$. The instantaneous centre of rotation of wheel A is the point p, and therefore the linear speed of the point a on the wheel A is

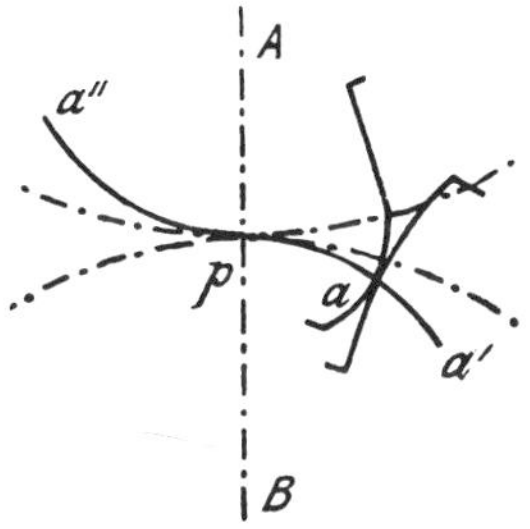

$$V\left(\frac{1}{r_1} + \frac{1}{r_2}\right) \times \text{chord } p\,a \quad \cdot \quad \cdot \quad \cdot \quad \cdot \quad \cdot \quad (9)$$

This is, of course, the same as the relative speed of rubbing of the teeth in contact at a. In particular, the speed of rubbing is greatest when the teeth are just coming into or just leaving contact, and is zero when the teeth are in contact at the pitch-point.

If the two wheels have internal contact, by the same reasoning their relative angular speed may be shown to be $V\left(\dfrac{1}{r_1} - \dfrac{1}{r_2}\right)$, and the speed of rubbing

$$V\left(\frac{1}{r_1} - \frac{1}{r_2}\right) \times \text{chord } p\, a \quad . \quad . \quad . \quad . \quad (10)$$

Thus, comparing two pairs of wheels with external and internal contact respectively, if the pitch-circles and arc of contact be the same in both, the wheels with internal contact have much less rubbing than those with external contact. If $r_2 = 3\,r_1$ the rubbing with external contact is twice as great as with internal.

Friction and Wear of Wheel-teeth.—The frictional resistance, and therefore the wear, of wheel-teeth will be proportional to the maximum speed of rubbing, and will therefore be greater the longer the path of contact. The arc of contact, therefore, should be chosen as short as possible ; the working length of the teeth will then be short, and it will be much easier to make the teeth accurately. The arc of contact must, of course, be at least equal to the pitch, so that one pair of teeth comes into contact before the preceding pair has left contact. It may be chosen a little greater, in order to allow a margin for the centres of the wheels being moved a little further apart than was intended. The rubbing of the teeth against each other during approach is said to be more injurious than during recess. In a pair of wheels in which the driver and driven are never interchanged (as in gear-wheels of cycles, which are always driven ahead and never backwards), the arc of recess may therefore be chosen a little larger than the arc of approach.

If c be the length of the arc of contact, the *average* speed of rubbing will be approximately (with external contact)

$$\frac{c}{4}\left(\frac{1}{r_1} + \frac{1}{r_2}\right) V.$$

If P be the average normal pressure on the teeth, and μ the co-efficient of friction, the work lost in friction will be

$$\frac{c}{4}\left(\frac{1}{r_1} + \frac{1}{r_2}\right)\mu\,P\,V.$$

The useful work done in the same time will be approximately $P\,V$, and the efficiency of the gear will be

$$\frac{P\,V}{P\,V + \dfrac{c}{4}\left(\dfrac{1}{r_1} + \dfrac{1}{r_2}\right)\mu\,P\,V} = \frac{1}{1 + \dfrac{c\,\mu}{4}\left(\dfrac{1}{r_1} + \dfrac{1}{r_2}\right)} \quad . \quad . \quad (11)$$

Example I.—In a pair of wheels with 12 and 24 teeth respectively, assuming $\mu = \cdot 08$, and $c = 1\cdot2\,p$, $\dfrac{p}{r_1} = \dfrac{2\,\pi}{12} = \cdot524$, and $\dfrac{p}{r_2} = \cdot262$, and the efficiency is

$$\frac{1}{1 + \cdot3 \times \cdot08\,(\cdot524 + \cdot262)} = \cdot982.$$

Example II.—In an internal gear with 12 and 36 teeth, with the same assumptions as above, the efficiency is

$$\frac{1}{1 + \cdot024\,(\cdot524 - \cdot175)} = \cdot992.$$

313. **Circular Wheel-teeth.**—Since only a small arc is used to form the tooth outline it is often convenient to approximate to the exact curve by a circular arc. Involute or cycloidal teeth are first designed by the above methods, then the circular arcs, which fit as closely as possible, are used for the actual tooth outlines. When this is done there will be a slight variation of the speed-ratio during the time of contact of a pair of teeth. The variation may be reduced to a minimum by (instead of proceeding as just described) finding the values of $A\,C$, $B\,D$, and $C\,D$ (fig. 474), such that the point e will deviate the smallest possible amount from p, the pitch-point. The author has investigated this subject in a paper on 'Circular Wheel-teeth,' published in the 'Proceedings of the Institution of Civil Engineers,' vol. cxxi. The analysis is too long for insertion here, but the principal results may be given :

For a given value of the speed-ratio $R = \dfrac{r_2}{r_1}$, three positions of the coupling-link $C D$ can be found in which it passes through the pitch-point p: let $C_1 D_1$, $C_2 D_2$ and $C_3 D_3$ (fig. 478) be these positions. The distance of C_2, the middle position of C,

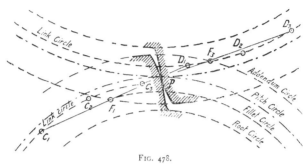

FIG. 478.

from the pitch-point p may be chosen arbitrarily ; but the greater this distance the less will be the speed variation and the greater the obliquity.

Let $$\frac{p\,C_2}{r_1} = m, \quad \frac{\text{arc of contact}}{r_1} = 2\,n\;;$$

then assuming that the other two positions of the equivalent link in which its intersection e with the line of centres coincides with the pitch-point p are at the beginning and end of contact of a pair of teeth, we may take approximately

$$p\,C_1 = (m + n)\,r_1, \quad p\,C_2 = m\,r_1, \quad p\,C_3 = (m - n)\,r_1.$$

Let $A\,C = l_1$, $B\,D = l_2$, and the length $C\,D$ of the equivalent coupling-link $= h$. The values of h, l_1, and l_2, for given values of R, m, and n, are given by the following equations :

$$\frac{h}{r_1} = \frac{3\,(R + 1)}{(R + 2)}\,m \quad . \quad . \quad . \quad . \quad . \quad . \quad . \quad (12)$$

$$\left(\frac{l_1}{r_1}\right)^2 = 1 - \frac{(R + 2)}{3\,R}\,(m^2 - n^2) \quad . \quad . \quad . \quad . \quad . \quad . \quad (13)$$

$$\left(\frac{l_2}{r_2}\right)^2 = 1 - \frac{(2\,R + 1)^3}{3\,R^2\,(R + 2)^2}\,m^2 + \frac{(2\,R + 1)}{3\,R^2}\,n^2. \quad . \quad (14)$$

Also V, the percentage speed variation above and below the average, is given by the equation

$$V = \frac{6\cdot415\,(R + 1)\,(R + 2)}{R^2} \cdot \frac{n^3}{m} \qquad . \quad . \quad . \quad . \quad (15)$$

from which, for a constant value of m, the variation is inversely proportional to the cube of the number of teeth in the smaller wheel. The values of $\frac{h}{r_1}$, $\frac{l}{r_1}$, $\frac{l_2}{r_2}$, and V, for $m = \cdot3$ and various values of R and n, are given in Table XVIII.

Having calculated, or found from the tables, the values of l_1, l_2, and h, the drawing of the teeth may be proceeded with as follows :

Draw the pitch-circles, with centres A and B, touching at the pitch-point p ; draw the link-circle $C_1\,C_2\,C_3$ with centre A and radius l_1 ; likewise draw the link-circle $D_1\,D_2\,D_3$. With centre p and radii equal to $(m + n)\,r_1$, $m\,r_1$, and $(m - n)\,r_1$ respectively, draw arcs cutting the link-circle C at the points C_1, C_2, and C_3, respectively. With centres C_1, C_2, and C_3, and radius equal to h, draw arcs cutting the link-circle D at D_1, D_2, and D_3 respectively. A check on the accuracy of the drawing and calculation is got from the fact that the straight lines $C_1\,D_1$, $C_2\,D_2$, and $C_3\,D_3$ must all pass through the pitch-point p.

Assuming that the arcs of approach and recess are equal, C_2 and D_2 will be the centres of the circular tooth outlines in contact at p. Mark off along $C_1\,D_1$ and $C_3\,D_3$ respectively, $C_1\,F_1$ and $C_3\,F_3$ each equal to $C_2\,p$. Then F_1 and F_3 will be the extreme points on the path of contact ; the addendum-circle of wheel B will pass through F_1, and the addendum-circle of wheel A through F_3. No working portion of the teeth will lie nearer the respective wheel centres than F_1 and F_3. Fillet-circles with centres A and B may therefore be drawn through F_1 and F_3.

The circular portion of the tooth will extend between the fillet- and addendum-circles ; the fillet, between the fillet- and root-circles, is designed as with involute or cycloidal teeth.

Internal Gear.—With internal gearing the radius of the larger wheel may be considered negative, and the value of R will also

TABLE XVIII.—CIRCULAR WHEEL-TEETH. External Gear. $m = 0\cdot30$.

R		$1\cdot0$			$1\cdot5$			$2\cdot0$			$3\cdot0$				$4\cdot0$			$5\cdot0$			$10\cdot0$			
h/r_1		$0\cdot600$			$0\cdot643$			$0\cdot675$			$0\cdot720$				$0\cdot750$			$0\cdot771$			$0\cdot825$			
n	Approximate number of teeth in small wheel	l_1/r_1	V Per cent.	Maximum obliquity	l_1/r_1	l_2/r_3	V Per cent.	l_1/r_1	l_2/r_3	V Per cent.	l_1/r_1	l_2/r_3	V Per cent.	Maximum obliquity	l_1/r_1	l_2/r_3	V Per cent.	l_1/r_1	l_2/r_3	V Per cent.	l_1/r_1	l_2/r_3	V Per cent.	Maximum obliquity
$0\cdot30$	10	$1\cdot000$	$3\cdot5$	$17\cdot4$	$1\cdot000$	$0\cdot992$	$2\cdot2$	$1\cdot000$	$0\cdot989$	$1\cdot7$	$1\cdot000$	$0\cdot980$	$1\cdot3$	$17\cdot4$	$1\cdot000$	$0\cdot989$	$1\cdot1$	$1\cdot000$	$0\cdot990$	$0\cdot97$	$1\cdot000$	$0\cdot994$	$0\cdot76$	$17\cdot4$
$0\cdot25$	13	$0\cdot986$	$2\cdot0$	$17\cdot4$	$0\cdot980$	$0\cdot984$	$1\cdot3$	$0\cdot991$	$0\cdot984$	$1\cdot0$	$0\cdot992$	$0\cdot985$	$0\cdot74$	$16\cdot7$	$0\cdot993$	$0\cdot987$	$0\cdot62$	$0\cdot994$	$0\cdot988$	$0\cdot56$	$0\cdot994$	$0\cdot993$	$0\cdot44$	$16\cdot5$
$0\cdot20$	16	$0\cdot975$	$1\cdot0$	$17\cdot4$	$0\cdot980$	$0\cdot977$	$0\cdot66$	$0\cdot983$	$0\cdot979$	$0\cdot51$	$0\cdot985$	$0\cdot982$	$0\cdot38$	$16\cdot0$	$0\cdot987$	$0\cdot985$	$0\cdot32$	$0\cdot988$	$0\cdot987$	$0\cdot29$	$0\cdot990$	$0\cdot992$	$0\cdot23$	$15\cdot6$
$0\cdot15$	20	$0\cdot966$	$0\cdot42$	$17\cdot4$	$0\cdot974$	$0\cdot971$	$0\cdot27$	$0\cdot977$	$0\cdot975$	$0\cdot21$	$0\cdot980$	$0\cdot980$	$0\cdot16$	$15\cdot4$	$0\cdot983$	$0\cdot983$	$0\cdot13$	$0\cdot984$	$0\cdot985$	$0\cdot12$	$0\cdot987$	$0\cdot991$	$0\cdot09$	$14\cdot7$
$0\cdot10$	31	$0\cdot960$	$0\cdot13$	$17\cdot4$	$0\cdot966$	$0\cdot968$	$0\cdot08$	$0\cdot973$	$0\cdot972$	$0\cdot06$	$0\cdot978$	$0\cdot978$	$0\cdot05$	$14\cdot8$	$0\cdot980$	$0\cdot982$	$0\cdot04$	$0\cdot981$	$0\cdot984$	$0\cdot04$	$0\cdot984$	$0\cdot991$	$0\cdot03$	$13\cdot8$
$0\cdot05$	63	$0\cdot956$	$0\cdot01$	$17\cdot4$	$0\cdot965$	$0\cdot965$	$0\cdot01$	$0\cdot971$	$0\cdot971$	$0\cdot01$	$0\cdot976$	$0\cdot977$	$0\cdot00$	$14\cdot1$	$0\cdot978$	$0\cdot981$	$0\cdot00$	$0\cdot980$	$0\cdot984$	$0\cdot00$	$0\cdot982$	$0\cdot990$	$0\cdot00$	$12\cdot9$
$0\cdot00$	∞	$0\cdot954$	$0\cdot00$	$17\cdot4$	$0\cdot954$	$0\cdot965$	$0\cdot00$	$0\cdot970$	$0\cdot970$	$0\cdot00$	$0\cdot975$	$0\cdot977$	$0\cdot00$	$13\cdot4$	$0\cdot977$	$0\cdot981$	$0\cdot00$	$0\cdot977$	$0\cdot984$	$0\cdot00$	$0\cdot982$	$0\cdot990$	$0\cdot00$	$12\cdot0$

TABLE XIX.—CIRCULAR WHEEL-TEETH. Internal Gear. $m = 0\cdot20$.

R			$2\cdot0$			$3\cdot0$			$4\cdot0$			$5\cdot0$			$10\cdot0$		
Maximum obliquity, for all values of R	h/r_1					$1\cdot200$			$0\cdot900$			$0\cdot800$			$0\cdot675$		
	n	Approximate number of teeth in small wheel	l_1/r_1	l_2/r_3	V	l_1/r_1	l_2/r_3	V	l_1/r_1	l_2/r_3	V	l_1/r_1	l_2/r_3	V	l_1/r_1	l_2/r_3	V
$14\cdot0$	$0\cdot30$	10		$A = 0\cdot000$		$1\cdot003$	$1\cdot081$	$0\cdot19$	$1\cdot004$	$1\cdot029$	$0\cdot32$	$1\cdot005$	$1\cdot016$	$0\cdot42$	$1\cdot007$	$1\cdot004$	$0\cdot62$
$12\cdot5$	$0\cdot25$	13		$l_1/r_1 = 1\cdot000,\ l_2/r_3 = \infty$		$1\cdot001$	$1\cdot083$	$0\cdot11$	$1\cdot002$	$1\cdot031$	$0\cdot19$	$1\cdot002$	$1\cdot018$	$0\cdot24$	$1\cdot003$	$1\cdot005$	$0\cdot36$
$11\cdot5$	$0\cdot20$	16				$1\cdot000$	$1\cdot085$	$0\cdot06$	$1\cdot000$	$1\cdot033$	$0\cdot10$	$1\cdot000$	$1\cdot019$	$0\cdot12$	$1\cdot000$	$1\cdot006$	$0\cdot18$
10	$0\cdot15$	20				$0\cdot999$	$1\cdot087$	$0\cdot02$	$0\cdot999$	$1\cdot034$	$0\cdot04$	$0\cdot998$	$1\cdot020$	$0\cdot05$	$0\cdot998$	$1\cdot006$	$0\cdot08$
9	$0\cdot10$	31				$0\cdot998$	$1\cdot088$	$0\cdot01$	$0\cdot998$	$1\cdot035$	$0\cdot01$	$0\cdot997$	$1\cdot021$	$0\cdot01$	$0\cdot996$	$1\cdot007$	$0\cdot02$
8	$0\cdot05$	63				$0\cdot998$	$1\cdot089$	$0\cdot00$	$0\cdot997$	$1\cdot035$	$0\cdot00$	$0\cdot996$	$1\cdot021$	$0\cdot00$	$0\cdot996$	$1\cdot007$	$0\cdot00$
7	$0\cdot00$	∞				$0\cdot998$	$1\cdot089$	$0\cdot00$	$0\cdot997$	$1\cdot035$	$0\cdot00$	$0\cdot996$	$1\cdot022$	$0\cdot00$	$0\cdot995$	$1\cdot007$	$0\cdot00$

be negative. In (12), (13), (14), and (15) substitute $R = -R$; they become respectively,

$$\frac{h}{r_1} = \frac{3\,(R - 1)\,m}{R - 2} \quad . \quad . \quad . \quad . \quad . \quad . \quad . \quad . \quad (16)$$

$$\left(\frac{l_1}{r_1}\right)^2 = 1 - \frac{(R - 2)}{3\,R}\,(m^2 - n^2) \quad . \quad . \quad . \quad . \quad . \quad (17)$$

$$\left(\frac{l_2}{r_2}\right)^2 = 1 + \frac{(2\,R - 1)^3\,m^3}{3\,R^2\,(R - 2)^2} - \frac{(2\,R - 1)}{3\,R^2}\,n^2 \quad . \quad . \quad (18)$$

$$V = \frac{6\cdot415\,(R - 1)\,(R - 2)}{R^2}\,\frac{n^3}{m} \quad . \quad . \quad . \quad . \quad (19)$$

Table XIX., with $m = \cdot2$, is calculated from these equations.

The values of $\dfrac{l_1}{r_1}$, $\dfrac{l_2}{r_2}$, in Tables XVIII. and XIX. change so slowly, that their values corresponding to any value of R and n not found in the tables can easily be found by interpolation.

314. **Strength of Wheel-Teeth.**—The mutual pressure F between a pair of wheels is sometimes distributed over two or more teeth of each wheel ; but when one of the pair has a small number of teeth it is impossible to have an arc of contact equal to twice the pitch, and the whole pressure will be borne at times by a single tooth ; each tooth must therefore be designed as a cantilever fixed to the rim of the wheel and supporting a transverse load F at its point. Let p be the pitch of the teeth, b the width, h the thickness of a tooth at the root, and l the perpendicular distance from the middle of the root of the tooth to the line of action of F. Then the section at the root is subjected to a bending-moment $F\,l$, while the moment of resistance of the section is $\dfrac{b\,h^2 f}{6}$. Therefore,

$$F\,l = \frac{b\,h^2\,f}{6} \quad . \quad . \quad . \quad . \quad . \quad . \quad . \quad (20)$$

The width of the teeth is usually made some multiple of the pitch ; let $b = k\,p$. The height of the tooth may also be expressed as a multiple of p ; it is often as much as $7\,p$, but since long teeth are necessarily weak, the teeth should be made as short as possible consistent with the arc of contact being at least equal to the pitch.

If the height be equal to ·6 p, the length l may be assumed equal to ·5 p. If there be no side-clearance the thickness at the pitch-line will be ·5 p, and with a strong tooth form the thickness at the root will be greater. Even with side-clearance, we may assume $h = \cdot 5\,p$. Substituting in (20) we have

$$F = \cdot 08333\,p^2\,k\,f \quad . \quad . \quad . \quad . \quad . \quad (21)$$

or, writing $p = \dfrac{\pi}{P}$, P being the diametral pitch-number,

$$F = \cdot 82246\,\frac{k\,f}{P^2} \quad . \quad . \quad . \quad . \quad . \quad (22)$$

The value that can be taken for f, the safe working stress of the material, depends in a great measure on the conditions to which the wheels are subjected. If the teeth be accurately cut and run smoothly, they will be subjected to comparatively little shock. For steel wheels with machine-cut teeth, 20,000 lbs. per sq. in. seems a fairly low value for f the safe working stress.

Table XX. is calculated on the assumptions made above.

TABLE XX.—SAFE WORKING PRESSURE ON TOOTHED
WHEELS.

Calculated from equation (22).

Pitch-number	Lbs. Pressure when $k = \dfrac{b}{p} =$				
	1	$1\frac{1}{2}$	2	$2\frac{1}{2}$	3
5	548	822	1097	1371	1645
6	381	571	761	952	1142
7	282	421	561	703	845
8	215	323	430	538	645
9	170	266	341	427	511
10	137	206	274	343	411
11	114	171	228	275	342
12	96	144	192	239	287
13	81	122	162	203	244
14	70	106	141	176	211
15	61	91	122	152	183
16	54	80	107	134	161
18	43	64	85	107	128
20	34	51	69	86	103
22	28	42	55	71	85
24	24	36	48	60	72

The arc of contact is sometimes made equal to two or three times the pitch, with the idea of distributing the total pressure over two or three teeth. But in this case, although the pressure on each tooth may be less than the total, they must be made longer in order to obtain the necessary arc of contact. It is therefore possible that when the pressure is distributed, the teeth may be actually weaker than if made shorter and the pressure concentrated on one.

In cycloidal teeth, for a given thickness at the pitch-line, the thickness at the root is greater the smaller the rolling-circle ; where strength is of primary importance, therefore, a small rolling-circle should be adopted. In involute teeth, the angle of obliquity influences the thickness at the root in the same manner ; the greater the angle of obliquity the greater the root thickness. In circular teeth, the greater m be taken, the thicker will be the teeth at the root.

315. **Choice of Tooth Form.**—It has already been remarked that involute toothed-wheels possess the valuable property that their centres may be slightly displaced without injury to the motion. Involute tooth outlines are simpler than cycloidal outlines, the latter having a point of inflection at the pitch-circle ; involute teeth cutters are therefore much easier to make to the required shape than cycloidal. With involute teeth the direction of the line of action is always the same, but with cycloidal teeth it continually changes, and therefore the pressure of the wheel on its bearing is continually changing. Taking everything into consideration, involute teeth seem to be preferable to cycloidal. The old millwrights and engineers invariably used cycloidal, but the opinion of engineers is slowly but surely coming round to the side of involute teeth.

316. **Front-driving Gears.**—Toothed-wheel gearing has been more extensively used for front-driving bicycles than for rear-drivers. A few special forms may be briefly noticed.

'*Sun-and-Planet*' *Gear*.—In the 'Sun-and-Planet' Safety (fig. 479), the pedal-pins are not fixed direct to the ends of the main cranks, but to the ends of secondary links, hung from the crank-pin. A small pinion is fastened to each pedal-link and gears with a toothed-wheel fixed to the hub. This is a simple form of epicyclic train, and can be treated as in section 306. If N_1 and N_2 be the numbers of teeth on the hub and pedal-link respectively, it can be shown (sec. 306) that the speed-ratio of the driving-wheel and main crank is $\dfrac{N_1 + N_2}{N_1}$.

FIG. 479.

If the driving-wheel be 40 in. diameter, and the pinion and hub have 10 and 30 teeth respectively, the bicycle is geared to $\dfrac{30 \times 10}{30} \times 40$ in. $= 53\cdot3$ inches. It should be noticed that the pedal-link will not hang vertically, owing to the pressure on the pinion. During the down-stroke the pedal will be behind the crank-pin; while on the up-stroke, if pressure be applied to the pedal, it will be in front of the crank-pin. The pedal path is therefore an oval curve with its longer axis vertical. If the pressure on the pedal be always applied vertically, the pedal path will be an ellipse, with its minor axis equal to the diameter of the toothed-wheel on the driving-hub.

This simple gear might repay a little consideration on the part of those who prefer an up-and-down to a circular motion for the pedals.

The 'Geared Facile' is a combination of the 'Facile' and 'Sun-and-Planet' gears, the lower end of the pinion-link of the latter being jointed to the pedal-lever of the former. In figure 124, the planet-pinion is 2 in. diameter, the hub-wheel 4 in. diameter, and the driving-wheel 40 in. diameter; the bicycle is therefore geared to $\dfrac{(4 + 2)}{4} \times 40 = 60$ in.

Perry's Front-driving Gear is similar in arrangement to the back gear of a lathe. The crank-axle (fig. 480) passes through the hub and is carried by it on ball-bearings. A toothed-wheel fixed to the crank-axle gears with a wheel on a short intermediate spindle, to which is also fastened a wheel gearing in turn with one fastened to the hub of the driving-wheel ; the whole arrangement being the same as diagrammatically shown in figure 470.

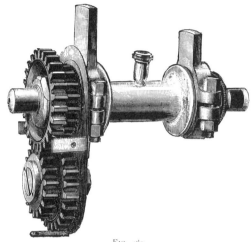

FIG. 480.

The mutual pressure between the wheels D and E (fig. 470) is equal to the tangential effort on the pedal multiplied by the ratio of the crank length to the radius of wheel D.

Example.—If the pedal pressure be 150 lbs., the crank length $6\frac{1}{2}$ in., and the radius of wheel D $1\frac{1}{2}$ in., the pressure on the teeth will be $\dfrac{6\frac{1}{2}}{1\frac{1}{4}} \times 150 = 780$ lbs.

The ' Centric' Front-driving Gear affords an ingenious example of the application of internal contact. A large annular wheel is fixed to the crank-axle and drives a pinion fixed to the hub of the driving-wheel, the arrangement being diagrammatically shown in figure 481 ; a and b being the centres of the crank-axle and the driving-wheel hub respectively. As the crank-axle has to pass right through the hub, the latter must be large enough to encircle the former, as shown in section (fig. 482). The hub ball-races are of correspondingly large diameter, the inner race being a disc set

eccentrically to the crank-axle centre. The central part of the
hub must be large enough to enclose the toothed-wheel on the

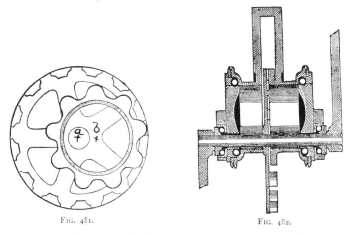

FIG. 481.   FIG. 482.

crank-axle. Instead of being made continuous and enclosing the
toothed-wheel completely, the hub is divided in the middle, and

FIG. 483.

the end portions are
united by a triangu-
lar frame.

From figure 482
it is evident that the
'Centric' gear can
only be used for
speed-ratios of hub
and crank-axle less
than 2.

*The   'Crypto'
Front-driving Gear*
is an epicyclic train,
similar in principle
to that shown in
figure 472. Figure

484 is a longitudinal section of the gear ; figure 483 an end view,
showing the toothed-wheels ; and figure 485 an outside view of the

hub, bearings and cranks complete. The arm C (fig. 472) in this case takes the form of a disc fastened to the crank-axle A, and carrying four wheels E, which engage with the annular wheel G, forming part of the hub of the driving-wheel, and with the small wheel D, rigidly fastened to the fork. The crank-axle is carried on ball-bearings attached to the fork, the hub runs on ball-bearings on

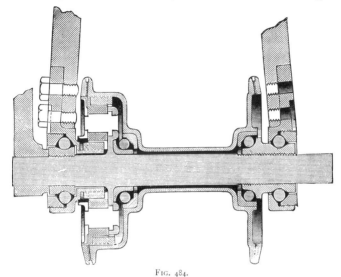

FIG. 484.

the crank-axle, while the small wheels E run on cylindrical pins B riveted to the disc C.

The pressures on the teeth of the wheels are found as follows : Considering the equilibrium of the rigid body formed by the pedal-pin, crank, crank-axle A, disc C, and pins B, the moment—about the centre of the crank-axle—of the tangential pressure P, on the pedal-pin is equal to that of the pressures of the wheels E on the pins B. Let l be the length of the crank, and r the distance of the centre of the pin B from the crank-axle ; the pressure of each wheel E on its pin will be

$$\tfrac{1}{4} \times \frac{l\,P}{r} = \frac{l\,P}{4\,r} \qquad \cdots \qquad \cdots \quad (23)$$

This pressure of the pin on the wheel E is resisted by the pressures of the wheels D and G; each of these pressures must, therefore, be equal to

$$\frac{l}{8}\frac{P}{r} \quad . \quad . \quad . \quad . \quad . \quad . \quad (24)$$

If N_1, N_2, and N_3 be the numbers of teeth on the hub-wheel G, fork-wheel D, and intermediate wheels E, respectively,

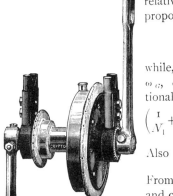

the speeds of the wheels and arm C, relative to the latter, are respectively proportional to

$$\frac{1}{N_1}, -\frac{1}{N_2}, \frac{1}{N_3}, \text{ and } \circ \quad . \quad . \ (25)$$

while, relative to the fork, the speeds ω_G, ω_D, ω_E, and ω_C are proportional to

$$\left(\frac{1}{N_1}+\frac{1}{N_2}\right), \circ, \left(\frac{1}{N_3}+\frac{1}{N_2}\right), \text{ and } \frac{1}{N_2}. (26)$$

Also $\qquad N_3 = \dfrac{N_1 - N_2}{2} \quad . \quad . \ (27)$

From (26) the speed-ratio of the hub and crank, relative to the fork, is

$$R = \frac{\left(\dfrac{1}{N_1}+\dfrac{1}{N_2}\right)}{\dfrac{1}{N_2}} = \frac{N_2}{N_1} + 1 \ . \ (28)$$

FIG. 485.

From (28) it is evident that when the hub speed is to be more than twice that of the crank, N_1 must be less than N_2; that is, the annular wheel must be fixed to the fork.

From (25) and (27) the speed-ratio of wheels E and D, relative to the disc C, is

$$-\frac{N_2}{N_3} = \frac{2N_2}{N_2 - N_1} = 2\frac{(R-1)}{(R-2)} \quad . \quad . \quad . \ (29)$$

But the wheel D makes -1 turn relative to the crank while the latter makes 1 turn relative to the fork. Therefore, for every turn of the crank, the wheels E make $2\dfrac{(R-1)}{(R-2)}$ turns in their bearings. Since these are plain cylindrical bearings, and the pressure on them is large, their frictional resistance will be the largest item in the total resistance of the gear.

Example. If $l = 6\frac{1}{2}$, $r = 1\frac{1}{4}$in., and $P = 150$ lbs. ; from (23) the pressure on the teeth is $\dfrac{6\frac{1}{2} \times 150}{8 \times 1\frac{1}{4}} = 97\cdot5$ lbs., and of the wheels E on their pins 195 lbs. Also if $R = 2\cdot5$, as in gearing a 28-inch driving-wheel to 70-inch, the wheels E each make $\dfrac{2 \times 1\cdot5}{\cdot5} = 6$ turns on their pins to one turn of the crank.

317. **Toothed-wheel Rear-driving Gears.**—A number of gears have been designed from time to time with the object of replacing the chain, but none of them have attained any considerable degree of success.

The 'Burton' Gear was a spur-wheel train, consisting of a spur-wheel on the crank-axle, a small pinion on the hub, and an intermediate wheel, gearing with both the former and running on an intermediate spindle on the lower fork. The intermediate wheel did not in any way modify the speed-ratio, so that the gearing up of the cycle depended only on the numbers of teeth of the wheels on the crank-axle and hub respectively. If r was the radius of the spur-wheel on the crank-axle and l the length of the crank, the upward pressure on the teeth of the intermediate wheel was $\dfrac{lP}{r}$, and therefore the upward pressure of the intermediate wheel on its spindle was $2\dfrac{lP}{r}$. This upward pressure was so great, that an extra bracing member was required to resist it.

Example.— If $P = 150$ lbs., $l = 6\frac{1}{2}$ in., $r = 4$, the pressure on the intermediate spindle $= \dfrac{2 \times 6\frac{1}{2} \times 150}{4} = 487\cdot5$ lbs.

The Fearnhead Gear was a bevel-wheel gear, bevel-wheels being fixed on the crank-axle and hub respectively and geared together by a shaft enclosed in the lower frame tube. If bevel-wheels could be accurately and cheaply cut by machinery, it is possible that gears of this description might supplant, to a considerable extent, the chain-driving gear ; but the fact that the teeth of bevel-wheels cannot be accurately milled is a serious obstacle to their practical success.

318. **Compound Driving Gears.**—For front-driving, Messrs. Marriott and Cooper used an epicyclic train (fig. 486), formed

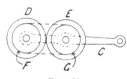

FIG. 486.

from a pair of spur-wheels and a pair of chain-wheels. Two spur-wheels, D and E, rotate on spindles fixed to the crank C. Rigidly fixed to E is a chain-wheel G, connected by a chain to a chain-wheel F, fixed to the fork. If the arm C be fixed and the pinion D be rotated, the chain-wheel F will be driven in the opposite direction. Let $-n$ be the speed-ratio of the wheels F and D relative to the arm C (in figure 486, $-n = -1$),then the angular speeds of F, C, and D are respectively proportional to n, o, and -1. If a rotation $+1$ be given to the whole system, their speeds will be proportional to $(n+1)$, 1 and o respectively. The wheel D is fixed to the fork, the wheel F to the hub of the driving-wheel, and C is the crank. The driving-wheel, therefore, makes $(n+1)$ turns to one turn of the crank. With this gear, any speed-ratio of driving-wheel and crank can be conveniently obtained.

A number of compound rear-driving gears have been made, some of which have been designed with the object of avoiding the use of a chain. In ' Hart's ' gear, a toothed-wheel was fixed on the crank-axle and drove through an intermediate wheel a small pinion ; a crank fixed on this pinion was connected by a coupling-rod to a similar crank on the back hub. In this gear, there was a dead-centre when the hub crank was horizontal, and when going up-hill at a slow pace the machine might stop. In ' Devoll's ' gear the secondary axle was carried through to the other side of the driving-wheel, two coupling-rods and pairs of cranks were used, and the dead-centre avoided.

The 'Boudard' Gear (fig. 487) was the first of a number of compound driving gears in which the chain is retained. An annular wheel is fixed near one end of the crank-axle and gears with a pinion on a secondary axle ; at the other end of the secondary axle a chain-wheel is fixed and is connected by a chain in the usual way to a chain-wheel on the back hub. A great deal of discussion has taken place on the merits and demerits of this gear ; probably its promoters at first made extravagant claims, and its opponents have overlooked some points that may be advanced in its favour. Of course, the mere introduction of an additional axle and a pair of spur-wheels is rather a disadvantage on account of the extra friction. In the chapter on Chain Gearing it has been shown that it is advantageous to make the chain run at a high speed ; this can be done with the ordinary chain gearing by making both chain-wheels with large numbers of teeth, but if the back hub chain-wheel be large, say with twelve teeth, that on the crank-axle must be so large as to interfere with the arrangement of the lower fork.

FIG. 487.

The 'Boudard' gear is a convenient means of using a high gear with a large chain-wheel on the back hub.

The 'Healy' Gear (fig. 488) is an epicyclic bevel gear having a speed-ratio of 2 to 1, which has the advantage of being more compact than the 'Boudard' gear, but has the disadvantage which applies to all bevel-wheels, viz. the fact that they cannot be cheaply and accurately cut.

Geared Hubs.—Compound chain gears have been used in which the toothed-wheel gearing is placed at the hub of the driving-wheel. In the 'Platnauer' gear (fig. 489) the small pinion is fixed to the hub and gears with a large annular wheel which runs on a disc set eccentrically to the hub spindle, a row of balls being introduced. The outer part of this wheel has projecting teeth to gear with the chain.

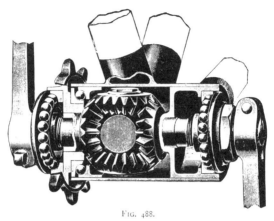

Fig. 488.

These hub gears, as far as we can see, have none of the advantages of the crank-axle gears to recommend them, since the speed of the chain cannot be increased unless a very large crank-

Fig. 489.

axle chain-wheel be used, and they possess the disadvantages of additional frictional resistance of the extra gear

319. **Variable Speed Gears.**—It has been shown, in Chapter XXI., that it is theoretically desirable to lower the gear of the cycle while riding up-hill.

In the '*Collier*' *Two-Speed Gear*, of which figure 490 is a section, and figure 491 a general sectional view, a stud-wheel *D* (that is, a wheel with pin teeth) fixed on the crank-axle gears with a toothed-pinion *P* attached to the chain-wheel *C*. The crank-axle *A* is carried on a hollow axle *B*, the axes of the two axles

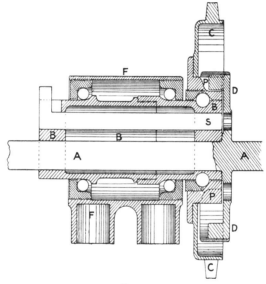

FIG. 490.

being placed eccentrically. The chain-wheel *C*, and with it the toothed-pinion *P*, revolves on a ball-bearing at the end of the hollow axle *B*. There are twelve and fifteen teeth respectively on the pinion and stud-wheel, so that the ratio of the high and low gears is 5 : 4. When the low gear is in use, the two axles are locked together by means of a slide bolt *S* in the hollow axle which engages with a hole in the stud-wheel *D*, the whole revolving together on ball-bearings in the bottom-bracket *F*. When the high gear is used, the bolt in the hollow axle is withdrawn from the hole in the stud-wheel and fits in a notch in the operating lever. The toothed-pinion, and with it the chain-wheel C, is then driven at a higher speed than the crank-axle.

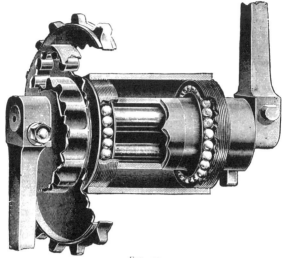

FIG. 491.

The arrangement of the two axles is shown diagrammatically in figure 492. When the high gear is in use the centre b of the crank-axle is locked in position vertically about the centre a of

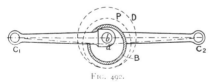

FIG. 492.

the hollow axle. If the cranks are exactly in line at high gear, the virtual cranks $a\,c_1$ and $a\,c_2$ will be slightly out of line at low gear. The pedals, however, describe practically equal circles with either gear in use.

The ' Eite and Todd' Two-Speed Gear (fig. 493) consists of a double-barrelled bracket carrying the crank-axle—on which is keyed a toothed-wheel—and a secondary axle, to which is fixed two small pinions at one end, and the chain-wheel at the other. The pinions on the secondary axle are in gear with intermediate pinions running on balls on adjustable studs attached to an arm which can swing round the secondary axle. One or other of the intermediate pinions can be thrown into gear with the spur-wheel on the crank-axle, by the shifting mechanism under the control of the rider, by means of a lever placed close to the handle-bar.

The Cycle Gear Company's Two-Speed Gear has an epicyclic train somewhat similar in principle to that of the 'Crypto' front-

FIG. 493.

driving gear. When high speed is required, the whole of the gear rotates as one rigid body; but when low speed is required the small central wheel is fixed and the chain-wheel driven by an epicyclic train.

The same Company also make a two-speed gear, the change of gearing being effected at the hub of the driving-wheel.

The 'J. and R.' Two-Speed Gear (fig. 494) consists of an epicyclic gear in the back hub; the central pinion of the gear is fixed to the driving-wheel spindle when the low gear is used, the wheel hub then rotating at a slower speed than the chain-wheel. When the high gear is used, the epicyclic gear—and with it, of course, the chain-wheel—is locked to the driving-wheel hub. C_1 is the main portion of the driving-wheel hub ; to this is fastened the end portion C_2, on which are formed a ball-race for the chain-wheel G, and an annular wheel D_2 in which the

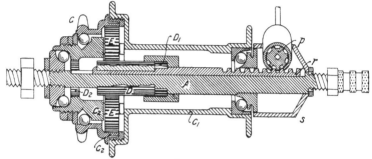

FIG. 494.

central pinion D can be locked. The intermediate pinions E, four in number, revolve on pins fastened to the hub C_1 and C_2. The annular wheel G_2, which gears with the intermediate pinions, is made in one piece with the chain-wheel G. When the low gear is in use the central pinion D is held by the axle-clutch D_1

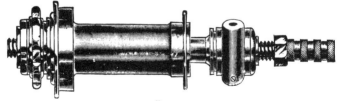

FIG. 495.

fastened to the spindle A. To change the gear, the central pinion D is shifted longitudinally out of gear with the axle clutch and into gear with the annular wheel D_2. This shifting is done by means of a rack r and pinion p ; the latter is supported in a shifter-case S fixed to the driving-wheel spindle, and is operated by the rider at pleasure.

Figure 495 shows an outside view of the hub with the spindle and shifter-case partially removed.

The 'Sharp' Two-Speed Gear (fig. 496) is an adaptation of the '.Boudard' driving gear.   On the crank-axle A the disc D_1 carries a drum D_2 on which are formed two annular wheels w_1 and w_2 which can gear with pinions p_1 and p_2 fastened to the secondary axle.   The secondary axle is in two parts ; the chain-wheel W is

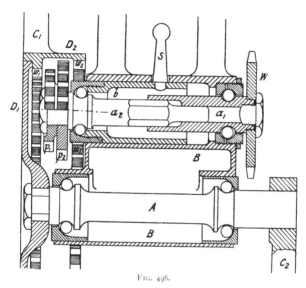

FIG. 496.

fixed to one part a_1, the pinions p_1 and p_2 to the other part a_2. The ball-bearing near the end of a_2 is carried by a secondary bracket b, which can be moved longitudinally in the main bracket B, so that the pinion p_1 may be moved into gear with the wheel w_1, or pinion p_2 into gear with wheel w_2 ; while in the intermediate position, shown in the figure, the crank-axle may remain stationary while the machine runs down hill.   A hexagonal surface on the portion a_2 fits easily in a hollow hexagonal surface on the portion a_1 of the secondary axle, so that the one cannot rotate without the other, although there is freedom of longitudinal movement.   The longitudinal movement is provided by a stud s, which passes through a small spiral slot in the main bracket B, and is screwed to the inner movable bracket b. The end of the stud can be raised or lowered, and the sliding bracket simultaneously moved longitudinally, by the rider, by means of suitable mechanism, and the gear changed from high to low, or *vice versa*.

The drum D_2 is wider than that on the ordinary 'Boudard' gear, the corresponding crank, C_1, may therefore be fastened to the outside of the drum instead of to the end of the crank-axle. The other crank, C_2, is fastened in the usual way to the crank-axle.

Linley and Biggs' Expanding Chain-wheel (fig. 465) provides for three or four different gearings, and though there is no toothed-wheel gear about it, it may be mentioned here, since it has the same function as the two-speed gears above described. The rim of the chain-wheel on the crank-axle can be expanded and contracted by an ingenious series of latches and bolts, so as to contain different numbers of cogs. When pedalling ahead the driving effort is transmitted direct from the crank-axle to the chain-wheel; but if the chain-wheel be allowed to overrun the crank-axle the series of changes is effected in the former. The right pedal being above, below, before, or behind the crank-axle, corresponds to one particular size of the chain-wheel ; if pedalling ahead be begun from one of these positions, the chain-wheel will remain unaltered. The length of chain is altered by the changes, therefore a loose pulley at the end of a light lever, controlled by a spring (fig. 465), is used to keep it always tight. Back-pedalling is impossible with this expanding chain-wheel, so a very powerful brake is used in conjunction with it.

A two-speed gear, with the gearing-down done at the hub, will be better than one with the gearing-down done at the crank-bracket, in so far that when driving with the low gear the speed of the chain will be greater, and therefore the pull on it will be less, presuming that the number of teeth on the back-hub chain-wheel is the same in both cases.

The frictional resistance of an epicyclic two-speed gear is probably much greater than that of an annular toothed-wheel gear, such as the 'Collier' or 'Sharp,' on account of the intermediate pinions revolving on plain cylindrical bearings under considerable pressure. The crank-axle of the former gear runs on plain cylindrical bearings when the gear is in action. The 'Sharp' and the 'Eite and Todd' two-speed gears have the disadvantage, compared with the others, that the additional gear and its consequent increased frictional resistance is always in action ; in this respect the former is exactly on a level with the ordinary 'Boudard' gear.

CHAPTER XXVIII

LEVER-AND-CRANK GEAR

320. **Introductory.**—A number of lever-and-crank gears have been used to transmit power from the pedal to the driving-axle of a bicycle ; the majority of them are based on the four-link kinematic chain. In general, a lever-and-crank gear does not lend itself to gear up or down ; that is, the number of revolutions made by the driving-axle is always equal to the number of complete up-and-down strokes made by the pedal. When gearing up is required, the lever-and-crank gear is combined with a suitable toothed-wheel mechanism, generally of the 'Sun-and-Planet' type. The four-link kinematic chain generally used for this gear consists of : (1) the fixed link, formed by the frame of the machine ; (2) the crank, fastened to the axle of the driving-wheel, or driving the axle by means of a 'Sun-and-Planet' gear ; (3) the lever, which oscillates to and fro about a fixed centre ; (4) the coupling-rod, connecting the end of the crank to a point on the oscillating lever.

Lever-and-crank gears may be subdivided into two groups, according as the pedal is fixed to the lever, or to the coupling-rod of the gear. In the former group, the best known example of which is the 'Facile' gear, the pedal oscillates to and fro in a circular arc, having a dead-point at the top and bottom of the stroke. In the latter group, of which the 'Xtraordinary' and the 'Claviger' were well-known examples, the pedal path is an elongated oval curve, the pedal never being at rest relative to the frame of the machine.

With lever-and-crank gears it is easy to arrange that the downstroke of the knee shall be either quicker or slower than the upstroke. In the examples analysed in this chapter, where a difference exists, the down-stroke is the quicker. Probably this is merely incidental, and has not been a result specially aimed at by the designers. Regarded merely as a mechanical question, it is immaterial whether the positive stroke be performed more quickly or slowly than the return stroke, though, possibly, physiological considerations may slightly modify the question.

321. **Speed of Knee-Joint with 'Facile' Gear.**—If the pedal be fixed to the oscillating lever, its varying speed can be found as in section 33, the speed of the crank-pin being considered constant. The speed of the knee-joint can be found as follows : Let $A\ B\ C\ D$ (fig. 497) be the four-link kinematic chain, $D\ C$ being the frame-link, $D\ A$ the crank, $C\ B$ the oscillating lever, and $A\ B$ the coupling-rod. Let the pedal be fixed to a prolongation of the oscillating lever at P. Let H and K be the rider's hip- and knee-joints respectively, corresponding to the points C and B of figure 21. In any position of the mechanism produce $D\ A$ and $B\ C$ to meet at I; I is the instantaneous centre of rotation of $A\ B$. Let $H\ K$ and

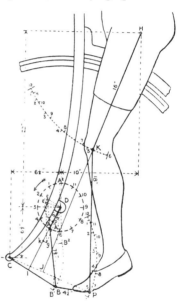

FIG. 497.

$P\ C$, produced if necessary, meet at J. Since P is at the instant moving in a direction at right angles to $C\ P$, it may be considered to rotate about any point in $I\ P$; for a similar reason, K may be considered to rotate about any point in $H\ K$; therefore J is the instantaneous centre of rotation of the rider's leg, $P\ K$, from the knee downwards. Let v_a, v_b, . . . be the speeds at any instants of the points A, B, . . . Draw $D\ e$, parallel to $B\ C$, meeting $B\ A$, produced if necessary, at e.

Then, since the points B and P are both rotating about the centre C,

$$\frac{v_p}{v_b} = \frac{C\,P}{C\,B} \qquad . \qquad . \qquad . \qquad . \qquad . \qquad . \qquad (1)$$

But from section 32,

$$\frac{v_b}{v_a} = \frac{D\,e}{D\,A} \qquad . \qquad . \qquad . \qquad . \qquad . \qquad . \qquad (2)$$

Draw $D\,e^1$ parallel to $P\,C$ and equal to $D\,e$. Draw $D\,g$ and $e^1\,g$, meeting at g, respectively parallel to $H\,K$ and $P\,K$. Since the triangles $J\,K\,P$ and $D\,g\,e^1$ are similar,

$$\frac{J\,K}{J\,P} = \frac{D\,g}{D\,e^1},$$

and

$$\frac{v_k}{v_p} = \frac{J\,K}{J\,P} = \frac{D\,g}{D\,e^1} \qquad . \qquad . \qquad . \qquad . \qquad (3)$$

Multiplying (1), (2), and (3) together, we get

$$\frac{v_p}{v_b} \cdot \frac{v_b}{v_a} \cdot \frac{v_k}{v_p} = \frac{C\,P}{C\,B} \cdot \frac{D\,e}{D\,A} \cdot \frac{D\,g}{D\,e^1},$$

that is, remembering that $D\,e$ and $D\,e^1$ are equal,

$$\frac{v_k}{v_a} = \frac{C\,P}{C\,B\,.\,D\,A} \cdot D\,g \qquad . \qquad . \qquad . \qquad . \qquad (4)$$

Therefore since the lengths $C\,P$, $C\,B$, and $D\,A$ are constant for all positions of the mechanisms, the speed of the knee-joint is proportional to the intercept $D\,g$. If $D\,k$ be set off along $D\,A$ equal to $D\,g$, the locus of k will be the polar speed-curve of the knee-joint.

322. **Pedal and Knee-Joint Speeds with 'Xtraordinary' Gear.**—If the pedal P be rigidly fixed to a prolongation of the coupling-rod $B\,A$, the construction is as follows : Produce $D\,A$ and $C\,B$, to intersect at I (fig. 498), the instantaneous centre of rotation of the coupling-rod $A\,B$. Draw $D\,e$, parallel to $I\,P$, cutting $A\,P$, produced if necessary, at e. [In some positions of the mechanism the instantaneous centre I will be inaccessible, and the direction of $I\,P$ not directly determinable ; the following

modification in the construction may be used : Draw $D\,e^1$ parallel to $B\,C$, meeting $A\,B$ at e^1 ; then draw $e^1\,e$ parallel to $B\,P$, meeting $A\,P$ at e.]

Then

$$\frac{v_p}{v_a} = \frac{I\,P}{I\,A} = \frac{D\,e}{D\,A} \quad (5)$$

or,

$$v_p = \frac{v_a}{D\,A}\,.\,D\,e \quad (6)$$

$D\,A$ is, of course, of constant length for all positions of the mechanism, and if the speed of the bicycle be uniform, v_a is constant, and therefore the speed of the pedal P along its path is proportional to the intercept $D\,e$. If $D\,p$ be set off along the crank $D\,A$, equal to this intercept, the locus of p will be the polar curve of the pedal's speed.

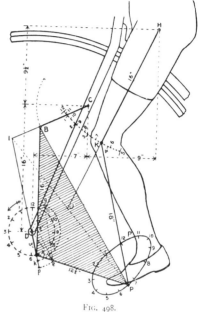

FIG. 498.

Produce $H\,K$ to meet $I\,P$ at J, then J is the instantaneous centre of rotation of the rider's leg $K\,P$ from the knee to the pedal. From D and e draw $d\,g$ and $e\,g$, meeting at g, respectively parallel to $K\,H$ and $P\,K$. Since the points K and P are at the instant rotating about the centre J,

$$\frac{v_k}{v_p} = \frac{J\,K}{J\,P} = \frac{D\,g}{D\,e} \quad\quad\quad (7)$$

Multiplying (5) and (7) together we get

$$\frac{v_k}{v_a} = \frac{D\,g}{D\,A}$$

or,

$$v_k = \frac{v_a}{D\,A}\,.\,D\,g \quad\quad\quad (8)$$

Therefore, since v_a and $D\,A$ are constant, the speed of the knee-joint is proportional to the intercept $D\,g$. If $D\,k$ be set off along the crank $D\,A$, equal to $D\,g$, the locus of k will be the polar curve of the speed of the knee-joint.

323. **Pedal and Knee-Joint Speeds with 'Geared-Facile' Mechanism.**—If toothed gearing be used in conjunction with a lever-and-crank gear, the motion of the mechanism is altered considerably. The toothed gearing usually employed in such cases is the well-known 'Sun-and-Planet' wheels, one toothed-wheel being fixed to the hub of the driving-wheel, the other centred on the crank-pin, and rigidly fixed to the coupling-rod of the gear. The driving-wheel will, as before, rotate with practically constant speed, since the whole mass of the machine and rider, moving horizontally, acts as a flywheel steadying the motion. Thus the sun-wheel of the gear moves with constant speed

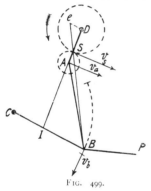

relative to the frame, but the speed of the crank is not constant, on account of the oscillation of the coupling-rod and planet-wheel.

Let $D\,A\,B\,C$ (fig. 499) be, as before, the lever-and-crank gear, and let the 'Sun-and-Planet' wheels be in contact at the point S, which must, of course, lie on the crank $D\,A$. Let I be the instantaneous centre of rotation of the coupling-rod $A\,B$, and planet-wheel; and let v_s be the speed, relative to the

FIG. 499.

frame, of the pitch-line of the sun-wheel; this will be, of course, the speed of the points of the wheel in contact at S. Draw $D\,e$ parallel to $C\,B$, meeting $B\,S$, produced if necessary, at e. Then,

$$\frac{v_b}{v_s} = \frac{I\,B}{I\,S} = \frac{D\,e}{D\,S},$$

or,

$$v_b = \frac{v_s}{D\,S} \cdot D\,e \quad . \quad . \quad . \quad . \quad . \quad (9)$$

That is, the speed of the pedal is proportional to the intercept

$D\,e$, since v_s and $D\,S$ are constant. Performing the remainder of the construction as in figure 497, we get

$$\frac{v_k}{v_s} = \frac{C\,P}{C\,B\,.\,D\,S} \cdot D\,g \quad . \quad . \quad . \quad . \quad (10)$$

The variation in the speed of the crank can easily be shown thus : The points A and S of the planet-wheel are at the instant rotating about the point I. Therefore,

$$\frac{v_a}{v_s} = \frac{I\,A}{I\,S} = 1 - \frac{A\,S}{I\,S} \quad . \quad . \quad . \quad . \quad (11)$$

$I\,S$ being considered negative when S lies between A and I.

324. **Pedal and Knee-joint Speeds with 'Geared Claviger' Mechanism.**—In this case the modification of the construction in figure 498 is the following : Let S (fig. 500) be the point of contact of the 'Sun-and-Planet' wheels. Join $P\,S$, and draw $D\,e$ parallel to $I\,P$, meeting $P\,S$ in e. Then, as in section 323,

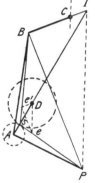

$$\frac{v_p}{v_s} = \frac{I\,P}{I\,S} = \frac{D\,e}{D\,S},$$

or,

$$v_p = \frac{v_s}{D\,S} \cdot D\,e \quad . \quad . \quad . \quad (12)$$

That is, the pedal speed is proportional to the intercept $D\,e$.

If the instantaneous centre I of the coupling-rod be inaccessible, the method of determining $D\,e$ may be as follows :—Join

FIG. 500.

$S\,B$, and draw $D\,e^1$ parallel to $C\,B$, meeting $S\,B$ at e^1. Draw $e^1\,e$ parallel to $B\,P$, meeting $S\,P$ at e.

325. **'Facile' Bicycle.**—Figure 497 represents the 'Facile' mechanism. From the centre of the driving-wheel D with radius $(D\,A + A\,B)$ draw an arc cutting the circular arc forming the path of B in the point B_1 ; from D with radius $(A\,B - D\,A)$ draw an arc cutting the path of B in B_2 ; then B_1 and B_2 will be the extreme positions of the pedal. The motion being in the direction of the arrow, and the speed of the machine being

uniform, the times taken by the pedal to perform its upward and downward movements are proportional to the lengths of the arcs A_1 9 A_2 and A_2 3 A_1. With the arrangement of the mechanism shown in the figure, the down-stroke takes a little less time than the up-stroke.

$p\,p$ (fig. 501) is the polar curve of pedal speed, found by the method of section 32, and $k\,k$, the polar curve of speed of knee-joint, found by the method of section 321, for the dimensions of the gear marked in figure 497. The speed of the knee-joint is greatest when the crank is about 30° from its lowest position, then very rapidly diminishes to zero, and rapidly attains its maximum speed in the opposite direction. It should be remembered that the speed curve, $k\,k$, is obtained on the assumption that the ankle is

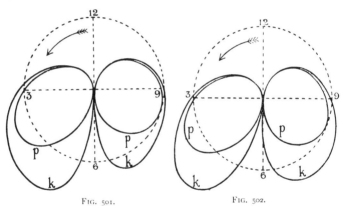

FIG. 501.   FIG. 502.

kept stiff during the motion. Using ankle action freely, the curve $k\,k$ may not even approximately represent the actual speed of the knee ; but the more rapid the variation of the radius-vectors to the curve $k\,k$, the greater will be the necessity for perfect ankle action. It should be noticed that with any mechanism a slight change in the position of the point H (fig. 497) may make a considerable change in the form of the curve $k\,k$ (fig. 501).

In some of the early lever-and-crank geared tricycles the pedal was placed at the end of a lever which, together with the oscillating lever of the four-link kinematic chain, formed a bell

crank (see fig. 146). The treatment of the pedal motion in this
case is the same as for the ' Facile' mechanism.

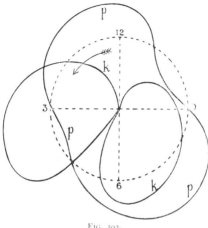

FIG. 503.

Geared Facile.—Figure 502 shows the polar curves of pedal
speed, *p p*, and of speed of knee-joint, *k k*, for a Geared Facile :

FIG. 504.

the dimensions of the mechanism being exactly the same as in figure 497, and the ratio of the diameters of the ' Sun-and-Planet ' wheels being 2 : 1.

326. **The 'Xtraordinary'** was, perhaps, the first successful Safety bicycle, the driving mechanism being arranged so that the

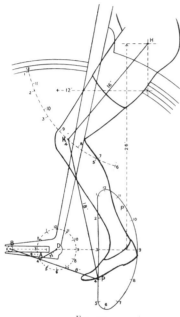

rider could use a large front wheel while sitting considerably further back and lower than was possible with an ' Ordinary.'

PP (fig. 498) is the pedal path in the 'Xtraordinary,' $p\,p$ (fig. 503) the polar curve of pedal speed, and $k\,k$ the polar curve of speed of the knee-joint. The down-stroke of the knee is performed much more quickly than the up-stroke, as is evident either from the polar speed curve, $k\,k$, or from the correspondingly numbered positions (fig. 498) of the knee and crankpin. During the down-stroke of the knee, the crank-pin moves in the direction of the arrow, from 12 to 5 ; during the up-stroke, from 5 to 12.

Fig. 505.

327. **Claviger Bicycle.**—In the Claviger gear, as applied to the ' Ordinary ' type of bicycle (fig. 504), the crank-pin was jointed to a lever, the front end of which moved, by means of a ball-bearing roller, along a straight slot projecting in front of the fork. At the rear end of the lever a segmental slot was formed to provide a vertical adjustment for the pedal, to suit riders of different heights. The mechanism is equivalent to the crank and connecting-rod of a steam-engine, the motion of the ball-bearing roller being the same as that of the piston or cross-head of the steam-engine. The mechanism may be derived from the four-link kinematic chain by

considering the radius of the arc in which the end, *B* (fig. 21), of the coupling-rod moves to be indefinitely increased.   The con-

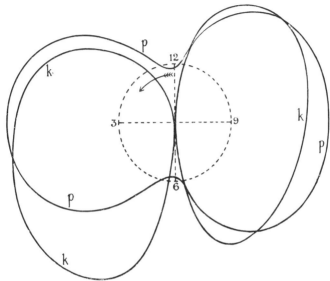

FIG. 506.

structions of figure 21 will be applicable, the only difference being that the straight line *B I* will always remain in the same direction,

FIG. 507.

that is, at right angles to the straight slot. By bending the pedal lever downwards as shown (fig. 504), the position of the saddle is further backward and downward than in the 'Ordinary.'

P P (fig. 505) is the pedal path, *p p* (fig. 506) the polar curve of pedal speed, and *k k* the polar curve of speed of knee-joint, for the mechanism to the dimensions marked in figure 505.

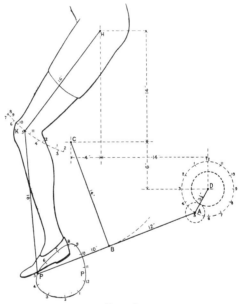

Fig. 508.

Geared Claviger.—*P P* (fig. 508) is the pedal path, *p p* (fig. 509) polar curve of pedal speed, and *k k* polar curve of speed of knee-joint, for a 'Geared Claviger' rear-driving Safety (fig. 507); the dimensions of the mechanism being as indicated in figure 508, and the ratio of the diameters of the 'Sun-and-Planet' wheels 2 : 1. The construction is as shown in figure 500.

A few peculiarities of the gear, as made to the dimensions marked in figure 508, may be noticed. The motion of the pedal in its oval path, is in the opposite direction to that of a pedal fixed to a crank. The speed of the pedal increases and diminishes three times in each up-and-down stroke ; the speed-curve, *p p* (fig. 509), shows this clearly.   The pedal path (fig. 508) also indicates the same speed variation ; the portions 2–3, 6–7, and 10–11, being

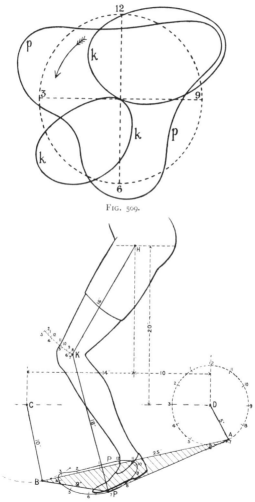

FIG. 509.

FIG. 510.

each longer than the adjacent portions, are passed over at greater speeds.

328. **Early Tricycles.**—In the *Dublin quadricycle* (fig. 117), and in some of the early lever-driven tricycles (fig. 142), the pedal was placed about the middle of the coupling-rod, one end of which was jointed to the crank-pin, the other to the end of the oscillating lever. The pedal path was an elongated oval, the vertical axis of which was shorter than the horizontal ; the early designers aiming

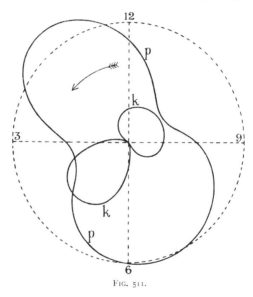

FIG. 511.

at giving the pedals a motion as nearly as possible like that of the foot during walking. *P P* (fig. 510) is the pedal path, *p p* (fig. 511) the polar curve of pedal speed, and *k k* the polar curve of speed of knee-joint, the dimensions of the mechanism being shown in figure 51c. The construction is as shown in figure 498.

It may be noticed either from figure 510 or the curve *k k* (fig. 511) that the down-stroke of the knee is performed in one-third the time of one revolution of the crank, the up-stroke in two-thirds. Also, the knee is at the top of its stroke, when the crank is nearing the horizontal position, descending.

CHAPTER XXIX

329. **The Tyre** is that outer portion of the wheel which actually touches the ground. The tyres of most road and railway vehicles are of iron or steel, and in the early days of the bicycle, when wooden wheels were used, their tyres were also of iron. The tyre of a wooden wheel serves the double purpose of keeping the component parts of the wheel in place, and providing a suitable wearing surface for rolling on the ground.

330. **Rolling Resistance on Smooth Surfaces.**—The rolling friction of a wheel on a smooth surface is small, and if the surfaces of the tyre and of the ground be hard and elastic the rolling friction, or tyre friction, may be neglected in comparison with the friction of the wheel bearings. This is the case with railway wagons and carriages. A short investigation of the nature of rolling friction has been given in section 78.

In Professor Osborne Reynolds' experiments the rolling took place at a slow speed. When the speed is great another factor must be considered. The tyre of a circular wheel rolling on a flat surface gets flattened out, and the mutual pressure is distributed over a surface. Let c (fig. 512) be the geometrical point of contact, a_1 and a_2 two points at equal distances in front of, and

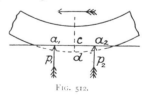

FIG. 512.

behind, c; p_1 and p_2 the intensities of the pressures at a_1 and a_2 respectively. The pressure p_1 opposes, the pressure p_2 assists, the rolling of the wheel. If the rolling takes place slowly, it is possible that p_2 may be equal to p_1, and the resultant reaction on the wheel may pass through the centre. But in all reversible dynamical actions which take place quickly, it is found that there is a loss of

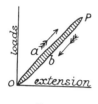

FIG. 513.

energy, which varies with the quickness of the action. The term 'hysteresis,' first used by Professor Ewing in explaining the phenomenon as exhibited in the magnetisation of iron, may be used for the general phenomena. In unloading a spring quickly, the load corresponding to a given deformation is less than when loading it ; more work is required to load the spring than it gives out during the removal of the load. If O a P (fig. 513) be the stress-strain curve during loading, that during unloading will be P b O, and the area O a P b O will be the energy lost by hysteresis. Thus, p_2 is less than p_1, the ratio $\frac{p_2}{p_1}$ lying between 1 and e, the index of elasticity. p_1, varies with the distance of a_1 from c, and is a maximum when a_1 coincides with c. Assuming the ratio $\frac{p_2}{p_1}$ to remain constant for all positions of a_1 and a_2 relative to c, we may say that the energy lost is proportional to

$$\frac{(p_1 - p_2)}{p_1} \cdot \overline{c\,d},$$

$\overline{c\,d}$ being the radial displacement of a point on the tread of the tyre.

Comparing three tyres of rubber, air, and steel respectively rolling on a *perfectly hard* surface, $\frac{p_1 - p_2}{p_1}$ will possibly be smallest for air, and largest for rubber ; while the displacement $\overline{c\,d}$ will be smallest for steel. The rolling resistance of the steel tyre will be least, that of the rubber tyre greatest.

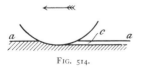

FIG. 514.

331. **Metal Tyre on Soft Road.**— The road surfaces over which cycles have to be propelled are not always hard and elastic, but are often quite the opposite. If a hard

metal tyre be driven over a soft road *a a* (fig. 514) it sinks
into it and leaves a groove *c* of quite measurable depth. The
resistance experienced in driving a cycle with narrow tyres
over a soft road is mainly due to the work spent in forming this
groove.

332. **Loss of Energy by Vibration.**—The energy lost on
account of the impact of the tyre on the ground is proportional
to the total mass which partakes of the motion of impact (see
chap. xix.). In a rigid wheel with rigid tyres, this will consist
of the whole of the wheel, and of that part of the frame which may
be rigidly connected to, and rest on, the spindle of the wheel.
If no saddle springs be used, part of the mass of the rider will also
be included. The energy lost by impact, and which is dissipated
in jar on the wheel of the machine, must be supplied by the
motive power of the rider ; consequently any diminution of the
energy dissipated in shock, will mean increased ease of propulsion
of the machine.

The state of the road surface is a matter generally beyond the
control of the cyclist or cycle manufacturer, and therefore so also
are the velocities of the successive impacts that take place.
However, the other factor entering into the energy dissipated, the
mass *m* rigidly connected with the tyre is under the control of
the cycle makers. In the first bicycles made with wooden wheels
and iron tyres, and sometimes without even a spring to the seat,
the mass *m* included the whole of the wheel and a considerable
proportion of the mass of the frame and rider ; so that the energy
lost in shock formed by far the greatest item in the work to be
supplied by the rider. The first improvement in a road vehicle is
to insert springs between the wheel and the frame. This practically means that the up and down motion of the wheel is performed to a certain extent independently of that of the vehicle
and its occupants ; the mass *m* in equation (2), chapter xix., is
thus practically reduced to that of the wheel. The effort required
to propel a spring vehicle along a common road is much less than
that for a springless vehicle.

333. **Rubber Tyres.**—If the tyre of the wheel be made elastic
so that it can change shape sufficiently during passage over an
obstacle, the motion of the wheel centre may not be perceptibly

affected, and the mass subjected to impact may be reduced to that of a small portion of the tyre in the neighbourhood of the point of contact. Thus, the use of rubber tyres on an ordinary road greatly reduces the amount of energy wasted in jar of the machine. Again, the rubber tyre being elastic, instead of sinking into a moderately soft road, is flattened out. The area of contact with the ground being much larger, the pressure per unit area is less, and the depth of the groove made is smaller; the energy lost by the wheel sinking into the road is therefore greatly reduced by the use of a rubber tyre.

Rolling Resistance of Rubber Tyres.—The resistance to rolling of a rubber tyre is of the same nature as that discussed in section 78, but the amount of compression of the tyre in contact with the ground being much greater than in the case of a metal wheel on a metal rail, the rolling resistance is also greater. This may appear startling to cyclists, but this slight disadvantage of rubber as compared with steel tyres is more than compensated by the yielding quality of the rubber, which practically neutralises the minor inequalities of the road surface.

334. **Pneumatic Tyres in General**.—The good qualities of a rubber tyre, as compared with a metal tyre for bicycles, are present to a still greater degree in pneumatic tyres. In a $\frac{3}{4}$-inch rubber tyre, half of which is usually buried in the rim of the wheel, the maximum height of a stone that can be passed over without influencing the motion of the wheel as a whole, cannot be much greater than a quarter of an inch. With a 2-inch pneumatic tyre, most of which lies outside the rim, a stone 1 inch high may be passed over without influencing the motion of the wheel to any great extent, provided the speed is great. The provision against loss of energy by impact in moving over a rough road is more perfect in this case. Again, the tyre being of larger diameter, its surface of contact with the ground is greater, and the energy lost by sinking into a road of moderate hardness is practically *nil*.

Rolling Resistance of Pneumatic Tyres.—Considering the tyre as a whole to be made of the material 'air,' and applying the result of section 194, if the material be perfectly elastic, there would be absolutely no rolling resistance. Now for all practical purposes air may be considered perfectly elastic, and there will be no dissipation of energy by the air of the tyre. The indiarubber tube in which the air is confined, and the outer-cover of the tyre, are, however, made of materials which are by no means perfectly elastic. The work done in bending the forward part of the cover will be a little greater than that restored by the cover as it regains its original shape. Probably the only appreciable resistance of a pneumatic tyre is due to the difference of these two forces. The work expended in bending the tyre will be greater, the greater the angle through which it is bent. This angle is least when the tyre is pumped up hardest; and therefore on a smooth racing track pneumatic tyres should be pumped up as hard as possible.

Again, the work required to bend the cover through a given angle will depend on its stiffness; in other words, on its moment of resistance to bending. For a tyre of given thickness d this resistance will be greatest when the tyre is of the single-tube type, and other things being equal, will be proportional to the square of the thickness d. If the cover could be made of n layers free to slide on each other, each of thickness $\dfrac{d}{n}$, the resistance of each layer to bending would be proportional to $\dfrac{d^2}{n^2}$, and that of the n layers constituting the complete covering to $\dfrac{d^2}{n}$. Thus for a tyre of given thickness its resistance is inversely proportional to the number of separate layers composing the cover. This explains why a single-tube tyre is slower than one with a separate inner air-tube; it also explains why racing tyres are made with the outer-cover as thin as possible.

Relation between Air Pressure and Weight Supported.—Let a pneumatic tyre subjected to air pressure p support a weight W. The part of the tyre near the ground will be flattened, as shown in figure 515. Let A be the area of contact with the ground, and let q be the average pressure per square inch on the ground. Then, if we assume that the tyre fabric is perfectly flexible, since the part in contact with the ground is quite flat, the pressures on the opposite sides must be equal. Therefore $q = p$. But the only external forces acting on the wheel are W and

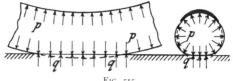

FIG. 515.

the reaction of the ground. These must be equal and opposite, therefore

$$A\, p = W \quad . \quad . \quad . \quad . \quad . \quad . \quad (1)$$

Let p_0 and V_0 respectively be the pressure per sq. in., and the volume of air inside the tyre, before the weight comes on the wheel ; and let p and V be these quantities when the tyre is deformed under the weight. The air is slightly compressed; *i.e.* V is slightly less than V_0, and p is a little greater than p_0. Now the pressure of a given quantity of gas is inversely proportional to the volume it occupies ; *i.e.*

$$\frac{p}{p_0} = \frac{V_0}{V} \quad . \quad . \quad . \quad . \quad . \quad . \quad (2)$$

p and p_0 being absolute pressures.

Example.—If a weight of 120 lbs. be carried by the driving-wheel of a bicycle, and the pneumatic tyre while supporting the load be pumped to an air pressure of 30 lbs. per sq. in. above atmosphere, the area of contact with the ground $= \dfrac{120}{30}$ $= 4$ sq. in.

If the diameter of the wheel be 28 inches and that of the inner tube be $1\frac{3}{4}$-inch, it would be easy by a method of trial and error to find a plane section of the annulus having the area required, 4 sq. in. If we assume that the part of the tyre not in contact with the ground retains its original form, which is strictly true except for the sides above the part in contact with the ground, the diminution of the volume of air inside the tyre would be the volume cut off by this plane section. In the above example this decrease is less than 1 cubic inch. The original volume of air is equal to the sectional area of the inner tube multiplied by its mean circumference. The area of a $1\frac{3}{4}$-inch circle is 2·405 sq. in., the circumference of a circle 26 inches diameter is 81·68 ins.

$$\therefore V_0 = 2·405 \times 81·68 = 196·5 \text{ cubic inches.}$$

V may be taken 195·5 cubic inches. Taking the atmospheric pressure at 14·7 lbs. per sq. in., $p = 30 + 14·7 = 44·7$. Hence, substituting in (2)

$$p_0 = \frac{44·7 \times 195·5}{196·5} = 44·47 \text{ lbs. per sq. in. absolute}$$

$$= 29·77 \text{ lbs. per sq. in. above atmosphere ;}$$

and therefore the pressure of the air inside the tyre has been increased by 0·23 lb. per sq. in.

335. **Air-tube.**—The principal function of the air-tube is to form an air-tight vessel in which the air under pressure may be retained. It should be as thin and as flexible as possible, consistent with the necessity of resisting wear caused by slight chafing action against the outer-cover. It should also be slightly extensible, so as to adapt itself under the air-pressure to the exact form of the rim and outer-cover. Indiarubber is the only material that has been used for the air-tube.

Two varieties of air-tubes are in use : the continuous tube and the butt-ended tube. The latter can be removed from a complete outer tube by a hole a few inches in length, while the former can only be removed if the outer-cover is in the form of a band with two distinct edges.

336. **Outer-cover.**—The outer-cover has a variety of functions to perform. Firstly, it must be sufficiently strong transversely and longitudinally to resist the air-pressure. Secondly, in a driving-wheel it must be strong enough to transmit the tangential effort from the rim of the wheel to the ground. Thirdly, the tread of the tyre should be thick enough to stand the wear and tear of riding on the road, and to protect the air-tube from puncture. Fourthly, though offering great resistance to elongation by direct tension, it should be as flexible as possible, offering very

little resistance to bending as it comes into, and leaves, contact with the ground, and as it passes over a stone.

Stress on Fabric.—We have already investigated (sec. 84) the tensile stress on a longitudinal section of a pneumatic tyre. We will now investigate that on a transverse section. Consider a transverse section by a plane passing through the axis of the wheel, and therefore cutting the rim at two places. The upper part of the tyre is under the action of the internal pressure, and the pull of the lower portion at the two sections. If we imagine the cut ends of the half-tyre to be stopped by flat plates, it is evident that the resultant pressure on the curved portion of the half-tyre will be equal and opposite to the resultant pressure on the flat ends. If d and t be respectively the diameter and thickness of the outer-cover, and p be the air-pressure, the area of each of the flat ends is $\dfrac{\pi d^2}{4}$, and therefore the resultant pressure on the curved surface is $2 \dfrac{\pi}{4} d^2 p$.

The area of the two transverse sections of the outer-cover is $2 \pi d t$; therefore the stress on the transverse section is

$$f = \frac{2 \dfrac{\pi}{4} d^2 p}{2 \pi d t} = \frac{p d}{4 t}. \qquad \ldots \ldots (3)$$

Comparing with section 84, the stress on a transverse section of the fabric is half that on a longitudinal section.

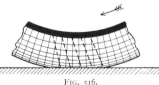

FIG. 516.

Spiral Fibres.—The first pneumatic tyres were made with canvas having the fibres running transversely and circumferentially (fig. 516). The fibres of a woven fabric, intermeshing with each other, are not quite straight, and offer resistance to bending as it comes into and leaves contact with the ground. Further, when the fibres are disposed transversely and circumferentially the cover cannot transmit any driving effort from the rim of the wheel to the ground, until it has been distorted through a considerable angle, as shown by the dotted lines.

In the ' Palmer ' tyre the fabric is made up of parallel fibres embedded in a thin layer of indiarubber, the fibres being wound

FIG. 517.

spirally (fig. 517) round an inner tube. Two layers of this fabric are used, the two sets of spirals being oppositely directed. When a driving effort is being exerted, the portion of the tyre between the ground and the rim is subjected to a shear parallel to the ground, which is, of course, accompanied by a shear on a vertical plane. This shearing stress is equivalent to a tensile stress in the direction $c c$ (fig. 518), and a compressive stress in the direction $d d$ (see sec. 105) ; consequently the fabric with spiral fibres is much better able to transmit the driving effort from the rim to the ground. This construction is

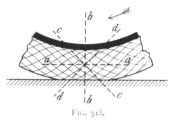

FIG. 518.

undoubtedly the best for driving-wheel tyres ; but in a non-driving wheel practically no tangential or shearing stress is exerted on the fabric of the tyre. Therefore, for a non-driving wheel the best arrangement is, possibly, to have the fibres running transversely and longitudinally ; the brake should then be applied only to the driving-wheel.

The tyre with spirally arranged fibres has another curious

property. It has been shown that the tensile stress on the transverse section bb of the tyre is half that on the longitudinal section aa. Let the stress on the section bb be denoted by p, that on aa by $2p$. This state of stress is equivalent to two simultaneously acting states of stress : the first, equal tensile stresses $\frac{3p}{2}$ on both sections ; the second, a tension $\frac{p}{2}$ on aa, and a compression $\frac{p}{2}$ on bb. The first system of stress tends to stretch the fibre equally in all directions ; the second state of stress is equivalent to shearing stresses $\frac{p}{2}$ on the planes cc and dd parallel to the spiral fibres. If the tyre be inflated free from the rim of the wheel, the fabric cannot resist the distortion due to this shearing stress, so that the tension $\frac{p}{2}$ on the section aa tends to increase the size of the transverse section of the tyre, and the compression $\frac{p}{2}$ on bb tends to shorten the circumference of the tyre. Thus, finally, the act of inflation tends to tighten the tyre on the rim.

337. **Classification of Pneumatic Tyres.**—Pneumatic tyres have been subdivided into two great classes : Single-tube tyres, in which an endless tube is made air-tight, and sufficiently strong to resist the air-pressure ; Compound tyres, consisting of two parts—an inner air-tube and an outer-cover. Quite recently, a new type, the 'Fleuss' tubeless tyre, has appeared. Mr. Henry Sturmey, in an article on 'Pneumatic Tyres' in the 'Cyclist's Year Book' for 1894, divides compound tyres into five classes, according to the mode of adjustment of the outer-cover to the rim, viz.: Solutioned tyres, Wired tyres, Interlocking and Inflation-held tyres, Laced tyres, and Band-held tyres.

A better classification, which does not differ essentially from the above, seems to be into three classes, taking account of the method of forming the chamber containing the compressed air, as follows :

Class I., with complete tubular outer-covers. This would include all single-tube tyres, most solutioned tyres, and some laced tyres. Tyres of this class can be inflated when detached from the rim of the wheel ; in fact, the rim is not an integral part of the tyre, as in the two following classes. This class may be referred to as *Tubular* tyres.

Class II., in which the transverse tension on the outer-cover is transmitted to the edges of the rim, so that the outer-cover and rim form one continuous tubular ring subjected to internal air-pressure. The 'Clincher' tyre is the typical representative of this class. With most tyres of this class the compression on the rim due to the pull of the spokes is reduced on inflation. This class will be referred to as *Interlocking* tyres.

Class III., in which the transverse tension on the outer-cover is transmitted to the edges of the latter, and there resisted by the longitudinal tension of wires embedded in the cover. This class includes most wired tyres. With tyres of this class the initial compression on the rim is increased on inflation. This class will be referred to as *Wired* tyres, and may be subdivided into two sections, according as the wire is endless, or provided with means for bringing the two ends together, and so adjusting the wire on the rim.

338. **Tubular Tyres.**—*Single-tube tyres*, which form an important group in this class, are made up of an outer layer of rubber forming the tread which comes in contact with the ground, a middle layer of canvas, or other suitable material, to provide the necessary strength and inextensibility, and an inner air-tight layer of rubber. The 'Boothroyd' and the 'Silvertown' were among the most successful of these single-tube tyres. The 'Palmer' tyre (fig. 517) was originally made as a single-tube.

Since a solid plate of given thickness offers more resistance to bending than two separate plates having the same total thickness, the resilience of a tyre is decreased by cementing the air-tube and outer-cover together.

Solutioned tyres.—The original 'Dunlop' tyre (fig. 519), which was the originator of the principle of air tyres for cycles, belongs to this class. The outer-cover consists of a thick tread of rubber A solutioned to a canvas strip B. A complete woven tube of canvas H, encircles the air-tube C, and is solutioned to the rim E, which is previously wrapped round by a canvas strip D ; while the flaps of the outer-cover are solutioned to the inner surface of the rim, one flap being lapped over the other, the side being slit to pass the spokes. A strip of canvas *F*, solutioned over the flaps, makes a neat finish.

In the Morgan and Wright tyre, the air-tube is butt-ended, or rather scarf-ended, the two ends overlapping each other about

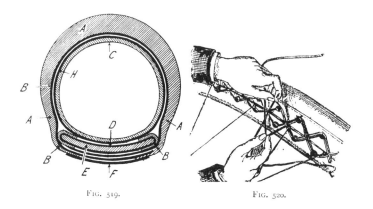

FIG. 519.   FIG. 520.

eight or ten inches. The outer-cover forms practically a tube slit for a few inches along its under side ; this opening serves for the insertion of the air-tube, and is laced up when the air-tube is in place. When partially inflated the tyre is cemented on to the rim.

Laced tyres.—In Smith's ' Balloon' tyre (fig. 520) the outer-cover was furnished with stud hooks at its edges, and enveloped the rim completely ; its two edges were then laced together.

339. **Interlocking Tyres.**—In this class of tyres the circumferential tension near the edge of the outer-cover is transmitted direct to the rim of the wheel, by suitably formed ridges, which on inflation are forced into and held in corresponding recesses of the rim.

Inflation-held Tyres.—In tyres which depend primarily on inflation for the fastening to the rim, the edge of the outer-cover is continued inwards forming a toe beyond the ridge or heel, the air-pressure on the toe keeping the heel of the outer-cover in close contact with the recess of the rim.

The 'Clincher' tyre was the first of this type. The 'Palmer' detachable tyre (fig. 521), so far as regards the fastening of the outer-cover to the rim, is identical with the 'Clincher.'

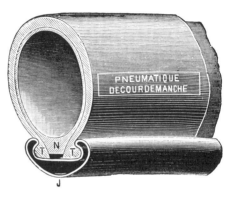

FIG. 521.   FIG. 522.

The 'Decourdemanche' tyre (fig. 522) is of the 'Clincher' type, but it has a wedge thickening N on the inner part of the air-tube, which on inflation is pressed between the ridges T of the outer-cover, and forces them into the recesses of the rim.

The 'Swiftsure' tyre differs essentially from those previously described. The outer-cover is furnished at the edges with circular ridges which lie in a central deep narrow-mouthed groove of the rim. The mouth of the groove is just large enough to admit the ridge of the cover, while the body of the groove is wide enough to let them lie side by side. On inflation, the tendency is to draw both ridges from the groove together, so that they lock each other at the mouth, and thus the tyre is held on the rim.

Hook-tyres.—In this subdivision the positive fastening of the outer-cover to the rim does not depend merely on inflation ; but the pull of the cover can be transmitted to the rim in the proper direction, even though there be no pressure in the air-tube.

In the original 'Preston-Davies' tyre eye-holes were formed near the edges of the outer-cover ; these were threaded on hooks turned slightly inwards, so that on inflation the cover was held securely to the rim.

The 'Grappler' tyre is a successful modern example of this same class. Near each edge of the outer-cover a series of turned-back hooks or *grapplers* are fastened. These engage with the in-turned edge of the rim, so that on inflation the tyre is securely fastened.

Band-held tyres.—In the 'Humber' pneumatic tyre (fig. 523) the outer-cover *A* is held down on the rim *D* by means of a lock-ing plate *C* on which the air-tube *B* rests.

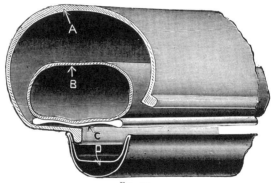

Fig. 523.

In the 'Woodley' tyre (fig. 524) it is possible that the flap acts in somewhat the same way as the plate in the 'Humber' tyre.

Fig. 524.

The 'Fleuss' tubeless tyre (fig. 525) is fixed to the rim on the 'Clincher' principle. The inner surface of the tyre is made air-tight, and thus a separate air-tube is dispensed with. A flap, per-manently fastened to one edge of the tyre, is pressed on the other edge, when inflation is com-pleted. The difficulty of keeping an air-tight joint between this loose flap and the edge of the tyre, right round the circumference (a length of over six feet), has been successfully overcome.

340. **Wire-held Tyres.**—The mode of fastening to the rim, used in this class of pneumatic tyre, differs essentially from that used in the other classes. Wires W (fig. 526) are embedded near the edges of the outer-cover C. On infla-

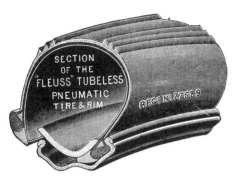

FIG. 525.

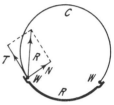

FIG. 526.

tion, a transverse tension T is exerted on the outer-cover, and transmitted to the wire W, tending to pull it out of the rim R. The wire is also pressed against the rim, the reaction from which N is at right angles to the surface. The resultant R of the forces T and N must lie in the plane of the wire W, and constitutes a radial outward force acting at all points of the ring formed by the wire. Thus, the chamber containing the air under pressure is formed of two portions : the outer-cover, subjected to tension T; and the rim R subjected to bending by the pressures N exerted by the wires W.

Let d be the diameter of the air-tube (not shown in figure 526), D the diameter of the ring formed by the wire W, and p the air-pressure. Then, by (7) chap. x., the force T per inch length of the wire is $\dfrac{p\,d}{2}$. The force R will be greater than T, depending on the angle between them. In the 'Dunlop' detachable tyre, this angle is about 30°, and therefore $R = 1\cdot155\ T$. The longitudinal pull P on the wire W is, by another application of the same formula,

$$P = \frac{1\cdot155\ T D}{2} = \cdot289\,p\,d\,D \quad . \quad . \quad . \quad (4)$$

Example.—A pneumatic tyre with air-tube $1\frac{3}{4}$ in. diameter is fixed by wires forming rings 24 ins. diameter ; and has an air pressure of 30 lbs. per sq. in. ; the pull on each wire is therefore $\cdot289 \times 30 \times 1\cdot75 \times 24 = 364$ lbs.

If the wire be No. 14 W. G. its sectional area (Table XII.), p. 346, is $\cdot00503$ sq. in., and the tensile stress is

$$\frac{364}{503} = 72300 \text{ lbs. per sq. in.}$$
$$\text{or } 32\cdot3 \text{ tons per sq. in.}$$

Wire-held tyres may be sub-divided into two classes ; in one the wire is in the form of an endless ring, and is therefore non-adjustable, in the other the ends of the wire are fastened by suitable mechanism, so that it can be tightened or released at pleasure.

The 'Dunlop' Detachable Tyre (fig. 527) is the principal representative of the endless wired division. In it two endless

FIG. 527.

wires are embedded near the edges of the outer-cover. These wires form rings of less diameter than the extreme diameter of the rim, and are lodged in suitable recesses of the rim. The rim is deeper at the middle than at the recesses for the wire. To detach the tyre, after deflation, one part of one edge of the outer-cover is depressed into the bottom of the rim, the opposite part of the same edge will be just able to surmount the rim, and one part of the wire being got outside the rest will soon follow.

The ' Woodley ' tyre (fig. 524) is formed from the ' Dunlop ' by adding a flap to the outer-cover, this flap extending from one of the main fixing wires to the other, and so protecting the air-tube from contact with the rim.

In the original ' Beeston ' tyre this flap was extended so far as to completely envelop the air-tube. In the newer patterns this wrapping has been discarded, and the ' Beeston ' is practically the same as the ' Dunlop ' detachable.

The '1895 Speed' tyre, made by the Preston-Davies Valve and Tyre Company, is fixed to the rim by means of a continuous wire of three coils on each side. At each side of the tyre a complete coil is enclosed in a pocket near the edge of the outer-cover ; one half of each of the other coils is outside, and the remaining halves inside, the pocket. By this device a wire, composed of two half coils, is exposed all round between the cover and the rim. When the tyre is deflated this exposed wire can easily be pulled up with the fingers, the detached coil is then brought over the edge of the rim, more of the slack pushed back into the pocket, enlarging the other coils, whereupon the outer-cover can be removed from the rim.

Tyres with Adjustable Wires.—The '1894 Preston-Davies' tyre was attached to the rim by means of a wire running through the edge of the outer-cover, one end of the wire having a knob which fitted into a corresponding slot in the rim, the other end having a screwed pin attached to the wire by an inch or two of a very small specially made chain. This chain was introduced to take the sharp bend where the adjusting nut drew up the slack of the wire in tightening it upon the rim.

In the 'Scottish' tyre (fig. 528) the ends of the adjustable wire are brought together by a right- and left-handed screw. A short wire, terminating in a loop, forms a handle for turning the screw. When in position this handle fits between the rim and outer-cover.

The 'Seddon' pneumatic tyre was the first successful wired tyre. Figure 529 is a view showing a portion of the

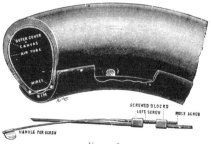

F I G. 528.

tyre with the fastening released. The ends of the wire were secured by means of a small screw which was passed through the rim and locked in place by a nut. The ends of the wire were pulled together by means of a special screw wrench.

In the 'Michelin' tyre the wires are of square tubular section.

The outer-cover, which is very deep, is provided at its edges with thick beads turned outwards, and each rested in the grooves of the specially-formed rim. A tubular wire is placed round these beads

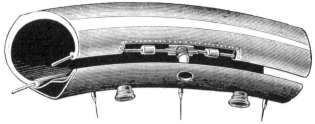

FIG. 529.

and its ends are secured in notches cut in the rim, a ⊤ bolt and screw securing the ends in position.

In the 'Drayton' tyre (fig. 530) the wires are tightened on the rim by a screw-and-toggle-joint arrangement.

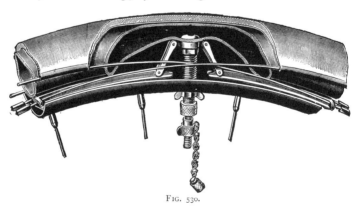

FIG. 530.

341. Devices for Preventing, and Minimising the Effect of, Punctures.—In the 'Silvertown Self-closure' tyre, which was of the single-tube variety, a semi-liquid solution of rubber was left on the inner surface of the tyre. When a small puncture was made, the internal pressure forced some of the solution into the hole, and the solvent evaporating, the puncture was automatically repaired.

In the 'Macintosh' tyre a section of the air-tube when deflated took the form shown in figure 531. On inflation the part of the air-tube at *S* was strongly compressed, so that if a puncture took place the elasticity of the indiarubber and the internal pressure combined to close up the hole.

In the 'Self-healing Air-Chamber' the same principle is made use of; the tread of an ordinary air-tube is lined inside with a layer of vulcanised indiarubber contracted in every direction. When the chamber is punctured on the tread, the lining of contracted indiarubber expands and fills up the hole, so preventing the escape of air.

FIG. 531.

In the 'Preston-Davies' tyre a double air-chamber with a separate valve to each was used. If puncture of one chamber took place it was deflated and the second chamber brought into use.

In the 'Morgan and Wright Quick-repair Tyre' (fig. 532) the air-tube is provided with a continuous patching ply, which normally rests in contact with that portion near to the rim. To repair a puncture a cement nozzle is introduced through the outer casing and

FIG. 532. FIG. 533.

tread of the air-tube (fig. 533), and a small quantity of cement is left between the tread and the patching ply. On pressing down the tread the patching ply is cemented over the hole, and the tyre is ready for use as soon as the cement has hardened.

A punctured air-tube is usually repaired on the outside, so that the air pressure tends to blow away the patch. In the 'Fleuss' tubeless tyre, on the other hand, a puncture is repaired from the inside, the tyre can be pumped up hard immediately, and the air pressure presses the patch closely against the sides of the hole.

342. **Non-slipping Covers** have projections from the smooth tread that penetrate thin mud and get actual contact with the solid ground (see sec. 170). These projections have been made diamond-shaped and oat-shaped, in the form of transverse bands, longitudinal bands (fig. 521), and interrupted longitudinal bands (fig. 527). They should offer resistance to circumferential as well as to side-slipping, though the latter should be the greater. Probably, therefore, the oat-shaped projections and the interrupted bands (fig. 527) are better than continuous longitudinal bands, and the latter in turn better than transverse bands.

343. **Pumps and Valves.**—Figure 534 shows diagrammatically the pump used for forcing the air into the tyre. The pump barrel B is a long tube closed at one end, and having a gland G screwed on to the other, through which a tubular plunger P works loosely. To the inner end of the plunger a cup-leather L is fastened. When the air-pressure in the inner part B^1 of the barrel is greater than in the outer part B, the edge of the cup-leather is pressed firmly against the sides of the barrel ; but when the pressure in the space B^1 is less than in the space B the cup-leather leaves the sides of the barrel and allows the air to flow past it from B into B^1. A valve V at the inner end of the plunger allows the air to flow from B^1 through the hollow plunger and connecting tube to the tyre, but closes the opening immediately the

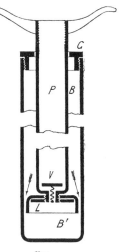

FIG. 534.

air tends to flow in the opposite direction. The action is as follows : The plunger being at the bottom, and just beginning the outward stroke, the volume B^1 is enlarged, the air-pressure in B^1 falls, and the valve V is closed by the air-pressure in the hollow plunger (the same as that in the tyre). The outward stroke of the plunger continuing, a partial vacuum is formed in B^1, the cup-leather leaves the sides of the barrel, and air passes from the space B to space B^1, until the outward stroke is completed. On beginning the inward stroke, the air in B^1 is compressed, forcing the edge of the cup-leather against the sides of the barrel, and so preventing any air escaping. The inward stroke continuing, the

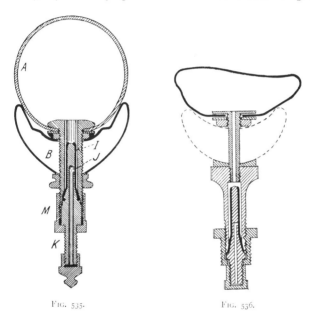

FIG. 535.   FIG. 536.

air in B^1 is compressed until its pressure reaches that of the air in the tyre, the valve V is lifted, and the air passes from B^1, along the hollow plunger, into the tyre. At the same time a partial vacuum is formed in the space B, and air passes into this space through the opening left between the plunger and the gland G.

A valve is always attached to the stem of the air-tube, so as to give connection, when required, between the pump and the interior of the tyre. A non-return valve is the most convenient, *i.e.* one which allows air to pass into the tyre when the pressure in the pump is greater than that in the tyre, and does not allow the air to pass out of the tyre. In the 'Dunlop' valve (fig. 535) the valve proper is a piece of indiarubber tube I, resting tightly on a cylindrical 'air-plug,' K. The air from the pump passes from the outside down the centre of the air-plug, out sideways at J, then between the air-plug and indiarubber tube I to the inside of the air-tube A of the tyre. Immediately the pressure of the pump is relaxed, the indiarubber tube I fits again tightly on the air-plug and closes the air-hole J. By unscrewing the large cap M, the tyre may be deflated.

Wood rims are seriously weakened by the comparatively large hole necessary for the valve-body B. Figure 536 shows a valve fitting, designed by the author, in which the smallest possible hole is required to be drilled through the rim.

CHAPTER XXX

344. **Pedals**.—Figure 537 shows the ball rubber pedal, as made by Mr. William Bown, in ordinary use up to a year or two ago. The thick end of the pin is passed through the eye of the crank and secured by a nut on the inner side of the crank. The pedal-pin is exposed along nearly its whole length, there are therefore four places at which dust may enter, or oil escape from, the ball-bearings.

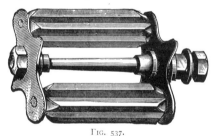

FIG. 537.

If the two pedal-plates be connected by a tube, a considerable improvement is effected, the pedal-pin being enclosed ; while if in addition a dust cap be placed over the adjusting cone at the end of the spindle,

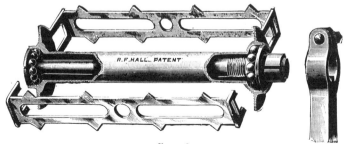

FIG. 538.

there is only one place at which dust may enter or oil escape from the bearings.

Figure 538 illustrates the pedal made by the Cycle Components Manufacturing Company, Limited, in which there are only three pieces, viz. : the pedal frame, pin, and adjustment cone. The adjustment cone is screwed on the crank end of the pedal-pin, a portion of the cone is screwed on the outside and split. The cone is then screwed into the eye of the crank, the pedal-pin adjusted by means of a screw-driver applied at its outer end ; then, by tightening up the clamping screw in the end of the crank, the crank, pedal-pin, and adjustment cone are securely locked together.

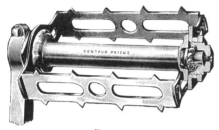

FIG. 539.

FIG. 540.

The 'Centaur' pedal (fig. 539) differs essentially from the others ; the arrangement is such that an oil-bath is possible for the balls, whereas in the usual form of pedal the oil drains out of the ball-bearings.

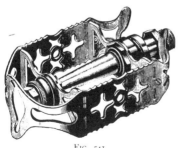

FIG. 541.

Recently, a number of new designs for pedals have been placed on the market, of which the 'Æolus Butterfly' (fig. 540), by William Bown, Limited, and that (fig. 541) by the Warwick and Stockton Company, Newark, U.S.A., may be noticed.

345. **Pedal-pins.**—The pedal-pin is rigidly fixed to the end of the crank; it may therefore be treated as a cantilever (fig. 542) supporting a load P, the pressure of the rider's foot. This load comes on at two places, the two rows of balls. One of these rows is close to the shoulder of the pin abutting against the crank, the other is near the extreme end of the pin. At any section between the balls and

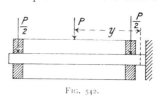

FIG. 542.

distant x from the outer row the bending moment is $\frac{1}{2} P x$. If d be the diameter of the pin at this section, and f the maximum stress on the material, we have, substituting in the formula $M = Z f$,

$$\tfrac{1}{2} P x = \frac{d^3 f}{10} \text{ or } d = \sqrt[3]{\frac{5 P x}{f}} \quad . \quad . \quad . \quad . \quad (1)$$

That is, for equal strength throughout, the outline of the pedal-pin should be a cubical parabola. On any section between the shoulder and the inner row of balls, and distant y from the centre of the pedal, the bending moment will be $P y$.

It will in general be sufficient to determine the section of the pin at the shoulder and taper it outwards.

Example.—If $P = 150$ lbs., $f = 20,000$ lbs. per sq. in., and the distance of P from the shoulder be 2 in., then

$$M = 150 \times 2 = 300 \text{ inch-lbs.} \quad Z = \frac{300}{20,000} = \text{·015 in.}^3$$

From Table III., p. 109, $d = \frac{9}{16}$ in.

346. **Cranks.**—Figure 543 is a diagrammatic view showing the crank-axle a, crank c, and pedal-pin p, the latter being acted on by the force P at right angles to the plane of the pedal-pin and crank. Introduce the equal and opposite forces P_1 and P_2 at the outer end of the crank, and the equal and opposite forces P_3 and P_4 at its inner end; P_1, P_2, P_3, and P_4 being each numerically equal to P. The forces P and P_1 constitute a twisting couple T of magnitude $P l_1$, acting on the crank, l_1 being the distance of P from the crank. The forces P_2 and P_3 constitute a bending couple M, of magnitude $P l$ at the boss of the crank. The force P_4

causes pressure of the crank-axle on its bearings. Thus the original force P is equivalent to the equal force P_4, a twisting couple $P\,l_1$, and a bending couple $P\,l$. No motion takes place along the line of action of P_4, nor about the axis of the twisting couple $P\,l_1$, the only work done is therefore due to the bending couple $P\,l$.

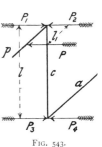

FIG. 543.

At any section of the crank distant x from its outer end, the bending-moment is $P\,x$. The equivalent twisting-moment T_e, which would produce the same maximum stress as the actual bending- and twisting-moments M and T acting simultaneously, is given by the formula $T_e = M + \sqrt{M^2 + T^2}$. Similarly, the equivalent bending-moment

$$M_e = \tfrac{1}{2}\,T_e = \tfrac{1}{2}\,(M + \sqrt{M^2 + T^2}).$$

Example.—If $l_1 = 2\tfrac{1}{4}$ in., $l = 6\tfrac{1}{2}$ in., $P = 150$ lbs., and $f = 20,000$ lbs. per sq. in.

$M = 150 \times 6\tfrac{1}{2} = 975$ inch-lbs., $T = 150 \times 2\tfrac{1}{4} = 337$ inch-lbs.

Then

$$T_e = 2007 \text{ inch-lbs., or } M_e = 1003 \text{ inch-lbs.}$$

Then

$$Z = \frac{M}{f} = \frac{1003}{20,000} = \text{·}0501 \text{ in.}^3$$

From Table III., p. 109, the diameter of a round crank at its larger end should be $1\tfrac{3}{16}$ in.

If the cranks are rectangular, and assuming that an equivalent bending-moment is 1000 inch-lbs., we get

$$Z = \frac{b\,h^2}{6} = \frac{1,000}{20,000}$$

$$\therefore b\,h^2 = \text{·}30.$$

If $b = \tfrac{1}{2}\,h$, *i.e.* the depth of the crank be twice its thickness, we get $\tfrac{1}{2}\,h^3 = \text{·}30$, $h^3 = \text{·}60$, and

$$h = \text{·}843 \text{ in., } b = \text{·}421 \text{ in.}$$

The cranks were at first fastened to the axle by means of a rectangular key, half sunk into the axle and half projecting into the boss of the crank. A properly fitted and driven key gave a very secure fastening, which, however, was very difficult to take apart, and detachable cranks are now almost invariably used. Perhaps the most common form of detachable crank is that illustrated in figure 544. The crank boss is drilled to fit the axle, and a conical

FIG. 544.

pin or cotter, flattened on one side, is passed through the crank boss and bears against a corresponding flat cut on the axle. The cotter is driven tight by a hammer, and secured in position by a nut screwed on its smaller end.

In the 'Premier' detachable crank made by Messrs. W. A. Lloyd & Co. a flat is formed on the end of the axle, and the hole in the crank boss made to suit. The crank boss is split, and on being slipped on the axle end is tightened by a bolt passing through it.

FIG. 545.

Figure 545 illustrates the detachable chain-wheel and crank made by the Cycle Components Manufacturing Company. A

long boss is made on the chain-wheel, over which the crank boss
fits. Both bosses are split and are clamped to the axle by means
of a screw passing through the crank boss. In addition to the
frictional grip thus obtained, a positive connection is got by means
of a small steel plate, applied at the end of the crank-axle and
wheel boss, and retained in position by the clamping-screw. The
pedal end of this crank is illustrated in figure 538.

The 'Southard' crank, which is round-bodied (fig. 544),
receives during manufacture an initial twist in the direction of the
twisting-moment due to the pressure on the pedal in driving ahead.
The elastic limit of the material is thus artificially raised, the
crank is strengthened for driving ahead, but weakened for back
pedalling ; as already discussed in section 123.

In the 'Centaur' detachable crank and chain-wheel, the crank
boss is placed over the chain-wheel boss. Both wheel and crank
are fixed to the axle by a tapered cotter, driven tight through the
bosses and retained in position by a nut.

It has been shown that near the boss of the crank the bending-
moment is greater than the twisting-moment. Round-bodied
cranks have the best form to resist twisting, rectangular-bodied to
resist bending. A crank rectangular towards the boss and round
towards the eye would probably be the best. Hollow cranks of
equal strength would of course be theoretically lighter than solid
cranks, but the difficulty of attaching them firmly to the axle has
prevented them being used to any great extent. In some of the
early loop-framed tricycles, the axle,
cranks, and pedal-pins were made of
a single piece of tubing.

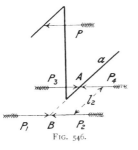

FIG. 546.

347. **Crank-axle.**—Figure 546 is
a sketch showing part of the crank-
axle a, the crank and pedal-pin, the
latter acted on by the force P. Introduce two equal and opposite forces P_3
and P_4 at the bearing A, and two equal
and opposite forces P_1 and P_2 at a
point B on the axis of the crank-axle,
the forces P, P_1, and P_2 lying in a plane parallel to the crank and
at right angles to the crank-axle. The forces P and P_1 constitute a

twisting couple of magnitude $P\,l$ acting on the axle. This twisting-moment is constant on the portion of the axle between the crank and the chain-wheel. The forces P_2 and P_3 constitute a bending couple of magnitude $P\,l_2$ at the point A, l_2 being the distance from the bearing to the middle of the pedal, measured parallel to the axis. The force P_1 produces a pressure on the bearing at A.

Example.—If l_2 be 3 ins., the other dimensions being as in the previous examples, $M = 150 \times 3 = 450$ inch-lbs., $T = 150 \times 6\frac{1}{2}$ $= 975$ inch-lbs., $T_e = 1524$ inch-lbs., $M_e = 762$ inch-lbs. The diameter d of the axle will be obtained by substituting in formula (15), chap. xii., thus :

$$\frac{d^3}{5} \times 20,000 = 1524$$

$$d^3 = \cdot381, \ d = \cdot725, \text{ say } \tfrac{3}{4} \text{ in.}$$

If the axle be tubular, $Z = \dfrac{762}{20,000} = \cdot0381$.

From Table IV., p. 112, a tube $\frac{7}{8}$ in. external diameter, 13 W. G., will be sufficient.

Comparing the hollow and solid axles, their sectional areas are ·226 and ·442 square inches respectively ; thus by increasing the external diameter $\frac{1}{8}$ inch and hollowing out the axle its weight may be reduced by one half ; while, if the external diameter be increased to 1 in., from Table IV., p. 112, a tube 16 W. G. will be sufficient, the sectional area being ·188 square inches ; less than 43 per cent. of that of the solid axle.

In riding ahead the maximum stresses on the axle, crank, and pedal-pins vary from zero, during the up-stroke of the pedal, to the maximum value *f*. If back-pedalling be indulged in, the range of stresses will be from $+ f$ to $- f$. The dimensions of the axle and crank above obtained by taking $f = 20,000$ lbs. per sq. in. are a little greater than those obtaining in ordinary practice. A total range of stress of 40,000 lbs. per sq. in. is very high, and cranks or axles subjected to it may be expected to break after a few years' working, unless they are made of steel of very good quality. It may be pointed out here that a pedal thrust of 150 lbs. will not be exerted continuously even in hard riding, though it may be exceeded in mounting by, and dismounting from, the pedal.

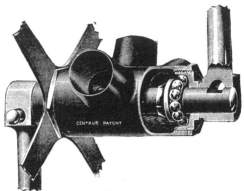

Fig. 547.

348. **Crank-brackets.**—The bracket and bearings for supporting the crank-axle form a kinematic inversion of the bearing shown in figure 404 ; the outer portion *H* forming the bracket is fastened

Fig. 548.

to the frame of the machine, while the spindle *S* becomes the crank-axle, to the ends of which the cranks are fastened. In the earlier patterns of crank-brackets, hard steel cups D were forced into the ends of the bracket, and cones C were screwed on the axle, the adjusting cone being fixed in position by a lock nut.

The barrel bottom-bracket is now more generally adopted ; being oil-retaining and more nearly dust-proof, it is to be preferred to the older pattern. The axle ball-races are fixed, and the adjustable ball-race can be moved along the bracket. In the 'Centaur' crank-bracket (fig. 547) the bearing *discs* or cups are screwed to the bracket, and secured by lock nuts. In the 'R. F. Hall' bracket (fig. 548) one cup is fastened to the bracket by a pin, and the other is adjusted by means of a stud screwed to the cup and working in a diagonal slot cut in the bracket. The pitch of this slot is so coarse that the adjustment is performed by pushing the stud forward as far as it will go, it being impossible to adjust too tightly. The cup is then clamped in place by the external screwed pin.

349. **The Pressure on Crank-axle bearings** is the resultant of the thrust on the pedals and the pull of the chain.

Example.—Taking the rows of balls $3\frac{1}{2}$ ins. apart, and the rest of the data as in the example of section 238, and considering first the vertical components due to the pressure P on the pedals, the condition of affairs is represented by figure 549. Taking moments about b, we get

$$3\tfrac{3}{4}P_1 = 3\tfrac{1}{2}P_3, \therefore P_3 = \frac{3\cdot75}{3\cdot5} \times 150 = 160\cdot7 \text{ lbs.}$$

In the same way, taking moments about c, we find $P_2 = 310\cdot7$ lbs.

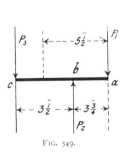

FIG. 549.

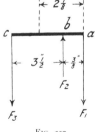

FIG. 550.

Consider now the horizontal forces. Fig. 550 represents the condition of affairs ; F_1, F_2, F_3 being respectively the horizontal components of the pull of the chain and of the pressure on the bearings. Taking moments about b, we get

$$\tfrac{3}{8}F_1 = 3\tfrac{1}{2}F_3, \text{ therefore, } F_3 = \frac{\cdot 375}{3\cdot 5} \times 340 = 36\cdot 4 \text{ lbs.}$$

In the same way, taking the moments about c, we find
$$F_2 = 376\cdot 4 \text{ lbs.}$$

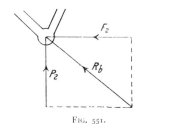

FIG. 551. FIG. 552.

The resultant pressures R_b and R_c on the bearings b and c can be found graphically as shown in figures 551 and 552, or by calculation, thus :

$$R_b = \sqrt{P_2{}^2 + F_2{}^2} = \sqrt{311^2 + 376^2} = 488 \text{ lbs.}$$
$$R_c = \sqrt{P_3{}^2 + F_3{}^2} = \sqrt{161^2 + 36^2} = 165 \text{ lbs.}$$

CHAPTER XXXI

SPRINGS AND SADDLES

350. **Springs under the Action of suddenly Applied Load.**—
We have already seen (sec. 82) that when a load is applied at the
end of a long bar, the bar is stretched, and a definite amount of
work is done. If the load be not too great, such a solid bar of
iron or steel forms a perfect spring. If a greater extension be
required for a given load, instead of a cylindrical bar a spiral
spring is used. The relation between the steady load and the
extension of a spiral spring is expressed by an equation similar to
(2), chap. x., and the stress-strain curve is, as in figure 74, a
straight line inclined to the axis of the spring.

Let a spiral spring be fixed at one end with its axis vertical
(fig. 553), and let A_0 be the position of its free end when supporting no load. Let A_1 be the position of
the free end when supporting a load W,
the ordinate $A_1 P_1$ being equal to W, to
a convenient scale. Let $A_0 P_1 P_2 P_3$ be
the stress-strain curve of the spring.
When this spring is supporting steadily
the load W, let an extra load w be suddenly applied. The end of the spring
when supporting the load $W + w$ will be
in the position A_2. The work done by
the loads in descending from A_1 to A_2
is $(W + w) x$, and is graphically represented by the area of the rectangle
$A_1 A_2 P_2 p_1$. The work done in stretch-
ing the spring is $W x + \frac{1}{2} w x$, and is represented by the area
$A_1 A_2 P_2 P_1$.

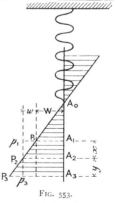

FIG. 553.

The difference of the quantities of work done by the falling weight and in stretching the bar is $\frac{1}{2} w x$, and is graphically represented by the triangle $P_1 p_1 P_2$. In the position A_2 of the end of the spring, this exists as kinetic energy, so that in this position the load must be still descending with appreciable speed. The spring continues to stretch until its end reaches a point A_3, where it comes to rest and then begins to contract. At the position of rest A_3, the work done by the loads in falling the distance $A_1 A_3$ must be equal to the work done in stretching the spring, since no kinetic energy exists in the position A_3. Therefore, area $A_1 A_3 p_3 p_1 = $ area $A_1 A_3 P_3 P_1$.

It is easily seen that this is equivalent to saying that the triangles $P_2 p_1 P_1$ and $P_2 p_3 P_3$ are equal, and therefore $y = x$; *i.e.* a load suddenly applied to a spring will stretch it twice as much as the same load applied gradually.

In the position A_3, the tension on the spring is greater than the load supported, and therefore the spring begins to contract and raise the load. If the spring had no internal friction it would contract as far as the original position A_1, and continue vibrating with simple harmonic motion between A_1 and A_3; but owing to internal friction of the molecules (or *hysteresis*) the spring will ultimately come to rest in the position of equilibrium A_2, and therefore the work lost internally is

$$P_1 p_1 P_2 = \tfrac{1}{2} w x \quad . \quad . \quad . \quad . \quad . \quad (1)$$

For a stiff spring the slope of $A_0 P_1 P_2 P_3$ is great, *i.e.* the extension x corresponding to a load w is small, and therefore the work lost is also small. For a weak spring the slope $A_0 p$ is small, and for a given load w the extension x, and therefore work lost, is large. But for a given extension x the work lost with a stiff spring is greater than with a weak spring.

351. **Spring Supporting Wheel.**—The function of a spring supporting the frame of a vehicle from the axle of a rolling wheel is to allow the frame to move along in a horizontal line without partaking of any vertical motion due to the inequalities of the road. This ideal motion would be attained if the stress-strain curve of the spring were a straight line parallel to its axis, and distant from it W; W being the steady load to be

supported. The wheel centre would then remain indifferently at any distance (within certain limits) from the frame of the vehicle, and since the pressure of the spring in all positions would be just equal to the weight supported, no vertical motion would be communicated to the frame. With this ideal spring the motion would be perfect until the spring got to one end or other of its stops, when a shock would be communicated to the frame. A better practical form of spring would be one having a stress-strain curve with a portion distant W from, and nearly parallel to, the axis ; the slope increasing at lower and higher loads, practically as shown in figure 554.

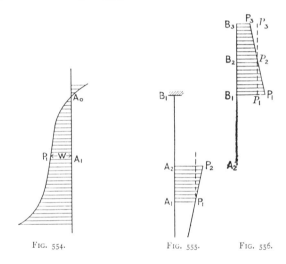

FIG. 554.   FIG. 555.   FIG. 556.

Let a cycle wheel running along a level road be supported by a spring under compression, the steady load on the latter being W, A_1 and B_1 (fig. 555) being the steady positions of the ends of the spring at the wheel axle and frame respectively. Let the wheel suddenly move over an obstacle so that its centre is raised the distance $A_1 A_2$, and the spring is further compressed. The frame end B of the spring may be considered fixed, while the wheel-centre is being raised. The work $A_1 P_1 P_2 A_2$ is expended in compressing the spring. The end A_2 may now be considered

fixed, and as the pressure on the spring is greater than the
load supported, the end B will rise and lift the frame. The
work $B_1 B_2 p_2 p_1$ (fig. 556) is expended in raising the frame
from B_1 to B_2, where static equilibrium takes place. If the
wheel-centre remain at the level A_2 the difference of energy
$P_1 p_1 P_2 = \frac{1}{2} w x$ is dissipated, the frame end of the spring vibrating between positions B_1 and B_3. If the wheel return quickly to
its former level A_1, little or no energy may be lost. The quantity
of lost energy is smaller the more nearly the stress-strain curve P
is parallel to the axis of the spring ; therefore a spring for a spring-
frame or wheel should be long, or the equivalent. An ideal spring
would have to be very carefully adjusted, as a small deviation from
the load it was designed for would send it to one end or other of
its stops.

352. **Saddle Springs**.—With a rigid frame cycle, the saddle
spring should perform the function above described, so that no
vertical motion due to the inequality of the road be communicated
to the rider ; practically, the vertical springs of saddles are
arranged so as to make as comfortable a seat as possible. It has
been shown (Chap. XIX.) that in riding over uneven roads, the
horizontal motion of the saddle is compounded of that of the
mass-centre of the machine, and a horizontal pitching due to the
inequalities of the road. If the saddle springs cannot yield
horizontally, the rider will slip slightly on his saddle.

A saddle, as in figure 557, with three vertical spiral springs
interposed between the upper and lower frames will yield horizontally more than one in which the frame and spring are merged
into one structure (fig. 560).

353. **Cylindrical Spiral Springs**.—Let d be the diameter of
the round wire from which the spring is made ; D the mean
diameter, and n the number, of the coils ; C the modulus of transverse elasticity ; δ the deflection, and q the maximum torsional
shear, produced by a load W. Then

$$\delta = \frac{8\,n\,D^3}{C\,d^4}\,W \quad . \quad . \quad . \quad . \quad . \quad . \quad (2)$$

$$q = \frac{8\,D}{\pi\,d^3}\,W \quad . \quad . \quad . \quad . \quad . \quad . \quad . \quad (3)$$

Mr. Hartnell says that a safe value for q for $\frac{3}{8}$-inch to $\frac{1}{4}$-inch wire, as used in safety-valve springs, is 60,000 to 70,000 lbs. per sq. in. Probably cycle springs have not such a large margin of strength as safety-valve springs. If q be taken slightly under 80,000 lbs. per sq. in., the greatest safe load, W, is given by the equation

$$W = 30,000 \frac{d^3}{D} \quad . \quad . \quad . \quad . \quad . \quad (4)$$

lbs. and inches being the units.

The value of C is between 12 and 14,000,000 lbs. per sq. in. If we take $C = 12,800,000$ the deflection is given by the equation

$$\delta = \frac{n\,D^3\,W}{1,600,000\,d^4} \quad . \quad . \quad . \quad . \quad (5)$$

Example.—A spiral spring $1\frac{1}{2}$-inch mean diameter, made from $\frac{1}{8}$-inch steel wire, will carry safely a load

$$W = \frac{30,000 \times 1}{1\cdot5 \times 8^3} = 40 \text{ lbs. nearly.}$$

The deflection per coil with this load will be

$$\delta = \frac{1\cdot5^3 \times 8^4 \times 40}{1,600,000 \times 1} = \cdot345 \text{ inch.}$$

Round wire is more economical than wire of any other section for cylindrical spiral springs.

354. **Flat Springs.**—The deflection of a beam of uniform section of span l, supported at its two ends and carrying a load W in the middle, is given by the formula

$$\delta = \frac{W\,l^3}{48\,E\,I} \quad . \quad . \quad . \quad . \quad . \quad (6)$$

E being the modulus of elasticity of the material, and I the moment of inertia of the section. For steel wire, tempered, $E =$ 13,000 to 15,000 tons per sq. in. If E be taken 33,600,000 lbs. per sq. in., substituting for I its value for a circular section $\frac{\pi}{64}\,d^4$, we get

$$\delta = \frac{W\,l^3}{80,000,000\,d^4} . \quad . \quad . \quad . \quad (7)$$

lbs. and inches being the units.

In many saddles the springs are made of round wire, and are subjected both to bending and direct compression. The deflection due to stress along the axis of the wire is very small in comparison with that due to bending, and may be neglected.

355. **Saddles**.—The seat of a cycle is almost invariably made of a strip of leather supported hammock fashion at the two ends,

FIG. 557.

the sides being left free. In the early days of the 'Ordinary' bicycle the seat was carried by a rigid iron frame, to which the peak and back of the leather were riveted. After being in use for some time such a seat sagged considerably, and the necessity for providing a tension adjustment soon became apparent. This tension adjustment is found on all modern saddles. The iron

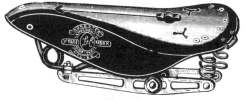

FIG. 558.

frame was itself bolted direct either to the backbone or to a flat spring, the saddle and spring were considered to a certain extent as independent parts, and were often supplied by different manufacturers. In modern saddles the seat, frame, and springs are so intimately connected that it is impossible to treat them separately.

One of the most comfortable types of saddles consists of the leather seat, the top-frame with the tension adjustment, an under-

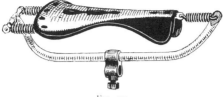

FIG. 559.

frame with clip to fasten to the **L**-pin of the bicycle, and three vertical spiral springs between the top- and under-frames.

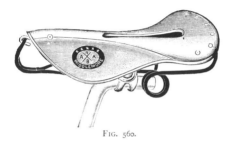

FIG. 560.

In the 'Brampton' saddle (fig. 557) the under-frame forms practically a double-trussed beam made of two wires. In

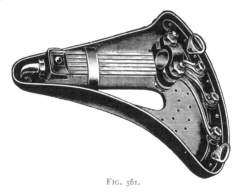

FIG. 561.

Lamplugh's saddle (fig. 558) the under-frame is made of two thin plates.

A simple hammock saddle with the seat supported by springs (fig. 559), made by Messrs. Birt & Co., consists of leather seat, tubular frame, and three spiral springs subjected to tension, no top-frame being necessary.

The springs, top- and under-frames, are often merged into one structure, as in the saddle shown in figure 560, made by

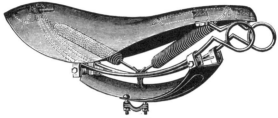

FIG. 562.

Mr. Wm. Middlemore, and that shown in figure 561, made by Messrs. Brampton & Co. In the former two wires, in the latter six wires, are used for the combined springs and frames.

All saddle-clips should be of such a form that the rider can adjust the tilt of the saddle so as to get the most comfortable

FIG. 563.

position. In the 'Automatic Cycle Saddle (fig. 562) the rider can alter the tilt while riding.

It may be noticed that the leather seats of the saddles illustrated above are slit longitudinally, the object being to avoid injurious pressure on the perineum.

The 'Sar' saddle (fig. 565), of the Cameo Cycle Company, is provided with a longitudinal depression, for the same purpose.

356. **Pneumatic Saddles.**—A number of pneumatic saddles have been made, in which the resilience is provided by com-

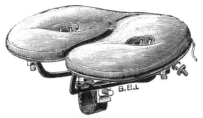

FIG. 564.

pressed air instead of steel springs. The 'Guthrie-Hall' saddle (fig. 563) is one of the most successful. The 'Henson Anatomic' saddle (fig. 564) is made without a peak, and consists of two air pads, each with a depression in which the ischial tuberosities

FIG. 565.

rest, the whole design of the saddle being to avoid perineal pressure. The 'Sar' saddle (fig. 565) is also provided with two depressions for the same reasons.

CHAPTER XXXII

357. Brake Resistance on the Level.—Let W be the total weight of machine and rider, W_b the load supported by the wheel to which the brake is applied, and μ_g the coefficient of friction between the ground and the tyre. If the brake be powerful enough, it may actually prevent the wheel from rotating, in which case the tyre will rub along the ground while the machine is being brought to a standstill. Then R, the greatest possible brake resistance, would be $\mu_g W_b$. The pressure applied at the brake handle should be, and usually is, less than that necessary to make the tyre rub on the ground; this rubbing might have disastrous results. Let v be the speed in feet per second, V in miles per hour, and l the distance in feet which must be travelled when pulling up under the greatest brake resistance. Then, since the kinetic energy of the machine and rider is expended in overcoming the brake resistance,

$$\frac{W v^2}{2 g} = \cdot 0334 \ W \ V^2 = \mu_g \ W_b \ l,$$

or

$$l = \frac{\cdot 0334 \ W \ V^2}{\mu_g \ W_b} \qquad . \quad . \quad . \quad . \quad . \quad (1)$$

Example.—Taking the data of the example in section 228, with the weight of the machine, 30 lbs., equally divided between the two wheels, speed 20 miles per hour, $\mu = 0\cdot4$, and the brake applied to the front wheel, we have $W = 180$ lbs., $W_b = 54\cdot3$ lbs., $R = 0\cdot4 \times 54\cdot3 = 21\cdot7$ lbs., and substituting in (1),

$$l = \frac{\cdot 0334 \times 180 \times 400}{0\cdot4 \times 54\cdot3} = 111 \text{ ft.}$$

If the brake be applied to the rear wheel, $W_b = 125\cdot7$ lbs., and

$$l = \frac{\cdot0334 \times 180 \times 400}{0\cdot4 \times 125\cdot7} = 48 \text{ ft.}$$

It should be noticed that the load W_b should be taken as that actually on the wheel while the brake is applied (see sec. 164).

358. **Brake Resistance Down-hill.**—If the machine be on a gradient of x part vertical to 1 on the slope, the force parallel to the road surface necessary to keep it from running downhill is $x\,W$ (see fig. 58). The brake resistance is $\mu_g\,W_b\,cos\,\phi = \mu_g\,W_b\sqrt{1-x^2}$, ϕ being the angle of inclination to the horizontal. For all but very steep gradients, $\sqrt{1-x^2}$ does not differ much from 1, and therefore the brake resistance is approximately $\mu_g\,W_b$, as on the level. Thus, if the brake be fully applied, the resultant maximum retarding force is $\mu_g\,W_b\sqrt{1-x^2} - x\,W$, and therefore, as in section 357, the distance which must be travelled before being pulled up is given by the equation

$$\cdot0334\;W\,V^2 = (\mu_g\,W_b\,\sqrt{1-x^2} - x\,W)\,l \quad . \quad . \quad (2)$$

or

$$l = \frac{\cdot0334\;W\,V^2}{\mu_g\,W_b - x\,W} \text{ approx.} \quad . \quad . \quad . \quad (3)$$

If

$$x\,W = \mu_g\,W_b \quad . \quad . \quad . \quad . \quad . \quad . \quad . \quad (4)$$

the machine cannot be pulled up by the brake, however powerful; while if $x\,W$ is greater than $\mu_g\,W_b$ the speed will increase, and the machine run away.

Example I.—With the data of the example of section 357, brake on the front wheel, running down a gradient of 1 in 10, $x = 0\cdot1$; substituting in (3),

$$l = \frac{\cdot0334 \times 180 \times 400}{0\cdot4 \times 54\cdot3 - 0\cdot1 \times 180} = 643 \text{ ft.}$$

Example II.—With the same data except as to gradient, find the steepest gradient that can be safely ridden down, with the brake.

Substituting in (4), $0\cdot1 \times 180 = 0\cdot4 \times 54\cdot3$; or $x = \cdot121$. That is, no brake, however powerful, can stop the machine on a gradient of 121 in 1,000, about 1 in 8.

If the brake be applied to the back wheel, the corresponding gradient is

$$x = \frac{0.4 \times 125.7}{180} = .279$$

i.e. about 1 in 4.

359. **Tyre and Rim Brakes.**—The brake is usually applied to the tyre of the front wheel, not because this is the best position, but on account of the simplicity of the necessary brake gear. In the early days of the 'Ordinary' a roller or spoon brake was sometimes applied to the rear wheel, a cord communicating with the handle-bar (fig. 338). The ordinary spoon brake (fig. 131) at the top of the front wheel fork is depressed by a rod or plunger operated by the brake-lever on the handle-bar, the leverage being about $2\frac{1}{2}$ or 3 to 1. If r be this leverage, and μ_s the coefficient of friction between the brake-spoon and the tyre, the pressure P on the brake-handle necessary to produce the maximum effect is given by the equation $\mu_s \, r \, P = \mu_g \, W_b$, or

$$P = \frac{\mu_g \, W_b}{\mu_s \, r} \qquad . \qquad . \qquad . \qquad . \qquad . \qquad . \qquad (5)$$

Example.—With the data of the example of section 357, $r=3$, and $\mu_s = 0.2$; substituting in (5), we get

$$P = \frac{0.4 \times 54}{0.2 \times 3} = 36 \text{ lbs.}$$

In the *pneumatic brake* the movement of the brake block on to the tyre is produced by means of compressed air, pumped by a rubber collapsible ball placed on the handle-bar, and led through a small india-rubber tube to an air chamber, which can be fastened to any convenient part of the frame. With this simple apparatus the brake can be as easily applied to the rear as to the front wheel.

360. **Band Brakes** are applied to the hubs of both the front and rear wheels, and have been occasionally applied at the crank-axle. The spoon brake, rubbing on the tyre, may possibly injure it ; the band brake is not open to this objection. Since a small drum fixed to the hub has, relative to the frame, a less linear speed than the rim of the wheel, to produce a certain effect the brake resistance must be correspondingly larger. One end of the band is fastened to the frame, the other can be tightened by means of the

brake gear. The gear should be arranged so that when the brake is applied the tension on the fixed end of the band is the greater. If t_1 and t_2 be the tensions on the ends of the band, the resistance at the drum is $t_1 - t_2$, and, as in section 251,

$$\log. \frac{t_1}{t_2} = \cdot 4343\, \mu\, \theta. \quad . \quad . \quad . \quad . \quad . \quad (6)$$

If D and d be respectively the diameters of the wheel and the brake drum, to actually make the wheel stop revolving we must have

$$(t_1 - t_2)\, \frac{d}{D} = \mu_g\, W_b \quad . \quad . \quad . \quad . \quad . \quad (7)$$

Example I.—Let the band have an arc of contact of three right angles with the drum, *i.e.* $\theta = \frac{3\,\pi}{2} = 4\cdot71$, let $\mu = \cdot15$, $D = 28$ in., $d = 5\frac{1}{4}$ in., and the rest of the data as in section 357, then, substituting in (6)

$$\log. \frac{t_1}{t_2} = \cdot4343 \times 0\cdot15 \times 4\cdot71 = \cdot3068.$$

Consulting a table of logarithms,

$$\frac{t_1}{t_2} = 2\cdot027 ;$$

and $t_1 - t_2 = 1\cdot027\, t_2$. Substituting in (7),

$$1\cdot027\, t_2 \times \frac{5\frac{1}{4}}{28} = \cdot4 \times 54,$$

or

$$t_2 = \frac{\cdot4 \times 54 \times 28}{5\frac{1}{4} \times 1\cdot027} = 112 \text{ lbs.}$$

Example II.—If a band brake of same diameter as in last example be applied at the crank-axle, the necessary tension t_2 will be $\frac{N_2}{N_1}$ times as great, N_1 and N_2 being the numbers of teeth in the chain-wheels on the driving-hub and crank-axle respectively. With $N_1 = 8$, $N_2 = 18$,

$$t_2 = \frac{18 \times 112}{8} = 252 \text{ lbs.}$$

This example shows the ineffectiveness of a crank-axle band brake, since the elasticity of the gear is such that the brake lever would be close up against the handle-bar long before the required pull was exerted on the band.

If oil gets in between the band and its drum, the coefficient of friction will be much less, and a much greater pull will be required, than in the above examples.

INDEX

(The figures indicate page numbers.)